ALSO BY RICHARD RHODES

Hedy's Folly

The Twilight of the Bombs

Arsenals of Folly

John James Audubon

Masters of Death

Why They Kill

Visions of Technology

Deadly Feasts

How to Write

Dark Sun

Nuclear Renewal

Making Love

A Hole in the World

Farm

The Making of the Atomic Bomb

Sons of Earth

The Last Safari

Looking for America

Holy Secrets

The Ozarks

The Ungodly

The Inland Ground

RICHARD RHODES

ENERGY

A HUMAN HISTORY

SIMON & SCHUSTER

NEW YORK LONDON TORONTO SYDNEY NEW DELHI

Simon & Schuster
1230 Avenue of the Americas
New York, NY 10020

First Simon & Schuster hardcover edition May 2018

Interior design by Ruth Lee-Mui

Manufactured in the United States of America

10 9 8 7 6 5 4 3 2 1

Library of Congress Cataloging-in-Publication Data is available.

ISBN 978-1-5011-0535-7
ISBN 978-1-5011-0537-1 (ebook)

For Isaac Rhodes

CONTENTS

A grant from the Alfred P. Sloan Foundation supported
the research and writing of this book.

FOREWORD

You're about to embark on a four-hundred-year journey with some of the most interesting and creative people who ever lived. Since they're scientists and inventors and engineers, their names don't always attach to their work. But they shaped the world we live in, for better and for worse. Mostly for the better, I believe. After you travel with them, I think you will too. At least you'll know more about what they did and why and how they did it. I was surprised and sometimes amazed at how many of their stories have been forgotten. Some of the references I use to tell those stories were histories and biographies that date back two hundred years or more. The books and documents are old, but the stories are new.

Who are these paragons? One writer at least: William Shakespeare, not as playwright but as part-owner of the Theatre, the first in London. He and his partners dismantled it (the landowner claimed they stole it) for the wood in a time when wood had become scarce around London. They carted it across the Thames River to build the larger Globe Theatre in naughty Southwark, next door to a bear-baiting arena.

A Frenchman, Denis Papin, concerned with feeding the poor, whose invention of the pressure cooker prepared the way for the steam engine.

James Watt, of course, the Scotsman who gave us the steam engine itself, but also Thomas Newcomen before him, whose great galumphing atmospheric steam machine preceded Watt's elegant elaboration.

I visited a replica Newcomen engine in England on one of the few days a year when its keepers fire it up. It was the size of a house and a champion coal hog. (Coal isn't cheap anymore, which is why they seldom fire it up.) I shoveled a scoopful of coal into the firebox and talked with the retired engineer who ran it. I asked him what equipment he needed to keep it running, and, with a chuckle, he hefted a big hammer. The Newcomen was all pipes and cranks and often out of whack, so he whacked it.

Newcomens squatted at the pithead—the surface opening into a mine—and pumped water out. They were too inefficient to be made portable. Watt's more efficient engine could be smaller—small enough to mount on wheels and rails to haul the coal from the pithead to the river in order to be barged down to London. Then someone realized you could haul people as well as coal, and the passenger railroad emerged and quickly branched out all over England. America too, but our engines burned wood through most of the nineteenth century, penetrating the wilderness far from any coal mine and then connecting the continent.

Among twentieth-century paragons, there's Arie Haagen-Smit, a Dutch specialist in essences who was teaching at the California Institute of Technology (Caltech), in Pasadena. One day in 1948, concerned government officials found him in a laboratory full of ripe pineapples, condensing their tropical aroma from the air. They asked him to do the same for the ghastly Los Angeles smog. He cleared out the pineapples, opened a window, and sucked in thousands of cubic feet of smoggy air. He ran the air through a filter chilled with liquid nitrogen, and scraped up a few drops of brown, smelly gunk. After he'd analyzed the gunk chemically, he announced it was automobile exhaust and the exhaust of nearby refineries. Unlike the old and often deadly smoke and fog (smoke + fog = "smog") that blighted cities where coal was burned, this new stuff compounded in the air like a binary poison gas. Catalyzed by sunlight, it turned the air sepia.

Oil companies didn't want to know that. Their chemists scoffed at Haagen-Smit's analysis. They found no such reaction, they told the world. Which fueled the stubborn Dutchman's anger. Back to the lab. He showed that the oil company chemists' fancy equipment couldn't distinguish the smog-forming process. For his part, Haagen-Smit used strips of old inner tube to measure how much smog ozone embrittled rubber and his pineapple-analyzing gear to sniff out the components that combined to blight the air. Government stepped in then and began the process of cleaning up Los Angeles.

This book is full of such stories. It's more than merely stories, however. Its serious purpose is to explore the history of energy; to cast light on the choices we're confronting today because of the challenge of global climate change. People in the energy business think we take energy for granted. They say we care about it only at the pump or the outlet in the wall. That may have been true once. It certainly isn't true today. Climate change is a major political issue. Most of us are aware of it—increasingly so—and worried about it. Businesses are challenged by it. It looms over civilization with much the same gloom of doomsday menace as did fear of nuclear annihilation in the long years of the Cold War.

Many feel excluded from the discussion, however. The literature of climate change is mostly technical; the debate, esoteric. It's focused on present conditions, with little reference to the human past—to centuries of hard-won human experience. Yet today's challenges are the legacies of historic transitions. Wood gave way to coal, and coal made room for oil, as coal and oil are now making room for natural gas, nuclear power, and renewables. Prime movers (systems that convert energy to motion) transitioned from animal and water power to the steam engine, the internal combustion engine, the generator, and the electric motor. We learned from such challenges, mastered their transitions, capitalized on their opportunities.

The current debate has hardly explored the rich *human* history behind today's energy challenges. I wrote *Energy* partly to fill that void—with people, events, times, places, approaches, examples, parallels, disasters, and triumphs, to enliven the debate and clarify choices.

People lived and died, businesses prospered or failed, nations rose to world power or declined, in contention over energy challenges. The record is rich with human stories, a cast of characters across four centuries that includes such historic figures as Elizabeth I, James I, John Evelyn, Abraham Darby, Benjamin Franklin, Thomas Newcomen, James Watt, George Stephenson, Humphry Davy, Michael Faraday, Herman Melville, Edwin Drake, Ida Tarbell, John D. Rockefeller, Henry Ford, Enrico Fermi, Hyman Rickover, the coal barons of old Pennsylvania, and the oil barons of California and Saudi Arabia—to name only some of the more obvious.

Whole oceans of whales enter the story, the oil of their bodies lighting the world. Petroleum seeps from a streambed, and a Yale chemistry professor wonders what uses it might have. Horses foul cities with their redolent manure, an increasing public health challenge, and when the automobile replaces them, rural populations no longer required to grow their feed fall into permanent decline. The development of arc welding paces the pipeline distribution of natural gas. Nuclear energy announces itself by burning down two Japanese cities, an almost indelible taint.

Global warming itself, the evidence slowly accumulating across a century of increasingly anxious observation, provokes a biblical-scale confrontation of ideologies and vested interests. Wind energy, the bountiful energy from sunlight, vast supplies of coal and natural gas compete for dominance in a turbulent world advancing toward a population of ten billion souls by the year 2100. Most of them are residents of China and India, the two most populous countries in the world, just now moving out of subsistence into prosperity and consuming energy supplies accordingly. The energy is there, but can the earth sustain the waste of its burning?

You will not find many prescriptions in this book. Every century had its challenges and opportunities—some intended, some unintended—but in any case, too complex, too rich in implication, for simple moralizing. What you will find are examples, told as fully as I am able to tell them. Here is how human beings, again and again, confronted the deeply human problem of how to draw life from the raw materials of the world. Each invention, each discovery, each adaptation brought further challenges in its wake, and through such continuing transformations, we arrived at where we are today. The air is cleaner, the world more peaceful, and more and more of us are prosperous. But the air is also warmer. In August 2015, for example, northern Iran suffered under a heat index of 165 degrees Fahrenheit (74 degrees Celsius). May all this curious knowledge from our history help us find our way to tomorrow. I have children and grandchildren. I hope and believe that we will.

PART ONE

POWER

ONE

NO WOOD, NO KINGDOM

A cold, gray day, and heavy snow billowing. Saturday, 28 December 1598, the forty-first year of the reign of Elizabeth Tudor, Queen of England and Ireland. On the edge of London Town, in the precinct of Holywell, workmen gather in the yard before the old Theatre, snow on their beards, stamping their boots and clapping their gloved hands to keep warm. Hailing each other with ale-warmed breath: work to do, and that quickly, shillings to earn even in holiday time. Wood was scarce in London, the forests that ringed the city stripped bare. The workmen had been hired to tear down the Theatre, the first of its kind, and move the salvaged framing to master carpenter Peter Street's Thames-side warehouse, hard by Bridewell Stairs. Steal a whole building, someone winked, right out from under the absent landlord's nose, though who rightly owned the Theatre would need years of litigation to decide.[1] The Burbage brothers, William Shakespeare's partners in the theater business, believed they did. They'd built it, in 1576. Let the landlord keep his land. They would dismantle their playhouse and raise it elsewhere.

Giles Allen, the landlord, away at his country house in Essex, would tell the court that men with weapons bullied aside the servants he sent with a power of attorney to stop them. With all the shouting, a crowd gathered. The Burbage brothers were there that day. So was Shakespeare. Moving the playhouse was urgent if their acting company would have a stage to perform on. Allen was threatening to pull it down himself and salvage the timbers to build tenements, as apartments were called in Shakespeare's day.

The Burbages' workmen dismantled the wooden building and carted the framing away. Two days earlier, the company had played before the Queen at Whitehall Palace. It was scheduled to play there again on New Year's night. The Theatre came down between the two performances.

It went up again in Spring 1599 across the Thames in bawdy Southwark, enlarged and renamed the Globe, a twenty-sided polygon three stories high and a hundred feet across, with a thatched ring of roof open to the sky above a wide yard. Peter Street probably cut the new timber for the enlargement in a forest near Windsor, west of London, lopped and topped and barked and shaped it there to avoid the cost of barging whole trees down the Thames. A Swiss tourist, Thomas Platter, attended a production of *Julius Caesar* in the new Globe on the afternoon of 21 September 1599, so it was up and running by then. He thought the play "quite aptly performed."[2]

Elizabethan England was a country built of wood. "The greatest part of our building in the cities and good towns of England," the Elizabethan observer William Harrison reported in 1577, "consisteth only of timber."[3] Even the country's implements, its plows and hoes, were wooden, if iron edged. London was a wooden city, peak-roofed and half-timbered, heating itself with firewood burned on stone hearths called reredos raised in the middle of rooms, the sweet wood smoke drifting through the house and out the windows.

But wood was growing dear, its price increasing as London's population increased and woodcutters carted firewood into the city from

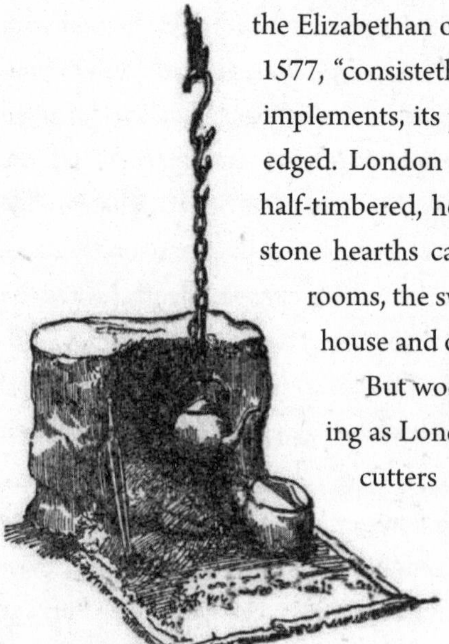

A reredos, with hook above for hanging a kettle.

farther and farther afield. Parliament provided a limited remedy in 1581: a law prohibiting the production of charcoal for iron smelting within fourteen miles of London, to reserve the nearby trees for domestic fuel. Even so, the cost of firewood delivered to the city more than doubled between 1500 and 1592, consistent with the burgeoning population, which quadrupled between 1500 and 1600, from 50,000 to 200,000.[4] (England's entire population increased across that century from 3.25 million to 4.07 million.[5])

Some economists today question if England was running out of wood. The Burbages and their company moved the Theatre's framing not only to save wood but also to save time and money putting up their new, enlarged Globe bankside. And wood, after all, is a renewable resource. Yet many seventeenth- and eighteenth-century government officials, parliamentarians, and private observers feared a wood shortage, especially of large oak trees suitable for ships' masts.

Warships were as valuable to national security in those days as aircraft carriers are today. About 2,500 large oak trees went into an average English ship of the line.[6] It was a beautiful wooden fighting machine, massive and solid, fifty feet wide and two hundred feet long. Two rows of cannon mounted on wooden trucks pierced its bulging yellow sides. Its decks were painted dull red to veil the blood that flowed in battle.[7] It carried its sails on no fewer than twenty-three masts, yards, and spars, from the forty-yard-long, eighteen-ton mainmast to the little fore topgallant-yard, a light seven-yard stick.[8] Patriots said the Royal Navy was England's "wooden walls," protecting it from invasion. The Admiralty built and maintained about one hundred ships of the line as well as several hundred smaller ships and boats. Battle and shipworms ravaged them; they needed replacing every decade or two.

But the great mast trees took 80 to 120 years to grow to sufficient diameter. A landowner who planted an acorn could hope his grandchildren or great-grandchildren might harvest it for profit—if the intervening generations could wait so long. Many could not; many did not. Selling timber was an easy means to raise cash; landowners from the king on down took advantage of the opportunity whenever their purses emptied. Wood, the dilettante second Earl of Carnarvon told a friend of the diarist Samuel Pepys, was "an excrescence of the earth provided by God for the payment of debts."[9]

Crooked hedgerow timbers—"compass timbers," the Admiralty called

The *Ark Royal*, built for Sir Walter Raleigh in 1587, carried fifty-five guns on two gun decks. In 1588 she chased the Spanish Armada into the North Sea.

them—were as important to ship construction as the straight forest timbers needed for the masts. These great bent oaks supplied curved and branched single pieces for the keel, the stern-post, and the ribs of the ship's hull. They were always scarce and priced accordingly, but with the enclosure movement of late-medieval England—the privatization and consolidation of communal fields into sheep pasture to benefit the manorial lords—most of the compass trees were cut down. Finding the right piece for a ship could take years.

The Royal Navy was not the only enterprise consuming the forests of England. By the 1630s, the country supported some three hundred iron-smelting operations, which burned three hundred thousand loads of wood annually to make charcoal, each load counting as a large tree.[10] Building and maintaining the more numerous ships of British commerce required three times as much oak as did navy shipping.

Timber, oak in particular, competed with grain for arable land. Great trees needed deep, rich soil, but it was more profitable to farm such land for feed. A Suffolk County official named Thomas Preston associated mighty forests with primitive conditions, "the past age" when the kingdom possessed "a great plenty of oak." The diminution of oak measured the kingdom's improvement, he argued, "a thousand times more valuable than any timber can ever be." Preston hoped the diminution would continue: "While we are forced to feed our people with foreign wheat, and our horses with foreign oats, can raising oak be an object? . . . The scarcity of timber ought never to be regretted, for it is a certain proof of national improvement; and for Royal navies, countries yet barbarous are the right and only proper nurseries."[11]

Those barbarous countries included North America, especially New England, where the colonists had just begun to harvest the primeval forest. There, from 1650 onward, the Admiralty sought the strong "single stick" masts its warships required, forty yards long and three to four feet in diameter. The colonists competed for the wood, however. The first American sawmill began operations in 1663 on the Salmon Falls River in New Hampshire, long before the English advanced from sawing board by hand to using water power. By 1747, there were 90 such water-powered mills along the Salmon Falls and the Piscataqua, with 130 teams of oxen working hauling logs. Among them, they cut about six million board feet of timber annually for sale in Boston, the West Indies, and beyond. England got her share. The eighteenth-century historian Daniel Neal, in his *The History of New-England,* noted that the Piscataqua was "the principal place of trade for masts of any of the king's dominions."[12]

Unfortunately for the Royal Navy, America's successful revolution three decades later cut off its supply of American white pine. It had to return to its earlier expedient of using "made masts": weaker composite masts of multiple trees strapped together around a central spindle.

Besides making charcoal to smelt iron, the English cut down timber to build houses, barns, and fences; to produce glass and refine lead; to build bridges, docks, locks, canal boats, and forts; and to make beer and cider barrels. More than one of these uses consumed as much wood as the navy. Even royalty was guilty of misusing the royal forests, while Parliament stood by. "The final failure of the woodlands," a historian concludes, "was the result of constant neglect and abuse."[13]

The Jacobean agriculturalist Arthur Standish was concerned less with the needs of the Royal Navy and more with what he called "the general destruction and waste of wood" when he published *The Commons Complaint* under King James I's endorsement in 1611, but he included "timber . . . for navigation" among the shortages that he foresaw. Paraphrasing one of the king's speeches before Parliament in his stark summary of the consequences, Standish concluded: "And so it may be conceived, *no wood, no kingdom*."[14]

A cheaper alternative was burning coal—sea coal or pit coal, the Elizabethans called it to distinguish it from charcoal. (A coal was originally any burning ember, thus char-coal for charred wood, and sea coal or pit coal for the fossil fuel, depending on whether it outcropped on the headlands above the beaches or was dug from the ground.) Harrison, in his 1577 contribution to the Elizabethan anthology *Holinshed's Chronicles*, had found the English Midlands already in transition to the fossil fuel: "Of coal-mines, we have such plenty in the north and western parts of our island as may suffice for all the realm of England."[15] Coal had served blacksmiths for hundreds of years. Soap boilers used it; so did lime burners, who roasted limestone in kilns to make quicklime for plaster; so did salt boilers, who boiled down seawater in open iron pans, a tedious process prodigal of fuel, to make salt for food preservation in the centuries before refrigeration.

But the acrid smoke and sulfurous stench of the Midlands's coal had not encouraged its domestic use in houses devoid of chimneys where meat was roasted over open fires. "The nice dames of London," as a chronicler called them, were unwilling even to enter such houses. In 1578 Elizabeth I herself objected to the stink of coal smoke blowing into Westminster Palace from a nearby brewery and sent at least one brewer to prison that year for his effrontery.[16] A chastened Company of Brewers offered to burn only wood near the palace.

Like nuclear power in the twentieth century, but justifiably, coal in the sixteenth and seventeenth centuries was feared to be toxic, tainted by its origins, diabolic: "poisonous when burnt in dwellings," a historian summarizes Elizabethan prejudices, "and . . . especially injurious to the human complexion. All sorts of diseases were attributed to its use."[17] The black stone found layered underground that burned like the stinking fires of hell—the Devil's very excrement, preachers ranted—suffered as well from its association with mining, an industry that poets and clergy had long condemned. Geoffrey Chaucer, in his short poem "The Former Age," written about 1380, set the tone:

But cursed was the time, I dare well say,
That men first did their sweaty business
To grub up metal, lurking in darkness,
And in the rivers first gems sought.
Alas! Then sprung up all the cursedness
Of greed, that first our sorrow brought![18]

The German humanist Georgius Agricola, a physician in the mining town of Joachimstal, paraphrased the arguments of mining's detractors in his 1556 work *De re Metallica* and quoted Ovid condemning mining in similar terms. The Roman poet, he wrote, had portrayed men as ever descending " 'into the entrails of the earth, [where] they dug up riches, those incentives to vice, which the earth had hidden and had removed to the Stygian shades. Then destructive iron came forth, and gold, more destructive than iron; then war came forth.' "[19] A century after Agricola, John Milton was still condemning mining, associating it with the fallen angel Mammon in the first book of *Paradise Lost*:

There stood a Hill not far whose grisly top
Belched fire and rolling smoke; the rest entire
Shone with a glossy scurf, undoubted sign
That in his womb was hid metallic ore,
The work of sulfur. Thither winged with speed
A numerous brigade hastened . . . .
Mammon led them on,
Mammon, the least erected spirit that fell
From Heaven, for even in Heaven his looks and thoughts
Were always downward bent, admiring more
The riches of Heaven's pavement, trodden gold,
Than aught divine or holy else enjoyed
In vision beatific: by him first
Men also, and by his suggestion taught,

Ransacked the center, and with impious hands
Rifled the bowels of their mother Earth
For treasures better hid.[20]

Impious hands or not, the Elizabethans were short of wood, so they began to dig coal and burn it. To do that without asphyxiating themselves, they needed chimneys to exhaust the smoke. Harrison, the chronicler, says old men in his village noticed the increase in chimneys, "whereas in their young days there was not above two or three." For Harrison, the development was doubtful, even hell-in-a-handcart:

> Now we have many chimneys, and yet our tenderlings complain of rheums, catarrhs, and poses [head colds]; then had we none but reredoses, and our heads did never ache. For as the smoke in those days was supposed to be a sufficient hardening for the timber of the houses, so it was reputed a far better medicine to keep the good man and his family from the quack [hoarseness] and the pose, wherewith as then very few were acquainted.[21]

Shipments of coal from Newcastle upon Tyne, an expanding coal port on the Tyne River in the northeast of England, increased accordingly from about thirty-five thousand tons in the midsixteenth century to about four hundred thousand tons by 1625. In two generations, the historian J. U. Nef concludes, "the coal trade from the Tyne had multiplied twelvefold."[22]

When Queen Elizabeth I died at sixty-nine in 1603, the king of Scotland, James VI, united the Scottish and English crowns as James I, moving in slow procession to London. The Scots had deforested their lands a century before the English. They were used to burning coal, and luckily for them, hard Scottish coal burned cleaner and brighter than soft Newcastle bituminous. Scottish anthracite's sulfur content was only 0.1 percent, compared with 1 percent to 1.4 percent for English bituminous.[23] Unfortunately, Scottish anthracite burned faster as well, which made it more expensive. Expense was no problem for the king; he had good Scottish coal shipped to Westminster to warm his palaces. Emulating the king, wealthy Londoners took up the custom. The middle classes began burning coal as well. Coal allowed Londoners to keep warm and feed themselves as the city's population increased rapidly, from roughly 200,000 in 1600 to 350,000 by 1650.[24]

Chimneys needed sweeping to prevent fires, a new and ultimately deadly trade for children apprenticed as young as five or six years old, who walked the streets crying "Sweep! Sweep!" to solicit work and crawled large-hatted and naked through the narrow chimneys like human brooms. In a 1618 "Petition of the Poor Chimney Sweepers of the City of London to the King," two hundred sweeps complained that the city was at risk of fire, and they "were ready to be starved for want of work" because people neglected to clean their chimneys. They asked that an overseer be appointed to enter houses and compel the owners to have their chimneys cleaned. The overseer and his deputies, the petition proposed, could be paid "by the delivery to them of the soot gathered," which they could sell for fertilizer. The king was sympathetic, but the Lord Mayor of London wasn't: there were already officers who oversaw the condition of London's chimneys, he claimed—and the poor chimney sweepers' petition was denied.[25]

Constant exposure to soot and creosote led to an epidemic of soot wart among chimney sweeps—squamous cell carcinoma of the scrotum—characterized by the English surgeon Percivall Pott in 1775, the first time a cancer was associated with an industrial occupation. The scrotum was the point of entry of the cancer into the body because that was where the sweeps' sooty sweat collected as they broomed their way up London's chimneys.

An engineer, Richard Gesling, invented a method of combining smoky Newcastle coal with common materials: chopped straw, sawdust, even cow manure. These coal balls, as he called them, were something like the charcoal briquettes of American backyard barbecues and burned more cleanly than coal alone. Gesling died before he could make his method public, but someone published an anonymous report of it, *Artificiall Fire, or, Coale for Rich and Poore,* in 1644.[26] Whoever it was, he did so for a reason: it was cold in London that winter, indoors as well as out. The Royalists were waging civil war against the Puritan Oliver Cromwell and his Parliamentarians, who had Scottish support. In 1644 the Scottish army besieged Newcastle, blocking coal shipments to the English capital. The author of *Artificiall Fire* writes contemptuously of "some fine Nosed City Dames [who] used to tell their Husbands; O Husband! We shall never be well, we nor our Children, whilst we live in the smell of this City's Seacoal smoke." But with Newcastle under siege and coal scarce in London, he continues, "how many of these fine Nosed Dames now cry, Would to God we had Seacoal, O the want of Fire undoes us! O the sweet Seacoal fire we used to have!"

As coal replaced wood, its denser and more toxic smoke became a pestilence. Between 1591 and 1667, coal shipments into London increased from 35,000 tons to 264,000 tons; by 1700, that tonnage had almost doubled to 467,000 tons.[27] An adequate supply of fossil fuel kept people warm and sustained the growth of English industry, but it also fouled the London air. John Evelyn, a wealthy diarist and horticulturalist who was one of the founders of the scientific Royal Society of London, condemned the city in his diatribe *The Character of England*, published in 1659.

London, Evelyn wrote, though large, was "a very ugly town, pestered with hackney coaches and insolent car men, shops and taverns, noise, and such a cloud of sea-coal [smoke], as if there be a resemblance of hell upon earth, it is in this volcano [on] a foggy day: this pestilent smoke . . . corrodes the very iron, and spoils all the movables, leaving a soot upon all things that it lights; and so fatally seizes on the lungs of the inhabitants, that the cough, and the consumption spare no man. I have been in a spacious church where I could not discern the minister for the smoke, nor hear him for the people's barking."[28]

A long-faced and solemn man, ambitious for laurels, Evelyn did more than complain. He also looked for ways to clear the air. He accepted appointment as one of London's commissioners of sewers. And since he was interested in gardening and in trees, his inventive mind turned to moving industry out of London and perfuming the city's precincts with flowering plants—reversing, as it were, at least locally, the transition from wood to coal. King Charles II had been restored to the throne on his thirtieth birthday, 29 May 1660, and the traitor Oliver Cromwell's head pickled and mounted on a pike on London Bridge after a seventeen-year interregnum bloodied with regicide and civil war; Evelyn's vision of a refreshed and healthier London drew as well on his renewed sense of public order.

Evelyn was walking in Whitehall one day, he told the king in the dedication that introduced his proposal, when "a presumptuous smoke . . . did so invade the court that all the rooms, galleries, and places about it were filled and infested with it; and that to such a degree [that] men could hardly discern one another for the cloud, and none could support [endure] without manifest inconveniency."[29] He had been thinking about the problem for some time, he added, but it was "this pernicious accident," and "the trouble that it must needs procure to Your Sacred Majesty, as well as hazard to your health," that inspired him to write his proposal. He titled it, grandly, *Fumifugium: or, the Inconvenience of the Aer, and Smoake of London Dissipated.* ("Fumi-," from Latin *fumus,* smoke, and "fuge," from Latin *fuge,* to drive away: approximately, *Fumigation.*) To pique the king's interest, Evelyn claimed that the project would render the palace and the whole city "one of the sweetest and most delicious habitations in the world, and this with little or no expense."[30]

Evelyn defined "pure air" expressively as "that which is clear, open, sweetly ventilated, and put into motion with gentle gales and breezes; not too sharp, but of a temperate constitution."[31] London should enjoy such air, he observed: it was built on high ground, its gravel soil "plentifully and richly irrigated . . . with waters which crystallize her fountains in every street." The city sloped down to "a goodly and well-conditioned river" which carried off industrial wastes to be dissipated by the sun.[32] He blamed home coal burning less for London's air pollution than coal burning in trade. The problem wasn't "culinary fires," he argued shrewdly. No, the truly destructive smoke came from the works of the "brewers, dyers, lime-burners, salt and soap-boilers, and some other private trades"—the same nuisances Londoners had decried all the way back to the Middle Ages. When they were belching coal smoke, "the City of London resembles the face rather of Mount Etna, the court of Vulcan, Stromboli, or the suburbs of Hell." Their pernicious smoke induced "a sooty crust or fur upon all that it lights, spoiling the movables, tarnishing the Plate, Gildings, and Furniture, and corroding the very Iron bars and hardest Stones with those piercing and acrimonious Spirits which accompany its Sulfur."[33]

Coal-smoke pollution not only damaged London's built environment, Evelyn insisted, but it also sickened and killed her citizens, "executing more in one year than exposed to the pure Air of the Country it could effect in some hundreds." People who moved to London found "a universal alteration in their Bodies, which are either dried up or inflamed, the humours being exasperated and made apt to putrefy, their sensories and perspiration . . . exceedingly stopp'd, with the loss of Appetite, and a kind of general stupefaction." Yet these same visitors were quickly restored to health when they returned home, evidence that it was London's pollution that sickened them. Evelyn added for good measure, "How frequently do we hear men say (speaking of some deceased neighbor or friend), 'He went up to London, and took a great cold . . . which he could never afterwards claw off again.' "[34]

How could an enlarging, increasingly industrial city—a city on the cusp of the industrial revolution—be purified? The first step, Evelyn argued, was to clear London of the polluters: Parliament should require them to remove five or six miles down the Thames below the Isle of Dogs, a square mile of reclaimed marshland around which the river made a winding, pear-shaped meander that might block their smoke.[35] Evelyn knew of it because in 1629 the several commissioners of sewers in London, he among them, had been assigned responsibility for its upkeep.

Siting coal-burning industry there, like siting factories in suburban industrial parks today, would help clear London's smoke-fouled air. It would also, Evelyn added, give employment to "thousands of able Watermen" delivering the products of industry upriver into the city, would free up "Places and Houses" within the city for conversion into "Tenements, and some of them into Noble Houses for use and pleasure" with attractive river views. (Urban renewal and gentrification have ancient antecedents.) Moving industry to the suburbs would help prevent fires as well, Evelyn concluded. He thought accidental fires originated in "places where such great and exorbitant Fires are perpetually kept going."[36] London in the year of *Fumifugium*'s first publication, 1661, was indeed only five years away from her Great Fire of 1666, which burned out all the city within the old medieval walls. That fire, however, started in a bakery.

Moving coal-burning industry out of London was only the first part of Evelyn's remedy for smoke pollution. The second reflected his experience designing gardens. He proposed that all the low grounds surrounding the city should be converted into fields planted with fragrant flowers and shrubs, including sweetbriar, honeysuckle, jasmine, roses, Spanish broom, bay, juniper, and lavender, "but above all, Rosemary," which was reputed to cast its scent a hundred miles out to sea.[37]

He would fill the spaces between the fields around the city with flowers as well, and with "Plots of Beans, Pease" but "not Cabbages, whose rotten and perishing stalks have a very noisome and unhealthy smell." Blossom-bearing grains would "send forth their virtue" and be marketable in London; "amputations and prunings" might be burned at appropriate times in the winter "to visit the City with a more benign smoke."[38]

But Evelyn's vision was not to be fulfilled. Charles II discussed it with its author on the royal yacht, the *Catherine*, during a yacht race on the Thames, telling Evelyn he was "resolved to have something done on it" and asking him to prepare a bill for Parliament. Evelyn did, but no action followed. The king was too busy selling monopolies to restore his fortunes to invest in rearranging his smoky capital.

The Royal Society of London had been founded in November 1660—Evelyn was a charter member—and honored the horticulturalist's work in 1662 by inviting him to write a report on the state of the kingdom's timber. The Royal Navy had requested it, anxious about the increasing scarcity of large trees for building and maintaining its ships. Published in February 1664, the report was to be Evelyn's best-known work: *Sylva: Or, a Discourse of Forest-Trees and the Propagation of Timber in his Majesty's Dominions.* It was the Royal Society's first published book.

For decades to come, the English would burn coal primarily for home heating. The new fuel had still to be adapted to perform useful work. Burning it at home was straightforward; adapting it to industrial production, challenging and complex. Homes needed only a hearth with a chimney. Industry needed changes in coal's very chemistry. In the meantime, increasing demand soon exhausted the superficial outcroppings of sea coal. Coal had been dug in pits open directly to the sky. Now it began to be excavated through tunnels from deepening mines. Digging deeper underground soon penetrated the water table. Some mines could be cleared with drains, but mines too deep for drainage filled up and had to be abandoned. Simple technologies had served to ease the transition from wood to coal as the English forests depleted. Coal made new demands. It would reward those who worked out how to meet them.

TWO

RAISING WATER BY FIRE

To dig coal, colliers had to find a coal seam. Mineral coal, the compressed and carbonized remains of ancient plants, lay in underground beds across much of the British Isles, densely in the English Midlands, most densely in the northeast around Newcastle upon Tyne. When a coal seam outcropped on a headland or a hillside, it could be dug out directly, but such accessible exposures soon depleted. Shallow seams were the next to be exploited, easily found and easily taken by trenching or skimming off the overburden of soil or by digging multiple bell-shaped pits.

As coal replaced wood in domestic heating and in industry, and as the British population grew, colliers sought deeper seams. A coal seam in Britain might range in thickness from a few inches up to a rare thirty feet. It might lay underground anywhere from a fathom or two—six to twelve feet—to eight hundred feet or more. It might run parallel to the surface or slant upward or downward. Water might flow through it or through porous strata above or below it. Often it harbored pockets or channels of noxious or explosive gas.

Exploring involved either sinking or boring, often both. Sinking meant digging a mine shaft six feet across with pick and shovel, with a windlass above ground to draw up the waste under a canvas to keep out the rain. Underground water and quicksand challenged the colliers, who might line the shaft with timbers sealed with earth or clay or packed with unshorn sheepskins.

Stone was a harder challenge. Stone required boring, which involved chiseling a three-inch hole through the earth using a chisel attached to the end of a string of wrought iron rods. A springing pole served as a lever: one end embedded in the ground and braced with a heavy stone, a forked fulcrum supporting the trunk halfway, and the upper end free to lift and drop the chisel string with the help of a collier's strong leg working a stirrup.

After each drop, the colliers rotated the chisel a quarter turn to make the hole round. Every six inches or so, the string had to be pulled to resharpen the chisel and check for traces of coal—an increasingly laborious chore as the hole deepened. If pulverized rock ("wreck," colliers called it[1]) clogged the hole, the men pulled the string and replaced the chisel with a screw-threaded auger to clear it. Through hard rock, a yard a day was considered good progress. Finding deep coal seams might require a year or more, the colliers paid at higher rates as the drill string lengthened and the work got harder.[2] If the borehole found a coal seam, it had to be enlarged to a mine shaft with pick and shovel.

A seventeenth-century record of a boring in Yorkshire reports the findings layer by layer: "in Earth 1 Yard, in yellow Clay 1 yard, in black Slate 1 Quarter

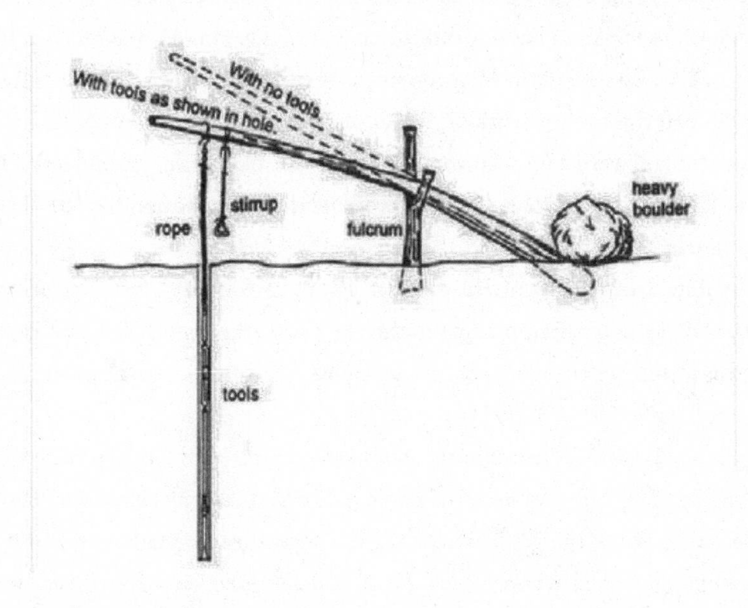

[that is, nine inches, or a fourth of a yard], in grey metal Stone two yards and two quarters, in black metal [stone] 2 quarters, in grey Stone 2 yards, in a Whinstone [a hard, dark-colored rock such as basalt] 1 qua[rter], in grey metal [stone] 2 qua[rters], in a Whinstone a Foot, in grey Metal [stone] a foot, in Iron-stone 6 Inches"—and on down through successive layers until the boring finally reached a coal seam a foot thick. "In all," the record concludes, "21 Fathom"—126 laborious feet of hammering a chisel down through dirt and rock.[3]

Once a mine was opened, it had to be kept dry. One Victorian expert calls water—from rain draining into the mine shaft and from underground flows—"the miner's first great enemy."[4] If nearby land sloped below a mine's working level, water could be drained by digging a narrow tunnel called an adit (from Latin *aditus,* entrance), which carried the water out to a natural drainage. Adits also delivered fresh air into a mine. In mines with gas pockets, such natural ventilation was controlled by a system of wooden doors. Since adits were typically no more than eighteen by eighteen inches square, children manned the doors, sitting in pitch darkness for up to twelve hours a day—saving the cost of a day's worth of candles or lamp oil. Until Parliament's reforming Mines Act of 1842, which prohibited women and children under ten from working the mines, whole families labored underground: the men hacking at the coal face with picks; the women hauling out the coal in wicker corves (baskets) on their backs or harnessed to iron or wooden tubs with belt and chain; the children helping haul the coal or working the doors. Families had to supply their own equipment and were paid according to the volume of coal they produced. Later, and in larger mines, ponies stabled permanently underground hauled out the coal in carts.

An illiterate seventeen-year-old girl, Patience Kershaw, testified before a parliamentary commission as late as 1841 about the conditions she experienced as a "hurrier" moving corves of coal from the pit face to the mine shaft:

> I go to a pit at five o'clock in the morning and come out at five in the evening; I get my breakfast of porridge and milk first; I take my dinner with me, a cake, and eat it as I go; I do not stop or rest any time for the purpose; I get nothing else until I get home, and then have potatoes and meat, not every day

> meat. I hurry in the clothes I have now got on, trousers and ragged jacket; the bald place upon my head is made by thrusting the corves; my legs have never swelled, but [my] sisters' did when they went to mill; I hurry the corves a mile and more underground and back; they weigh three hundredweight; I hurry eleven a day; I wear a belt and chains at the workings, to get the corves out; the getters sometimes beat me, if I am not quick enough, with their hands; they strike me upon my back; the boys take liberties with me; sometimes they pull me about; I am the only girl in the pit; there are about twenty boys and fifteen men; all the men are naked [to endure the heat and humidity]; I would 'ather work in mill than in coal pit.[5]

The gases in a coal mine could kill. Miners called them damps, from Middle Low German *dampf,* vapors. German miners first brought their skills and their terminology to England in medieval times. Damps formed underground from natural chemical and biochemical processes. Miners identified five kinds: suffocating chokedamp (mixed nitrogen and carbon dioxide); explosive firedamp (methane); explosive and suffocating stinkdamp, with a smell like rotten eggs (hydrogen sulfide); suffocating whitedamp (carbon monoxide); and suffocating afterdamp (a mixture of gases: carbon monoxide, carbon dioxide, nitrogen, and other products of explosions of firedamp or coal dust).[6] As mines lengthened and deepened, natural air circulation no longer sufficed to clear them. One solution was to maintain a fire at the bottom of the central mine shaft—the eye, as such shaft openings were called—which would draw air through the mine and out the eye like a chimney. But explosions were common and sometimes gruesome.

"The phenomenon of people being shot out of pits," writes the Victorian mining engineer Robert Galloway, ". . . was a frequent, indeed almost regular concomitant of early colliery explosions of any magnitude."[7] One of the more spectacular occurred in 1675 in Mostyn, Wales, on the Dee River, southeast of Liverpool. When the mine opened in 1640, the miners worked out a system for suppressing the firedamp at the beginning of each workday by sending one of their number ahead with a cluster of lit candles mounted on the end of a long pole to fire the night's accumulation. They called him the fireman. He wore old sackcloth overclothes, water-soaked for protection. "As the flame ran along the roof," Galloway writes, "the fireman lay flat on the floor of the mine till it passed over him."[8] Ventilation prevented methane from accumulating during the day, and the next morning the fireman repeated his risky detonation.

By 1675, the Mostyn mine had been worked for more than three decades. Then the owners decided to sink a pit into a parallel coal seam lower down. This fifty-foot blind pit filled with firedamp. Firing it, Galloway reports, produced an "alarmingly violent" explosion.[9] Worse was yet to come.

After a three-day work stay, a steward descended to the mouth of the pit to devise a way to move enough air to clear the pit of gas. He took two miners with him. The others who had dug the new pit followed. "One of them," says a contemporary account, "more indiscreet than the rest, went headlong with his candle over the eye of the damp pit, at which the damp immediately catched, and flowed over all the hollows of the work, with a great wind and a continual fire, and a prodigious roaring noise." The miners dove for cover in the loose slack on the floor or dodged behind one of the posts that shored up the roof. The blast roared out to the ends of the mine, reflected and roared back: "It came up with incredible force, the wind and fire tore most of their clothes off their backs, and singed what was left, burning their hair, faces, and hands, the blast falling so sharp on their skin as if they had been whipped with rods." Miners who hadn't found cover were blown through the mine tunnel and bashed against the roof or wrapped around posts and knocked senseless.[10]

One miner was standing near the eye of the upper shaft when the blast caught him. It carried him along as it roared up through the shaft, bursting from the eye with a crack like cannon fire, flinging the miner's body well above the treetops. The unlucky man had been fired from the mine shaft like a cannonball.

The hardest challenge of early coal mining was drainage. Rainwater flows through rills and streams into brooks and brooks into rivers, drawn always downward by gravity to the sea. About a third of any rainfall soaks into the soil and percolates downward into the earth. Eventually it encounters impermeable layers of rock. There it spreads out and flows along the rock layer until it finds cracks or permeable rock, when it continues percolating down to the next impermeable layer. Thus soaking, filtering, spreading, it saturates the permeable rock to form a subterranean lake: an aquifer. To create a water well, dig a hole far enough into the ground to penetrate below the surface of this aquifer; your hole will fill to the level of that surface—the water table—and refill as water is withdrawn.

Mines on high ground could be drained with adits, but as superficial coal seams depleted, owners opened deeper seams that extended below the water table. Then water had to be pumped out or the flooded mine abandoned. Many were, adding to the accumulating reward for finding a method of draining them and keeping them drained so that the coal could be wrought. Mine drainage was what Galloway called "the great engineering problem of the age."[11]

Windmills wouldn't do for pumping in the uncertain English weather. Waterwheels worked when there was sufficient water, but flows tended to be seasonal. Nor were many flooded mines located near streams of adequate volume. Mine owners turned first to horses harnessed to gins: raised horizontal drums large as waterwheels, which the animals worked by walking in circles, the rotary power winding up and unwinding a strong rope that turned through a pulley down a mine shaft.

Horse gins hauled water up the mine shaft in buckets. They hauled corves of coal as well. Galloway says the system was both limited and expensive: horses had to be bought or bred and raised, fed, and maintained. "In some instances, as many as fifty horses were employed in raising water at a single colliery"—at an expense, Galloway estimates, of not less than £900 a year (today £113,600, or $169,000). Deeper mines, impossible to drain with horsepower alone, had to be abandoned. Drowned mines, lost capital, lost work opened a space for invention.

Horse gin.

Discoveries in science prepared the way. That the atmosphere has weight had been known since the 1643 experiments of Galileo's protégé Evangelista Torricelli. Torricelli's experiments led to the invention of the mercury-column barometer, which responds to changes in air pressure—that is, to changes in the density of the column of air above the instrument. The Prussian engineer Otto von Guericke demonstrated the force of the atmosphere in 1654 in a famous public exhibition before Emperor Ferdinand III at Regensburg. Von Guericke pumped out the air in two copper hemispheres and mounted the resulting evacuated sphere between eight teams of horses. Only atmospheric pressure held the two hemispheres together, but the straining teams of horses could not pull them apart.

A friend of Von Guericke's, the Jesuit mathematician Kaspar Schott, added a report of the event (and a vivid engraving depicting the scene) to a book he published in 1657. In England, the wealthy Irish natural philosopher* Robert Boyle, a duke's son, read of Von Guericke's experiments and demonstration just as Boyle was trying to work out how to make a vacuum on a larger scale than within the narrow glass tube of Torricelli's barometer.[12] Boyle was impressed with Von Guericke's demonstration, less so with his laboratory vacuum system. Von Guericke made a vacuum in the laboratory by pumping the air from a jar inverted in a bowl of water. Boyle wanted to experiment with a vacuum—to

*Scientists were called natural philosophers until 1833, when William Whewell, the master of Trinity College, Cambridge, coined their modern name.

Von Guericke's demonstration of air pressure against a vacuum.

see, for example, what happened to a burning candle enclosed in a vacuum jar as the air was pumped out—and that wasn't something he could do with a chamber that had to be accessed underwater.

Though he was living by then in Oxford, Boyle turned to Ralph Greatrex, a London instrument maker of reputation. Greatrex proved unable to construct a workable air pump. An Oxford don who lectured in chemistry pointed Boyle to the young but ingenious Robert Hooke, twenty-three years old in 1658 and the don's laboratory assistant. Boyle hired Hooke to help him, and after several unsuccessful attempts using other people's designs, Hooke designed a vacuum pump that worked. It was a first-generation instrument, leaky and slow, but for Boyle it served to begin experimenting.

Boyle's pump and his subsequent vacuum experiments not only demonstrated that a vacuum could

Hooke and Boyle's first air pump. After withdrawing the stopper *K* at the top of the globe, inserting test materials through the opening, and reinserting the stopper, cranking down the plunger *C* in the cylinder *A* withdrew air from the globe. Closing the valve *L* kept the air from refilling the globe while cranking up the plunger. With the plunger fully inserted into the cylinder, opening valve *L* again allowed more air to be withdrawn from the globe, progressively improving the vacuum.

be created and studied and had distinctive properties (extinguishing candles, transmitting light but not sound). It also revealed the force of air pressure: the weight of the atmosphere above and surrounding us as we go about our lives. "There is a Spring, or Elastical power," Boyle wrote, "in the Air we live in."[13] The question then became how to harness such a powerful force at larger scale, outside the laboratory.

Men had been experimenting with using heat to make partial vacuums since at least the beginning of the seventeenth century. A Dutchman, Cornelius Drebbel, invented a simple mechanism for applying fire to draw water in 1604, one he later illustrated in a book.

Drebbel described hanging a retort—a gourd-shaped metal container—over a fire with its mouth submerged in a bucket of water. As the fire heated the retort, the air inside would expand and bubble out through the water. Withdraw the fire, and the air remaining in the retort would cool, contracting and forming a partial vacuum. Ambient air pressure would then drive water from the bucket into the underwater mouth of the retort. Drebbel's simple pump had potential. Enlarged and engineered further, it might draw water from a river, for example, to supply a community.

Drebbel, "a very light-haired and handsome man," according to one courtier who met him, "and of very gentle manners,"[14] produced other inventions as well: from fountain mechanisms to a barometric "perpetual-motion" display popular with royalty.[15] In 1605 he traveled to London as a tutor to Henry Frederick, Prince of Wales, James I's eldest son. Word of his inventive gifts spread across Europe, bringing Continental nobles to London to observe him at work. When the Holy Roman Emperor, Rudolf II, invited Drebbel to Prague, though he might have preferred to remain in England, he had little choice but to accept the invitation. Rudolf's death in 1612

Drebbel's simple pump.

liberated him. Unfortunately, the Prince of Wales died that year as well, at eighteen, of typhoid fever. Drebbel finally returned to James I's service in England in 1613.

There were those who laughed at King James for sponsoring Drebbel, "saying that this everlasting inventor has never achieved anything the cost of which has been covered by its usefulness." One who defended the Dutch inventor was a young Dutch diplomat and poet named Constantyn Huygens, who first met Drebbel in London in 1621. Huygens ranked Drebbel with the great English genius Francis Bacon. "By the aid of penetrating knowledge," Huygens praised him, "he has contributed remarkable mechanical instruments."[16]

The most remarkable may have been Drebbel's submarine, the first of its kind, an elongated diving bell that he demonstrated on the Thames to the Royal Navy in 1620. He had a rowboat with its bottom knocked out fitted with a domed wooden deck, the oarlocks and rudder sealed with leather gaskets and the entire boat covered with waterproofed leather. It could remain underwater for hours at a time, and there is reason to believe Drebbel knew how to generate oxygen chemically from saltpeter—potassium nitrate—to refresh the boat's air supply. (Nitrate is a compound of nitrogen and oxygen.)[17]

Later in the 1620s, Drebbel made mines and rockets for the Royal Navy, which was attempting to relieve the Protestant Huguenots, whom the French

Drebbel's submarine on the Thames, 1620.

were besieging at La Rochelle.[18] Huygens's son Christiaan, born at the end of that decade, would become one of the great natural philosophers of the seventeenth century. Drebbel died in 1633, but through his friendship with Christiaan's father, the inventive Dutchman influenced the boy's development.

Christiaan Huygens came to attention as a mathematician and astronomer. After studying law and mathematics at the University of Leiden, he published his first book, on the branch of mathematics called quadrature—finding the area of a geometric figure such as a circle—in 1651, when he was twenty-two. In the 1650s, Huygens learned to grind lenses and invented the first compound telescope eyepiece. In 1656 he correctly identified as a ring system the "ears" that other astronomers had seen jutting from Saturn. That year, he invented the pendulum clock as well.

These and other contributions prepared this brilliant and inventive young man to be selected as the first director of a new French academy of sciences, a project of King Louis XIV's finance minister, Jean-Baptiste Colbert, modeled on Britain's Royal Society. Colbert hoped that such an institution, established in 1666, might generate knowledge that could be turned into industry to increase the king's revenues. Huygens summarized his project for the newly appointed members of the academy:

> There is no better topic for research, and nothing is more useful to know, than the origins of weight, warmth, cold, magnetism, light, colors, the compounds of air, water, fire and all established matter, the breathing of animals, the development of metals, stones, and plants, all matters that man knows little or nothing of.[19]

Among the practical technologies Huygens thought worth pursuing, he included two possible methods of generating motive force: "Research into the power of gunpowder, of which a small portion is enclosed in a very thick iron or copper case. Research also into the power of water converted by fire into steam."[20]

Huygens pursued his gunpowder project in 1672, when the twenty-six-year-old German polymath Gottfried Leibniz arrived in Paris to ask Huygens to help him improve his knowledge of mathematics. Huygens agreed and set Leibniz

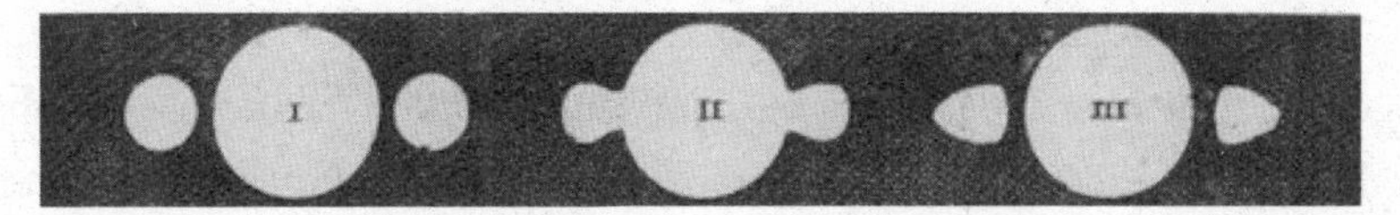

Early versions of Saturn: (1) Galileo, 1610; (2) Christoph Scheiner, 1614; (3) Giovanni Battista Riccioli, 1641. Huygens published these and others in the 1659 book where he correctly identified the "ears" as rings.

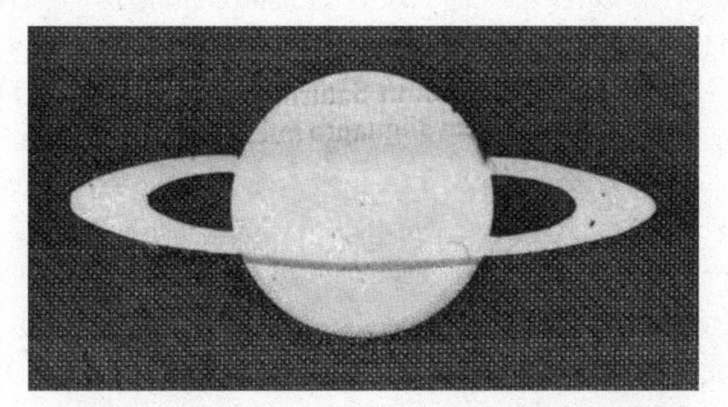

Huygens's version, in the same 1659 book, *Systema Saturnium.*

to work studying quadrature and calculating the value of pi. The other partner in what would be Huygens's gunpowder-engine adventure was Denis Papin, a physician a year younger than Leibniz who had forsaken medicine for engineering. Huygens met Papin in 1671 at Versailles, Louis XIV's great palace chateau twelve miles southwest of Paris, where the young engineer managed the windmill pumping system that powered the fountains in the extensive palace gardens. Papin's work so impressed Huygens that he hired him as his assistant.

The task Huygens set his two protégés in 1672 was to develop an engine powered by gunpowder. This curiosity was evidently conceived and perhaps prototyped by Caspar Kalthof, another Dutch engineer and gunmaker. Kalthof had worked for many years for the English Crown at Vauxhall, an experimental ordnance works in the London borough of Lambeth, the shop where Drebbel had developed his submarine. Huygens had met Kalthof on a visit to London and come away with some idea of how a gunpowder engine might work. Kalthof's death in 1667 or 1668 left the project open to whomever might choose to pursue it.

Leibniz, who had corresponded with Von Guericke, had already written a report on the Prussian engineer's vacuum demonstration for another member of the French academy.[21] The gunpowder engine that he and Papin now constructed for Huygens represented another way to turn atmospheric pressure into mechanical work: a small charge of gunpowder exploded below a piston inside a thick-walled metal cylinder that blew some of the air out of the cylinder through flap valves, creating a partial vacuum. Outside air pressure on the open-ended piston then drove it farther into the cylinder. If the piston was attached to a rod or a cable, objects connected to those extensions would move.

Huygens demonstrated a model of the engine to Colbert. It had already lifted "four or five footmen . . . with ease," he said, the footmen presumably standing on a platform connected to the piston cable.[22] Huygens imagined that his gunpowder engine "could be applied to raise great stones for building, to erect obelisks, to raise water for fountains, or to work mills to grind grain." The Dutch engineer predicted "new kinds of vehicles on land and water" and even "some vehicle to move through the air."[23]

But the gunpowder engine didn't work well. Not all the gases from the explosion left the cylinder, limiting the vacuum; the powder residue excoriated the cylinder walls; and, as designed, the explosions were one-off, requiring the piston to be withdrawn to insert another powder charge. No mills would grind grain under its power, nor would much water be raised.

Huygens moved on to inventing the mainspring-driven pocket watch and, a few years later, postulating the finite velocity of light. Leibniz crossed to London, where, despite election to the Royal Society, he continued his frustrating search for a secure position that might allow him time to work on philosophy. Papin, the physician-turned-engineer, recognizing the increasing risk of living as a Huguenot in Catholic France, moved to London in 1675. A letter of endorsement from Huygens recommended him to Robert Boyle, who had lost Hooke's services to London's Gresham College and to the Royal Society. Never a hands-on experimenter, Boyle hired Papin as his laboratory assistant.

In London, Papin applied his growing experience with steam to inventing a device for rendering palatable tough vegetables, tougher meat, and even bones: the pressure cooker. He called it his "New Digester for softening bones, etc."[24] He demonstrated it to the Royal Society in 1679. This innovative kitchen appliance might seem far removed from development toward a steam engine, but it incorporated a crucial feature that would be required later to make such engines safe: a self-regulating safety valve. Much like a modern pressure cooker safety valve, Papin's valve placed a levered weight over a small pipe that exited the lid of the cooker; steam from the cooker sufficient to lift the weight, release some of the steam, and reduce the internal pressure protected the machine from exploding.

Papin moved to Venice in 1681 to work as director of experiments at a new scientific academy that Venice's ambassador to England, Ambrose Sarotti, organized there in emulation of the Royal Society. He returned to England as temporary curator of experiments to the Royal Society in 1684 for a modest £30 annual salary (today £4,000, or $6,000), evidently hoping to be appointed secretary of the society. Demonstrators were not accorded scientific authority in eighteenth-century England, however, no matter how exceptional their gifts—they were closer in standing to servants, who were expected to represent their masters' opinions, right or wrong.[25] The appointment went instead to the astronomer Edmond Halley, and in 1687 Papin crossed Europe again to a mathematics professorship among fellow Huguenots at the University of Marburg, in Hessen.

At Marburg, Papin continued his experiments. In the late 1680s, observing that water occupies more than a thousand times its previous volume when it turns to steam, he decided that steam rather than gunpowder was the better working medium for his engine.

"Since it is a property of water," he wrote in 1690, "that a small quantity of it, converted into steam by the force of heat, has an elastic force like that of the air"—that is, expands and pushes against the walls of its container—"but, when cold supervenes, is again resolved into

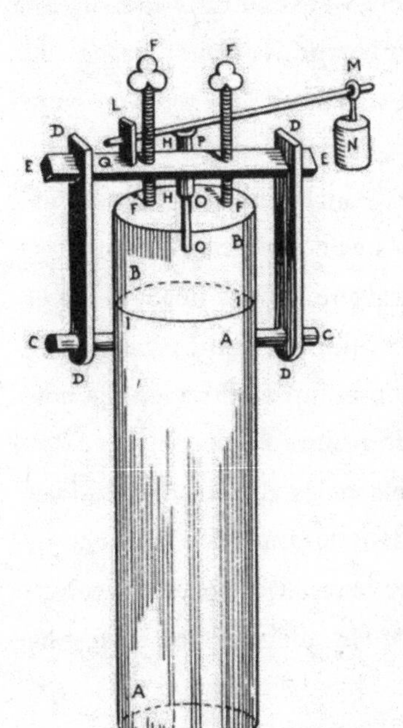

Papin's 1679 digester, with weighted safety valve *L-M-N.*

water, so that no trace of the said elastic force remains; I felt confident that machines might be constructed, wherein water, by means of no very intense heat, and at small cost, might produce that perfect vacuum which had failed to be obtained by aid of gunpowder."[26]

Papin was proposing to use steam to fill the cylinder, pushing up the piston to its highest position, and then hold the piston there with a latch while the cylinder cooled and the steam condensed back into water, losing most of the volume it had previously occupied. If he kept the cylinder sealed off from the air, the condensing steam would leave a vacuum in its place. Release the latch, and the full force of the atmosphere would push down the piston to fill the vacuum, pulling along whatever might be attached to it. Several such units operating together, like the cylinders in a modern automobile engine strung along a crankshaft, might produce a steady output of power.

Papin thought his "tubes," as he called them, could be "applied to draw water or ore from mines, to discharge iron bullets to a great distance, to propel ships against the wind, and to a multitude of other similar purposes." Of these possibilities, he was most interested in "moving vessels at sea. . . . My lightweight tubes would not slow down the ship; would take up little room; might also be readily manufactured in volume if a factory were built and fitted up for that purpose; and, lastly [unlike animal power or manpower], the tubes would consume no fuel except during operation; while in harbor, they would require no expenditure."[27]

The source of power in Papin's engine wasn't steam but the weight of the atmosphere acting on the vacuum the condensing steam left behind. So increasing the power of his engine required using a larger volume of steam in larger cylinders that could entrain a larger column of atmosphere. At the time, no one knew how to manufacture such large-scale machinery. Papin hoped his new engine might be a major inducement to its development.

While teaching at Marburg, Papin married his widowed cousin and added responsibility for her extended family to his burdens. He turned to Huygens for help in finding a better-paying position. Perhaps as a result, in 1695 he received appointment as a counselor to the landgrave* of Hesse-Kassel, the closest he

*In English, a count.

ever came to acquiring a noble patron. Unfortunately, Landgrave Moritz wasn't interested in financing an iron foundry or a factory for Papin's atmospheric engine. Instead, he wanted fountains in his gardens like those at Versailles.

For that project, Papin designed and had built a steam pump that moved water to an elevated storage tank from which it could flow by gravity to the landgrave's fountains. The system took a year to design and build. It worked, but only briefly: one of the pipes burst. Papin had another made. That burst too. Pipe construction wasn't yet up to the challenge of containing high-pressure steam.[28]

The landgrave then conceived a program of his own, Papin wrote Leibniz in April 1698, "a new plan, very worthy of a great Prince, to attempt to discover where the salt in salty springs comes from." To do that, he needed a way to draw out "a great quantity of water. . . . I've made many tests to try to usefully employ the force of fire for this task." He was building a new furnace to make large retorts of forged iron and had designed a new kind of bellows to blow the furnace fire. "And thus one thing leads to another," Papin concluded. He had to devise new infrastructure as he went along, that is, slowing and complicating each project.[29]

Leibniz responded immediately to Papin's letter, asking if his system for raising water was based on rarefaction, meaning condensing steam to make a vacuum. Papin replied that it was, but it *also used steam pressure directly.* "These [direct] effects are not bounded," he told Leibniz, "as is the case with suction."[30] Papin meant that his engine had two modes of action: (1) the pressure of expanding steam; and (2), rarefaction or suction—harnessing the power of atmospheric pressure to fill a partial vacuum. In the engine's direct-action phase, a small quantity of water was poured into a cylinder, a piston was inserted and pushed down until it contacted the water, a ported lid was screwed onto the cylinder, and a fire built under it. When the water turned to steam, it pushed up the piston, which a spring-loaded rod then pinned into place. Removing the fire and allowing the cylinder to cool caused the cooling steam inside to condense back into water, creating a vacuum where steam had been before. Removing the rod holding up the piston allowed the piston, in Papin's words, to be "pressed down by the whole weight of the atmosphere," forcing it down to fill the cylinder again.[31] With the piston connected to a crank, both the upward push of the steam and the downward push of the atmosphere could be applied to do useful work such as pumping water or turning a boat paddlewheel.

Papin understood that his move to using steam pressure directly was revolutionary. The condition of the roads in that era, he thought, probably foreclosed operating a steam carriage, "but in regard to travel by water, I would flatter myself that I could reach this goal quickly enough if I could find more support."[32]

Sadly, Papin was in no position to build even a model of his double-acting steam engine. The landgrave wasn't interested in investing in the project, and Papin had no money of his own. The best he could do, in 1695, was to publish a book about his inventions, *Collection of Various Letters Concerning Some New Machines.*[33] In that book, he described his Hessian bellows—a fan that rotated inside a casing, something like a large-scale version of a modern hair dryer without the heating element—and suggested its use, with air, replacing ordinary bellows for iron smelting or, with water, pumping fountains or putting out fires. More radically, Papin proposed using his steam engine to drain mines. His book was reviewed in a 1695 issue of the *Philosophical Transactions of the Royal Society of London,* so its basic ideas would have circulated at least among those members of the society who read the journal, and almost certainly more widely.[34]

Writing again to Leibniz in late 1698, Papin reported that he had been able to use steam pressure "to raise water up to 70 feet." That was a considerable achievement, since an atmospheric engine using steam only to create a vacuum was limited to raising water about 33 feet, the maximum lift that the pressure of the atmosphere—14.7 pounds per square inch at sea level—could produce. He discovered in the process that heating the steam above the boiling point greatly increased its power. Which meant, he told Leibniz, that steam was a better work agent than gunpowder.[35] Papin was right, as future developments would show, but the technology of the day, particularly the low melting temperature of the solder used to hold together the plates of steam boilers, wasn't adequate to allow the use of hotter, high-pressure steam: at higher pressure, solder softened, and steam boilers tended to blow apart.

In 1698 Papin's priority as an inventor and a pioneer came under challenge. While Papin was corresponding with Leibniz, an English engineer named Thomas Savery was patenting "A new invention for raising of water and occasioning motion to all sorts of mill work by the impellent force of fire, which will be of great use and advantage for drayning mines, serveing townes with water, and for the working of all sorts of mills where they have not the benefitt of water nor constant windes [for water- or windmills]." Like Papin's, Savery's engine combined both atmospheric and direct steam systems to pump water. The scale model he demonstrated before the Royal Society on 14 June 1699, impressed the members, but Savery, like Papin. would have difficulty making a full-sized engine work, and even more difficulty winning it a fair hearing.[36]

THREE

A GIANT WITH ONE IDEA

Denis Papin was an honest man. He might have claimed that Thomas Savery had stolen his idea for a double-acting steam engine, but he knew that ideas for raising water by fire were thick in the Enlightenment air. "I do not doubt that the same thought may have occurred to [Mr. Savery]," he responded to the news of Savery's 1698 patent, "as well as to others, without his having learnt it elsewhere."[1]

In 1704, however, when Leibnitz sent Papin a sketch of Savery's engine that he had acquired from contacts in London, Papin realized that it was grossly inefficient, perhaps even unworkable: Savery's engine, at one phase in its pumping cycle, used steam to blow cold water from a tank, with no intervening piston to prevent the cold water from condensing the steam. A much larger volume of steam would have to be injected into the tank to blow out the water.[2] That waste of energy reduced the engine's efficiency to less than 1 percent.[3] Nor was Savery's steam boiler protected with a safety valve, something Papin had invented and applied decades earlier when he developed his pressure cooker and now routinely included in his boiler designs.

Yet Papin's engine as well as Savery's suffered from design flaws that neither man found occasion to correct: both engines were designed to be controlled by hand. Their complicated sequences required a tireless operator to open and close their various valves several times a minute. Leibniz wrote to Papin as late as 1707 recommending he refine his design so that his engine's valves would be "alternately opened and closed *by the machine* without having to use a man for this purpose."[4]

But Papin was in no position to work on refinements. Falling out of favor with the landgrave, his Hessian patron, he had decided in 1706 to return to England. He carried with him plans for a steam-powered paddlewheel boat to present to the Royal Society, hoping for sponsorship. In London, unfortunately, his alignment with Leibniz worked against him. Isaac Newton had been elected president for life of the Royal Society in 1703. Both Leibniz and Newton had independently formulated the powerful mathematical system known as the calculus, and Newton and his followers were fighting a priority battle with Leibniz.

Papin described his steamboat to the Royal Society in February 1707. He asked the society to support building an eighty-ton prototype, which he estimated would cost £400 (today £57,000, or $84,000).[5] No such funds were forthcoming, nor was the society under Newton prepared to hire Papin again as a curator. It offered instead to pay him for any experimental demonstrations he gave, provided that he submitted his ideas in advance for approval. In 1708 the society leadership referred his steam engine plans to Savery, his primary competitor, for assessment. Unsurprisingly, Savery disparaged Papin's design. If Papin had dismissed Savery's grossly inefficient use of unshielded steam to pump cold water, Savery condemned Papin's cylinder and piston, claiming they wouldn't work "because the friction would be too great."[6]

Across the next four years, a desperate Denis Papin offered to demonstrate various inventions—a fuel-saving furnace; a method of purifying and heating room air—none of which the Royal Society cared to view or support. Down and out in Stuart London, he disappeared from history in 1712.[7]

Thomas Savery fared only a little better. "A small-scale Savery engine could be made to work most convincingly," writes the historian of technology Richard L. Hills. "No doubt, when Savery demonstrated his models, he stirred up great enthusiasm for his project."[8] Along with demonstrations, Savery published a

The Miner's Friend

promotional book called *The Miner's Friend; or, An Engine to Raise Water by Fire,* decorated with drawings of little workmen putti. [9]

But the larger the engine Savery constructed, the less efficiently it worked. It functioned adequately as an atmospheric engine, creating a partial vacuum by pouring cold water over a cylinder flushed with steam to draw water about twenty feet up a pipe. For lack of a safety valve, however, it could pump that water only a little higher by direct steam injection without risking blowing up its boiler. "The steam when too strong tore it all to pieces," one observer reported.[10] Savery's engines served as water pumps for the York Buildings water tower along the Thames in London as well as for a royal residence in Kensington, Queen Anne's Palace.[11]

"Savery vastly overestimated the capabilities of his engine," Galloway concludes, "and underrated the drawbacks to its use. He erected several which carried water very well for gentlemen's seats; but as an engine for draining mines, it proved an absolute failure."[12] Clearing a deep, drowned mine would have required five to ten of Savery's engines, shelved into the mine shaft every thirty feet, one above the other. Given their prodigious appetite for coal, ten such engines in a stack might have consumed most of the coal being mined, not to

York Buildings water tower.

mention exhausting the men required to operate them. Nor did mine owners welcome fire engines into their mines, given the risk of igniting any lingering methane. After 1705, having sold only two, Savery gave up trying to peddle his engines for draining mines.[13] He continued manufacturing them for town and estate waterworks.

When new technologies falter, reverting to earlier, more dependable systems can sometimes ease the transition, combining the old with the new. The earlier, commercially successful steam engine for mine drainage succeeded by retreating from such ambitious designs as those of Papin and Savery. If the craft skills of the day were inadequate to produce boilers capable of containing high-pressure steam, part of the answer to raising water by fire was to make use of steam only at atmospheric pressure, condensing it to create the partial vacuum of the atmospheric engine. Thomas Newcomen, a Devonshire ironmonger, pursued that path beginning around 1700.

Ironmongers in that era not only sold hardware but also crafted it—tools in particular. Making and selling tools carried Newcomen into the tin mines of Devon and Cornwall. Like England's coal mines, its tin mines were being extended deeper underground as more superficial veins of tin ore were exhausted.

By the turn of the eighteenth century, flooded mines had become a serious problem, and pumping them dry with horse sweeps was expensive. The anonymous author F. C., a miner, estimated in his *The Compleat Collier,* published in 1708, that "dry collieries would save several thousand pounds per annum which is expended in drawing water hereabouts."[14] Newcomen was responding to a lucrative opportunity.

Thomas Newcomen was a descendant of impoverished nobility in southwest England, born in Dartmouth in early 1663 and probably apprenticed to an Exeter ironmonger to learn his craft and trade. He finished his apprenticeship and returned to Dartmouth around 1685, when he was twenty-two, and established himself in business. A devout Baptist, Newcomen married late, at forty-one. In 1707 he leased a large house in Dartmouth for his family, which the Baptist congregation he led also used as a place of worship. A fellow Baptist, John Calley, had partnered with him at some earlier time and shared the work of invention.[15]

Not much is known about how Newcomen developed the engine that would bear his name, nor even how aware he may have been of the inventions of Papin and Savery. His most credible witness is a Swedish engineer named Marten Triewald, a founder of the Royal Swedish Academy of Sciences, who worked in England from 1716 to 1726. Triewald helped build a Newcomen engine there, built another one in Sweden after he returned, and knew Newcomen personally.[16] He wrote in 1734 that the English ironmonger had invented his engine "without any knowledge whatever of the speculations of Captain Savery." Instead, Triewald argues, Newcomen saw opportunity in "the heavy cost of lifting water by means of horses."[17] That may be so, but it's unlikely that Newcomen undertook a project that would occupy more than ten years of his life without some knowledge of previous efforts at harnessing fire to the problem of clearing flooded mines. Who except a fool reinvents the wheel?

Newcomen's engine borrowed the best features of its predecessors and incorporated new features of its own. It borrowed Huygens's cylinder and piston but followed Papin in substituting steam for gunpowder. It borrowed from Savery the idea of condensing steam to make a vacuum. The Newcomen engine, however, unlike Papin's or Savery's, heated water to steam in a large, separate boiler, then piped the steam through a flap valve up to an open-ended cylinder mounted overhead. Instead of using steam pressure to push up the piston, as Papin had, Newcomen hung the piston from a massive wooden rocking beam so that the weight of the beam as it rocked pulled up the piston to open the cylinder between cycles. Newcomen initially jacketed his brass cylinder with a lead casing into which cold water could be poured to condense the steam and create the vacuum that allowed atmospheric pressure to push down the piston, pulling down the rocking beam along with it.

Newcomen's design unquestionably improved on its predecessors. The piston separated the steam from the water it was lifting, reducing the volume of steam required and thus saving on coal. Cooling the exterior of the cylinder by pouring cold water over it condensed the steam more quickly, allowing the engine to pump faster. Since the steam was used only to create a vacuum, the system could operate at atmospheric pressure. And because the capacity of atmospheric engines depends on increasing the working area of their pistons, the piston and cylinder could be built larger or smaller to match the expected load.[18]

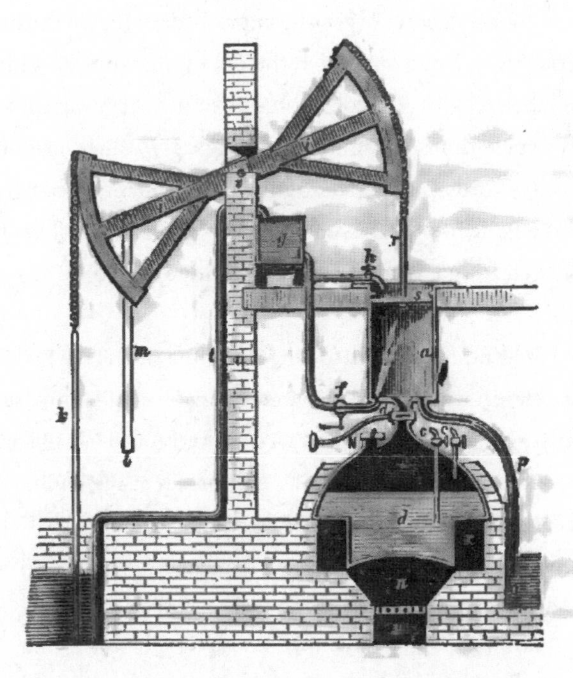

Boiler *d*, cylinder *a*, piston *s*, rocking beam *v*, pump rod *l*.

So far, so good. But pouring water over the cylinder to cool it from the outside was still relatively slow, limiting the power output. Newcomen's early design was much less efficient than it could be, though more efficient than Savery's.

A transformative breakthrough came by accident when Newcomen was still working with models and had not yet built a full-sized engine. According to Triewald, an "imperfection" in the brass cylinder—a hole that had been mended with solder—gave way, and the cold water pouring over the outside "rushed into the cylinder and immediately condensed the steam, creating such a vacuum that . . . the air . . . pressed with tremendous power on the piston, caused its chain to break and the piston to crush the bottom of the cylinder as well as the lid of the small boiler. The hot water which flowed everywhere thus convinced . . . the onlookers that they had discovered an incomparably powerful force which had hitherto been entirely unknown in nature—at least, no one had ever suspected that it could originate in this way."[19]

This accidental discovery of cold-water injection was the key to the success of the Newcomen engine. The inventor had to add a cold-water holding tank to his engine's machinery (*g* in the cutaway drawing on the preceding page, under the rocking beam, from which the cold water sprays inside the cylinder from a pipe labeled *f*).* With cold-water injection, a Newcomen engine could cycle about twelve times a minute, pumping up water from hundreds of feet below. Dorothy Wordsworth, sister of the poet William Wordsworth, observed a Newcomen engine pumping even more slowly on a tour of Scotland with her brother and their friend Samuel Taylor Coleridge in 1803:

> When we drew nearer, we saw, coming out of the side of the building, a large machine or lever, in appearance like a great forge-hammer, as we supposed for raising water out of the mines. It heaved upwards once in half a minute with a slow motion, and seemed to rest to take breath at the bottom, its motion being accompanied with a sound between a groan and a "jike." There would have been something in this object very striking in any place, as it was impossible not to invest the machine with some faculty of intellect; it seemed to have

*A second pipe, *h*, poured water onto the top of the piston to seal it from steam leakage.

> made the first step from brute matter to life and purpose, showing its progress by great power. William made a remark to this effect, and Coleridge observed that it was like a giant with one idea.[20]

Newcomen also had to devise a system of levers and valves to make his engine operate automatically rather than by hand, as Savery's did. Though these were crude by modern standards, requiring frequent adjustment, they worked well enough. With coal cheap at the pithead, Newcomen engines pumped water from British coal mines for more than two hundred years.

Unfortunately for Newcomen, Thomas Savery had written his 1698 patent so broadly that it covered all engines that raised water by fire, and Parliament in 1699 had extended the Savery patent for an additional twenty-one years beyond the original fourteen, to 1733. Having no other choice, Newcomen partnered with Savery, an arrangement that continued after Savery died in 1715 with a joint-stock company formed to exploit the Savery patent, the Proprietors of the Invention for Raising Water by Fire.[21] The proprietors issued eighty shares, of which Newcomen was awarded twenty.

Newcomen built his first full-scale commercial engine within sight of Dudley Castle, near Birmingham, in 1712. This Dudley Castle engine's cylinder, made of cast brass, was 21 inches in diameter and almost 8 feet long; it raised water from within a coal mine 153 feet below, and because it was built above the mine, at ground level, it risked no mine fires.[22] Other Newcomen engines followed across Britain. One behemoth with a 47-inch cylinder, built in Cornwall in 1720, raised water 360 feet.[23]

If draining a colliery with horsepower cost not less than £900 a year, a Newcomen engine did the same work at an annual cost of only £150, one-sixth as much.[24] With only marginally better efficiency than Savery's engine, however, and literally large as a house, the Newcomen engine was a transitional technology, limited almost entirely to pumping water from mines. "It takes an iron mine to build a Newcomen engine," contemporaries said, "and a coal mine to keep it going."[25]

Despite their drawbacks, Newcomens revitalized the mining industry in north-central England.[26] Between 1710 and 1733, when the patent expired, no fewer than 104 Newcomen engines were built in Britain and abroad.[27] Many more would follow—550 or more by 1800—but coal's industrial uses were still limited.[28] No one had yet devised a process for smelting good iron with coal; its primary market was still for home heating. As that market glutted, coal prices plummeted. "Between long adits, and wet pits, and hostmen's monopolies,"* Galloway writes of a slightly earlier period, "not to speak of the primitive methods of conveying coal both underground and on the surface then in vogue, the lessees of collieries seem to have been in quite a sorry plight."[29] The Newcomen engine would reduce the long adits and wet pits, but a use for the coal thus liberated and a more efficient means of transporting it remained to be found.

Early English roads were terrible. The Crown required landholders to maintain local roads at their own expense, one of three ancient obligations—to keep roads and bridges in repair, to build and maintain fortifications, and to serve in the militia—exacted to facilitate the kingdom's defense.[30] Roads for ordinary communication and commerce were effectively orphans. "That the ways, in winter, must be impassable for wheel traffic was habitually taken for granted," write historians Sidney and Beatrice Webb. They were hardly more than tracks until the middle of the eighteenth century, traversed on foot or on horseback when summer dust and winter mire allowed. "Coaches are not to be hired anywhere but at London," an early-seventeenth-century travel writer reports, "and although England is for the most part [a] plain, or consisting of little pleasant hills, yet the ways far from London are so dirty that hired coachmen do not ordinarily take any long journeys."[31]

The bulk of road traffic, the Webbs report, winter and summer, was animals on foot: a hundred thousand head of cattle and three quarters of a million sheep annually to Smithfield for fattening, vast droves of cattle to London for slaughter, legions of ducks, geese, and turkeys, numberless hogs. "For the further supplies of the markets of London," a writer noted in 1748, "they have within these years found it practicable to make the geese travel on foot too, and prodigious numbers are brought to London in like droves from the farthest parts of Norfolk." The flocks were large: one or two thousand birds driven together in a noisy, belligerent mass. "They begin to drive them generally in

*Hostmen were the protected middlemen whose river keelboats transferred coal from the riverbank to the coasting colliers that ferried it to London.

August, when the harvest is almost over, that the geese may feed on the stubble as they go. Thus they hold on to the end of October, when the roads begin to be too stiff"—sticky, viscous—"and deep for their broad feet and short legs to march in."[32] Moving animals on foot worked against road improvement; farmers wanted soft roadbeds for their herds and flocks, not hard surfaces that might lame them.

Coal moved locally in carts or in panniers on packhorses, but it was shipped by river and near shore to London. As mine owners devised methods of finding coal and then of draining flooded mines, output increased accordingly. Mines adjacent to rivers depleted, and new pits had to be opened farther inland. Two new problems arose: negotiating wayleave fees with landowners for crossing their property in a countryside of limited and primitive public roads; and moving the coal from the pithead to the river.

Wayleave was straightforward, if often costly. Francis North, Charles II's Lord Keeper of the Great Seal, found the wayleaves of Newcastle landowners "remarkable," his brother remembers: "For when men have pieces of ground

A packhorse convoy on an eighteenth-century English road.

between the colliery and the river, they sell leave to lead coals over their ground, and so dear that the owner of a rood of ground will expect £20 per annum for this leave."[33] A rood, an old English area measure, was only a quarter of an acre, or about a fifth of an American football field; in the midseventeenth century, £20 was the equivalent of about £2,500 today, or $3,700. A rood was 104 feet on a side; at 51 roods to a mile, the expense of wayleave could be prohibitive. A witness testified in Parliament as late as 1738, "There are fifty to sixty collieries unlet around Newcastle, caused partly by waterlogging [but partly by] prohibitive wayleaves."[34]

If wayleave could be negotiated, transport was the next challenge. Average cost per load across expensive wayleaves declined with volume. Packhorses and horse- or ox-carts served at first; in 1696 an estimated twenty thousand carts and cart horses moved coal from the collieries of the Tyne River and Wear River areas alone.[35] But as coal production increased, wagons supplanted them. In the second half of the seventeenth century, the Durham and Northumberland coalfields alone produced 1.2 million tons of coal annually, and the total for all of Britain was almost 3 million tons; packs and carts were inadequate to transport such quantities.[36] (To move that much coal today would require 260 one-hundred-car coal trains.) Britain, despite its roads, began shifting to wheeled transport early in the seventeenth century. The forty thousand Thames watermen serving London resented the competition; in 1623 the waterman poet John Taylor condemned the new "rattling, rowling, rumbling age" when "the World runnes on Wheeles."[37]

Larger mines with direct access to the surface had long been laid with wooden rails to make coal and ore carts easier to move; moving a cart on rails required about one-sixth the effort needed to haul a sled or a cart on a dirt path.[38] Moving coal to water on such rails—wagonways, they were called—would save money, time, and wear and tear.

The earliest known English wagonway dates from 1604. Huntingdon Beaumont, the son of a knight and an innovative coal entrepreneur, invented it or adapted it from the mining cart railing. Sir Percival Willoughby, Lord of the Manor of Wollaton, in Nottinghamshire in the East Midlands, was Beaumont's business partner in pit-mining coal. "News has reached me of Master Beaumont's efforts to move coal from Strelley to Wollaton Pits," Sir Percival wrote in

Mining cart on wooden rails.

1603. "His new invention will carry coal with wagons, with small wheels made from a single slice of oak, running on wooden rayles. I return home enlightened by this insight and possible cure for heavy loads; our roads are yet in unmade condition."[39]

In the next hundred years, wooden wagonways diffused across England. Sir Thomas Liddell laid one from his Ravensworth Colliery to the Wear River in 1671, the first section of what became an extensive system. Sir Humphrey Mackworth, a lawyer and early industrialist, commissioned a wagonway sometime after 1704 to move coal and copper ore from his estate in Neath, Wales, to the Neath River. Someone challenged Sir Humphrey's innovations; a 1706 legal document defends them with the claim that they were already in wide use: "These waggon-ways are very common, and frequently made use of about Newcastle, and also about Broseley, Bentall, and other places, in Shropshire, and are so far from being Nuisances, that they have ever been esteemed very useful to preserve the roads, which would be otherwise made very bad and deep by the carriage of coal in common waggons and carts."[40] Sir Humphrey pioneered horseless locomotion as well, outfitting his wagons with sails: "I believe he is the first gentleman in this part of the world," a countryman wrote of him, "that hath set up sailing engines on land, driven by the wind, not for any curiosity, but for real profit."[41]

Wagons that moved themselves would be even cheaper for hauling coal than horse-drawn wagons. Daniel Defoe, a prolific journalist as well as a novelist, described that development in his 1726 handbook *The Compleat English Tradesman.* "[Coals] are then loaded again into a great machine," Defoe wrote, "call'd a Waggon, which by the means of an artificial road, call'd a Waggon-way, goes with the help of but one horse, and carries two Chaldron or more [of coal] at a time."[42] (The weight of a chaldron of coal, 36 bushels, was fixed by law in 1678 at 5,880 pounds.)

Below the Tyne River in County Durham, the land slopes from west to east toward the sea, and from south to north into the Tyne Valley. Gravity would move a wagonload of coal down to the water, with a horse led along to haul the unloaded wagon back uphill. Flanges on the wagon wheels held the wagons on track so that they effectively steered themselves.

On longer slopes, Galloway reports, a wheeled platform attached to the coal wagon conveyed the horse downhill, "thus making the cart carry the horse"—putting the cart before the horse, that is. Horses adjusted quickly to the arrangement, he noticed, and seemed to enjoy the ride.[43]

But not all intervening landscapes sloped conveniently toward the nearest river. By 1725, the Liddell family had spent thousands of pounds improving

Coal wagon descending to river; note large brake lever. The horse hauled the wagon back uphill to be reloaded.

its wagonways. William Stukeley, an antiquarian and Royal Society fellow who later wrote an early study of Stonehenge, visited the Liddell works that year and inspected the wagonway improvements:

> We saw Col. Lyddal's coal-works at Tanfield, where he carries the road over valleys [he has] filled up with earth, 100 foot high, 300 foot broad at the bottom; other valleys as large have a stone bridge built across, and in other places hills are cut through for half a mile together; and in this manner a road is made and frames of timber laid, for five miles to the river side, where coals were delivered at 5*s*. the chaldron.[44]

Wooden rails broke down after a year or two of bearing such heavy loads and had to be replaced. Fortuitously, another technology developed during the same decades answered the rail problem: iron smelting with coal.

Iron had long been reduced from ore and purified with wood charcoal rather than mineral coal—pit coal—because the sulfur in coal embrittles iron. But the continuing scarcity of wood had begun to restrict the growth of British industry. At the same time, the market was glutted with coal because industrial coal use was limited to boiling operations such as salt boiling and cloth dyeing. As British industry expanded, there simply wasn't enough wood in all of Britain to meet the demand for iron.[45] Cracking the hard nut of smelting iron with coal had challenged British inventors and entrepreneurs for a hundred years.

A diligent Quaker entrepreneur named Abraham Darby, a farmer's son born in a village near Dudley, accomplished the long-sought breakthrough. Darby apprenticed with a malt-mill maker before establishing his own business in Bristol in 1700. (Malt, a key ingredient in beer making, is made by soaking barley in water until it sprouts, activating enzymes that convert the grain's starches into sugars. To interrupt the process, the sprouted grain is then kiln dried, and the resulting malt milled into powder to mix with unsprouted grain for fermentation.) In 1702 Darby also established a brass works in Bristol, where he pioneered smelting brass with coal. At about the same time, he opened a foundry there for casting iron pots in sand, a technological achievement for which he was awarded a patent in 1707.[46]

In 1709 Darby moved his foundry enterprise to Coalbrookdale, a village along the Severn River, about eighty miles north of Bristol, near Dudley. There he began developing a method of preparing coal for iron smelting by coking it—baking it in a kiln under low-oxygen conditions to drive out the sulfur and other impurities that would otherwise embrittle the iron. Writing about the invention later, his son's widow, Abiah Darby, would compare it to drying malt.[47]

As they mastered the technology, Darby and his descendants gradually substituted coke for charcoal. Smelting iron with coked coal then enabled British industry to bypass the bottleneck of wood scarcity, Abiah Darby noted in 1763: "Had not these discoveries been made, the Iron trade . . . would have dwindled away, for woods for charcoal became very scarce, and landed gentlemen [who owned the forests] rose the prices of Cord Wood exceeding high—indeed it would not have been to be got. But from pit coal being introduced in its stead, the demand for wood charcoal is much lessened and in a few years I apprehend will set the use of that article aside."[48] By the beginning of the nineteenth century, iron had largely replaced wood in manufacture and construction.[49]

Coalbrookdale cast the first iron wagonway wheels in 1729.[50] These damaged the wooden wagonway rails even worse than wheels of wood. Cast-iron plates soon followed, two inches wide and a half inch thick, nailed to the wooden rails to protect them.[51] Next came cast-iron rails, eventually a major Coalbrookdale product.[52] "Long lines of [wagonways] were now being constructed," Galloway writes, "with great care and cost, to collieries placed at greater distance from the river."[53]

In the meantime, the Coalbrookdale Company had begun casting iron cylinders for Newcomen engines to replace the more expensive brass cylinders of previous models. In the early 1740s, Coalbrookdale replaced its own horse-driven pumps with a Newcomen engine, the first time a steam engine was used to make iron, and a major reduction in expense.[54] From this point forward, mining advanced rapidly, as an increasing number of steam engines restored old mines previously drowned and kept new mines dry.[55]

Newcomen engines met the needs of coal mining, but away from the mines, their disadvantages limited their usefulness. They were too prodigal of coal, inefficient, and immovably bulky, and they could only, as Galloway writes, "perform this simple sea-saw motion of pumping."[56] If coal was to support Britain's accelerating industrialization, someone would have to invent a better engine.

FOUR

TO MAKE FOR ALL THE WORLD

James Watt, a Scotsman born near Glasgow in 1736, learned to craft mathematical and nautical instruments from a master instrument maker in London when he was twenty years old.[1] Watt's grandfather was a teacher of mathematics and navigation, his father a shipbuilder and chandler, his mother's cousin a professor of Latin at the University of Glasgow.[2] His twelve months' training in making "rules, scales, quadrants, etc.," as he listed them in a letter, cost his father 20 guineas (today £2,900, or $4,200).[3] A year in a London workshop hardly qualified as an apprenticeship, but Watt had grown up working with his father and knew how to handle tools. By the end of his training, he judged that he "should be able to get my bread anywhere as I am now able to work as well as most Journeymen tho I am not so quick as many."[4]

Watt found his first opportunity to get his bread in Glasgow in October 1756, when a shipment of astronomical instruments reached the university* from Jamaica. A wealthy Glasgow graduate and Jamaican merchant who had died the previous year had bequeathed them. Though they were rusted from

*Glasgow University was called the College in those days. It no longer is, and avoiding the anachronism here would be confusing.

the sea voyage and needed repair, the Astronomer Royal had exclaimed them worthy of "the Observatory of a Prince."[5] They would found a new observatory at Glasgow. Watt worked on them through the early winter. He finished in December and received £5 (today £700, or $1,000) for the work.[6] He spent the next six months at home in Greenock, downriver from Glasgow, acquiring tools and merchandise to establish himself as a maker, repairer, and seller of instruments. By July 1757, he had returned to Glasgow and set up shop within the university precincts. Laboratory instruments and demonstration models were not off-the-shelf purchases in the eighteenth century: they were made and maintained by hand. The university allowed Watt to take up residence and sell his goods as part of a larger service. It named him its "Mathematical-instruments maker" and paid him to keep its collections in repair.[7]

A recent graduate who visited Watt's shop on the university grounds, John Robison, found the young proprietor's scientific knowledge unsettling. "At first feasting my eyes with the view of fine instruments," Robison testified, "and prying into everything, I conversed with Mr. Watt. I saw a workman and expected no more—but was surprised to find a philosopher, as young as myself; and always ready to instruct me. I had the vanity to think myself a pretty good proficient in my favorite study and was rather mortified at finding Mr. Watt so much my superior." They became friends nonetheless. "I lounged much about him, and I doubt not, was frequently teasing him. Thus our acquaintance began."[8]

Nor was Robison the only Glaswegian who befriended Watt. "All the young lads of our little place that were any way remarkable for scientific predilection," Robison says, "were acquaintances of Mr. Watt, and his was a rendezvous for all of this description. Whenever any puzzle came in the way of any of us, we went to Mr. Watt."[9] So, it seems, did members of the faculty. Robison recalled finding Glasgow professor Joseph Black on Watt's premises. "Dr. Black used to come in, and, standing with his back to us, amuse himself with Bird's quadrant, whistling softly to himself, in a manner that thrilled me to the heart."[10] Bird's quadrant was an eight-foot brass quarter-circle astronomical instrument that mounted on the wall; John Bird, a fellow mathematical-instruments maker who worked in London, had made the original for the Royal Observatory at Greenwich in 1753.

Robison had taken his Glasgow master of arts degree in 1756 and then tried his hand at inventing. Watt credited Robison with turning his attention to the

Newcomen engine, "a machine of which I was then very ignorant, & suggested that it might be applied to giving motion to wheeled carriages, & that for that purpose it would be most convenient to place the cylinder with its open end downwards, to avoid the necessity of using a working beam." Watt evidently partnered with Robison in this early attempt to remove the tree-trunk-sized rocking beam from atop the Newcomen engine, liberating it for locomotion. He recalled building a model of Robison's engine out of tin plate with two inverted cylinders that would crank an axle to turn the carriage wheels. The hastily built model worked badly. Neither man had time to improve it nor yet what Watt called "any idea of the true principles of the machine." In 1759, when Robison enlisted in the Royal Navy to serve during the Seven Years' War between Prussia and Britain on the one hand, and Austria, France and Russia on the other, they set it aside.

Watt continued living on the Glasgow campus and operating his shop while learning and experimenting. If he had little knowledge of steam machinery in 1758, he soon had reason to acquire more: in 1760, Glasgow's newly appointed professor of natural philosophy, John Anderson, eight years Watt's senior, hired him to maintain the laboratory instruments and models that Anderson would be using to teach physics and chemistry.[11] (Anderson was an enthusiastic teacher; his students called him "Jolly Jack Phosphorus."[12]) He paid Watt five pounds five shillings in June of that year, probably as a retainer. The day prior to that payment order, another had been issued for £2 to cover the cost of picking up a model Newcomen engine from a London instrument maker, Jonathan Sisson.[13] Watt served Anderson not only by maintaining instruments and models but also as a demonstrator. He probably operated the Newcomen model for lecture hall demonstrations.

Many years later, Watt recalled that he "went on with some detached experiments on steam until 1763."[14] He said that differently in 1769, when he wrote that between 1761 and 1763, he "tried some experiments on the force of steam in a Papin's digester."[15] Watt described only one of his experiments in detail, but it's clear that he conducted a number of steam experiments during the period when he was working with Anderson.

The one experiment Watt described involved using a Papin's digester to construct what he calls a "species of steam engine." Denis Papin's 1679 pressure cooker survived his disappearance into anonymity in 1712 to become a standard laboratory device for experimenting with high-pressure steam. Unlike the riveted and soldered boilers of the day, the digester was sufficiently thick walled to hold together under pressures considerably higher than atmospheric. In Watt's experiment, the digester served as a boiler for generating and confining high-pressure steam. For a cylinder, he connected the digester to a pencil-sized metal syringe with a piston inside and with taps between syringe and digester to turn the steam on and off. By injecting high-pressure steam into this cylinder, Watt could raise a fifteen-pound stack of metal weights.[16] Opening one of the taps then allowed the steam to escape into the air and the cylinder to return to its starting position, lowering the weights again. "It was easy to see how [controlling the tap] could be done by the machine itself," Watt writes, "and . . . make it work with perfect regularity." He gave up the idea of building a full-sized engine based on the model, he says, because he understood that people would object to it for the same reasons they had objected to Savery's engine: "The danger of bursting the boiler, and the difficulty of making the joints tight, and also that a great part of the power of the steam would be lost" because the piston had to work against atmospheric pressure to return to its starting position.[17]

Robison's efforts to invent a steam-powered road vehicle had encouraged Watt to explore the possibilities of steam. He had almost immediately begun modeling a steam engine that worked with high pressure rather than indirectly by condensing steam to create a vacuum as the Newcomen engine did. High-pressure steam carried more energy per unit volume than did steam at atmospheric pressure, which meant a high-pressure engine could be smaller—small enough to mount on wheels. Though Watt never built one at full size, he would specify his "species of steam engine" in two patent applications several decades later, one of which included "a mode of applying it to the moving of wheel carriages."[18] The demands of his business forced him to set aside this early effort of invention, but by 1763, James Watt had learned enough about existing steam engines to understand something of how they worked.[19]

That winter, Anderson asked Watt to repair the model Newcomen engine that the university had purchased from Sisson in 1760. It had never worked very well, cycling a few times and then wheezing to a stop. Watt first set about repairing what he calls "the very bad construction of some of its parts." That done, he adjusted it to reduce the amount of cold water it injected into the cylinder. Once he had it working regularly, he studied its operation. Watt was surprised at the "immense" quantity of fuel it consumed in proportion to its size—its brass cylinder was only two inches in diameter but thick-walled.[20] Watt concluded that the cylinder was radiating away too much heat, overcooling the system and wasting energy.

At that point, John Robison returned to Glasgow from navy duty, and the two friends picked up where they had left off.[21] Discussing the faulty Newcomen model, Watt recalls, they concluded that if the cylinder were made of wood rather than metal, the device would lose much less heat. Watt had resolved by then "to improve the machine," he says—not only Anderson's model but also, after three years of informal study and experiment, the Newcomen engine itself. It was grossly inefficient, and there was wide scope for improvement. "I always thought the machine might be applied to other as valuable purposes as drawing water," Watt testifies in his notebook.[22] Robison had ventured one of those purposes: transportation. The numerous British iron foundries and woolen mills presently dependent upon unreliable water power might also operate better on steam.

In late 1763 Watt moved from the university premises into Glasgow, opening a larger store and workshop on Trongate, near the salt market, a few blocks north of the River Clyde and west of Glasgow Green.[23] He had taken partners by then, including his father, James Watt Senior, and his professorial friend Joseph Black. Their business had outgrown Watt's university rooms. By 1764, they employed some sixteen journeymen and clerks in an operation expanded to include retail merchandise, from telescopes, to musical instruments, to toys. Their gross sales that year would total about £600 (today £70,000, or $100,000).[24] Lodgings fit for a wife were another reason for the move. Watt had recently become engaged to a maternal cousin named Margaret Millar—Peggy—whom he had known since childhood. They would marry in July.[25]

Despite these preoccupying changes, Watt still found time to experiment with fire-engine improvements. Robison would testify to their extent:

> He greatly improved the boiler by increasing the surface to which the fire was applied; he made flues through the middle of the water; he placed the fire in

> the middle of the water; and made his boiler of wood. . . . He cased the cylinder, and all the conducting-pipes, in materials which conducted heat very slowly; he even made them of wood. After much acquaintance with his models (for he had now made others), he found that there was still a prodigious and unavoidable waste of steam and fuel, arising from the necessity of cooling the cylinder very low at every effective stroke; and he was able to show that more than three-fourths of the whole steam was thus condensed and wasted during the ascent of the piston. . . . This great cause of loss seemed to be unavoidable.[26]

More experiments then, into the summer of his marriage. Watt measured how much steam it took to heat a volume of cold water to boiling. To his surprise, he discovered that "water converted into steam can heat about six times its own weight of well-water to 212°. . . . Being struck with this remarkable fact, and not understanding the reason of it, I mentioned it to my friend Dr. Black, who then explained to me his doctrine of latent heat."[27]

Joseph Black, whose research field we would today call physical chemistry, was the discoverer of latent heat. It had been taken for granted for centuries that adding heat to a substance steadily increases its temperature. To the contrary, Black noted, boiling off a small quantity of water takes about six times as long as does bringing it to a boil in the first place.[28] In 1757 Black had begun measuring this curious lag. He found that heating a pot of water increased its temperature until the pot began to boil, at which point the boiling water's temperature remained the same—212°F—despite the continuing addition of heat, until all the water boiled away. Similarly, heating hard-frozen ice increased its temperature until it began to melt, at which point the temperature of the melting ice remained the same—32°F—despite continuing to add heat, until all the ice had melted into water. The heat that measurably increased the temperature, Black called "sensible heat." The heat that seemed to disappear into the boiling water or melting ice without changing its temperature, he named "latent heat."

What happened to the heat? Black experimented with these phenomena across the next five years. In 1762 he came to the conclusion, as he reported to the university Philosophical Club that April, that in the case of the ice, the heat goes to changing the ice to water; and in the case of boiling, to changing the water to steam.[29] (Today these changes are called changes of state. When heat is added to a material, the material either changes temperature or changes state—from ice to water, from water to steam, and generally from solid, to liquid, to gas—and the "missing" heat represents the work necessary to overcome the atomic forces resisting the change. Steam at atmospheric pressure occupies a volume some 1,700 times as great as the water it came from, for example: latent heat represents the energy—the work—necessary to drive such an expansion.)

Watt, with his after-hours experiments, had rediscovered latent heat: "I thus stumbled upon one of the material facts by which [Black's] beautiful theory is supported."[30] Watt took from the theory the information he needed: that water absorbed a great deal of heat in changing into steam and lost a great deal of heat changing back into water. If he wanted to make a more efficient steam engine, he reasoned, one that used less coal and therefore cost less to operate, then "it was necessary that the cylinder was always as hot as the steam that entered it, and that the steam should be cooled down below 100° (Fahrenheit) [when injected with cold water to condense it to make a vacuum] to exert its full powers."[31]

But these two requirements were mutually exclusive. How could he keep the cylinder hot while injecting it with cold water to condense the steam? He puzzled over that question for several months. On a Sunday afternoon in April 1765, the answer came to him as a classic stroke of insight.[32] Watt was walking in Glasgow Green, thinking about his problem, when it occurred to him that since steam expands to fill a vacuum, if he attached a separate, connected tank to the cylinder, pumped out its air and then opened the connection, the steam would rush into the separate tank—and that would empty the cylinder.[33]

The solution, that is, was not to cool the main cylinder to condense the steam. Doing so wasted the heat necessary to reheat the cylinder for the next cycle. Heat was coal; coal was money. Better to save it. "His mind ran upon making engines *cheap* as well as *good*," Watt would write in his third-person patent application.[34] The simplest way to save heat, the Scotsman had realized, was to move the condensation process out of the cylinder entirely and into a separate tank, which he called a separate condenser. Then the cylinder could be kept hot throughout its work cycle, saving energy. So: connect the cylinder and the smaller condenser through a pipe, with a valve to control the flow. When the steam from the boiler has filled the main cylinder, inject cold water *into the condenser only* and open the valve. The steam in the cylinder would rush into the condenser, condensing there and leaving behind a vacuum in the cylinder as well. The vacuum in the cylinder would then allow the pressure of the atmosphere to push up the piston, performing work.

But with a vacuum now in both cylinder and condenser, how were the cold injection water, the air that came in with it, and the condensed steam to be drained for the next cycle? Watt thought of two ways, one involving a long drain pipe sunk into the ground below the engine. But the method he chose was to pump out the dregs with an air pump. A long drainpipe would limit where the engine could be sited. A pump could go anywhere.

With great excitement, Watt built a model of his separate condenser: a tall, narrow tin cylinder with a smaller cylinder fitted inside for the pump.

He was sitting in his parlor examining this model one day in May 1765 when John Robison turned up. Presumably a servant had let Robison in; his entry surprised Watt. Robison began talking about steam. But Watt was too excited to contain himself. "You need not *fash* [trouble] yourself any more about that, man," he told Robison. "I have now made an engine that shall not waste a particle of steam. It shall all be boiling hot, aye, and cold water injected if I please." At which point, Watt noticed that Robison was staring at the model condenser; he kicked it under the table with his foot.[35]

Robison asked Watt about the tube he had just tried to hide. Watt answered unhelpfully. Robison was chagrined. He says he had recently "blabbed" to a competitor an idea Watt had shared with him, and word had found its way back to Watt. He slunk off, prepared to return to the country that evening. Robison walked to the nearby riverside to wait for his ride and encountered a mutual friend, Alexander Brown, out for a walk.

"Well," Brown asked him, "have you seen Jamie Watt?" Robison said he had. Brown said Watt must be in high

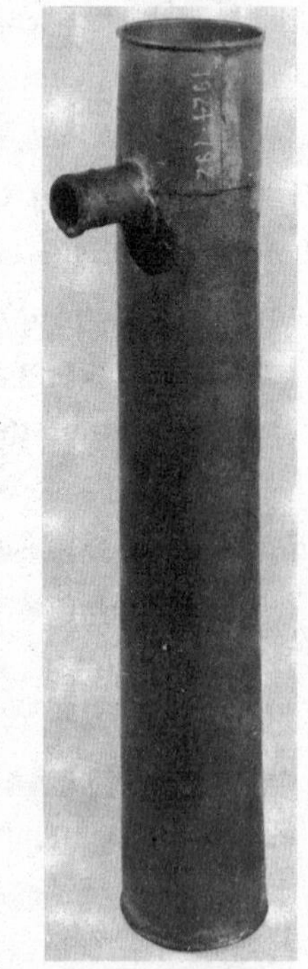

Watt's original model separate condenser.

spirits. Robison agreed that he was. "Gad," Brown went on, eager to share the news, "the condenser's the thing: keep it but cold enough, and you may have a perfect vacuum, whatever be the heat of the cylinder." Then Robison got it: keeping the cylinder hot, exhausting the steam, the separate condenser: "The whole flashed on my mind at once."[36]

Robison pumped Brown for more information. He learned that Watt was having trouble preventing the cylinder from leaking steam. Watt had tried both leather and felt, but neither could withstand the heat. That was clearly a problem yet to be solved. The important point, Robison decided, was that Watt was using steam to work his engine, not relying simply on air pressure.

Back home the next day, Robison ordered up a model himself. When it arrived from the shop of the craftsman who made it for him, he excitedly tried it out. Within minutes, he had repeatedly pumped most of the air out of a large tank. He had no doubt then "that Mr. Watt had really made a perfect steam engine."[37]

Watt's steam engine was far from perfect. The first engines he built, though much less wasteful of coal than the Newcomen engines that preceded them, were still only 2 percent efficient. Significantly, when he applied for a patent, awarded in 1769, he proposed to secure his right of invention under the rubric "Methods of Lessening the Consumption of Steam, and, consequently, of Fuel, in Fire-Engines."[38] If it was more efficient, it was still an atmospheric engine, operating not by direct steam pressure but by condensing steam to create a vacuum.

Nor was Watt able to pursue engine improvements full-time. For most of the ten years after his flash of insight on Glasgow Green, he worked as a surveyor and civil engineer, laying out canals for the new wave of canal development sweeping Britain and building conventional Newcomen engines. In 1775 he found a dynamic, supportive, well-funded partner in the successful industrialist Matthew Boulton. Where Watt was querulous, Boulton was bold; he famously told Watt, who had initially proposed selling Boulton the exclusive right to produce steam engines only for the counties of Warwickshire, Staffordshire, and Derbyshire, "It would not be worth my while to make for three Countys only, but I find it very well worthwhile to make for all the World."[39] Boulton lobbied Parliament for an extension of Watt's patent, which was granted out to 1800.

Boulton & Watt engines followed that harnessed steam directly, mounted automatic throttles, produced rotary motion, measured output and more. From only pumping mine water, the new steam engines came to blow smelting furnaces, turn cotton mills, grind grain, strike medals and coins, and free factories of the energetic and geographic constraints of animal or water power. At the beginning of the eighteenth century, coal already supplied half of Britain's energy; at the beginning of the nineteenth century, the proportion rose above 75 percent and continued to increase.[40]

The other face of these bounties, their unintended consequence, was increasing air pollution: the pollution of domestic and industrial coal smoke and pollution from the processes the steam engine drove. Already in Ireland as early as 1729, Jonathan Swift wrote in the *Dublin Weekly Journal,* "The physicians in Dublin make it their constant practice to remove their patients to some purer air, near the suburbs, out of the smoke of the city, which in winter is so thick, and cloudy enough to stifle men and beasts, so great an influence that it affects even the blossom and bloom of the flowers in the spring."[41] Adding to the smoke of burning coal, from about the middle of the eighteenth century, the "dark Satanic mills" of William Blake's 1808 poem "Jerusalem" began strewing their blight across England's green and pleasant land.

"In London," writes John Farey in his 1827 treatise on the steam engine, "all the large breweries and distilleries were in a few years furnished with engines. . . . An iron forge was built at Rotherhithe, about 1787, for making up scraps of old iron into bars. Also a paper-mill with an atmospheric engine. A fulling mill and logwood-mill was set up in the Borough [of London] in 1792, at a large dyehouse for woolen cloth, with a 20-horse engine. An oil mill was begun in the Borough soon after, and then a mill for grinding apothecaries' drugs; also a mill for calendaring, glazing, and packing cloths for exportation.

"These first establishments in each place," Farey concludes, "were greatly multiplied and extended in the course of 10 years, particularly the steam cotton mills at Manchester and Glasgow, and steam corn mills in every large town; the extension has been still more rapid since the expiration of Mr. Watt's patent in 1800."[42]

Also with the expiration of Watt's patent, it became possible for other inventors to explore the use of steam in transportation. First had come coal, to replace increasingly costly wood for residential heating. Demand for coal close to waterways had then driven mining deeper underground, intersecting the water table and flooding many of the mines. Pumping water from flooded mines had soon exceeded the capabilities of men and animals. Atmospheric engines invented for the purpose had then cleared the mines while revealing the power of steam to substitute for animal and human labor. The advent in 1781 of Boulton & Watt's double-acting steam engine—steam alternately admitted to opposite ends of the piston, pushing as well as pulling—made rotary motion possible, allowing manufacturers to site factories away from the rivers that had powered them previously. "During the first seventy years of its history," Galloway confirms, evoking Coleridge's metaphor, "the sole accomplishment of the steam-giant had been to draw a straight line. It had now been taught a new idea, *viz.*, how to describe a circle—an addition to its repertoire which was of infinite value to the world."[43]

On 10 July 1776, in the newly declared United States of America, newspapers published the Declaration of Independence, as yet unsigned, as British ships fired on New York from the Hudson River. The Continental Congress, meeting in session in Philadelphia, resolved to advance one month's pay to the Pennsylvania militia, established rules of order for its own proceedings, and reviewed a battle report from Brigadier General Benedict Arnold.

On the same day, away in England, the ironmaster John Wilkinson welcomed his friend Samuel More to Birmingham. More, the secretary of the Royal Society for Arts, Manufacture and Commerce, had traveled up from London to review the impressive new technologies of steam power and iron making. The two friends visited a canal, spent the evening talking, and slept that night at an inn. The next day, Josiah Wedgwood, the master potter, joined them, as did Matthew Boulton, who showed them through the Boulton & Watt factory at Soho. They dined there that afternoon. After dinner, Boulton reviewed the history of the steam engine, and they examined an early-model engine improved with a new Wilkinson cylinder.

"Then an astonishing thing happened," writes Wilkinson's biographer. "The four men took off their coats, rolled up their sleeves, and dismantled the engine piece by piece to understand the detail of its working. They then reassembled it and had it running satisfactorily again by evening."[44] If Americans had rolled up their sleeves to learn the workings of a political revolution, Englishmen had rolled up their sleeves to learn the workings of a technical revolution.

At least, that was how an astute Irish politician, John Baker Holroyd, Lord Sheffield, saw the equivalency ten years later, after the British had lost the American war. "If Mr. Cort's very ingenious and meritorious improvements in the art of making and working iron," Sheffield wrote, "the steam engine of Boulton and Watt, and Lord Dundonald's discovery of making coke [from coal] at half the present price, should all succeed, it is not asserting too much to say that the result will be more advantageous to Great Britain than the possession of the thirteen colonies."[45] Perhaps, but the first steam engine in America had been pumping water from a New Jersey copper mine since 1755, and the Connecticut Yankee inventor John Fitch was busy testing a steamboat on the Delaware River.[46]

FIVE

CATCH ME WHO CAN

British canals date their rise to the retreat of a heartbroken duke from London.[1] Francis Egerton, the third Duke of Bridgewater, born in 1736, was the youngest of eleven children and the only one of seven males to survive into adulthood. After his father died when he was nine, he was neglected and grew into something of a rake. A grand tour of Europe at seventeen with the scholar Robert Wood as his tutor hardly cured him. In Paris and Lyons, Egerton cavorted with a French actress so publicly that Wood threatened to resign. They repaired their relationship while young Bridgewater studied engineering at the Lyons Academy. The expensive works of art that Wood recommended he buy in Italy he shipped home but never bothered to uncrate. Returned to England and living in London, with an income of £30,000 a year (today £420,000, or $600,000), he drank, wenched, gambled, and raced horses.

In 1751 two Anglo-Irish beauties, the Gunning sisters, arrived in London. Elizabeth, the younger sister, married the Duke of Hamilton in 1752. Across the next six years, she bore him three children, only to lose him in 1758 to a fatal infection following a cold. Bridgewater promptly fell in love with the young widow. She accepted his proposal of marriage but broke off the engagement to marry the future Duke of Argyll. Permanently embittered in matters of the heart and never to marry, the twenty-three-year-old duke gave a final grand ball and cantered off to his rural estate in Worsley, northwest of Manchester, to mine coal and build a canal.

So the story goes. But on his grand tour, young Bridgewater had been smitten with the Languedoc Canal in southwestern France before he was smitten with the Duchess of Hamilton. The Languedoc Canal (the Canal du Midi today), opened in 1681, crosses France at the French end of the neck it shares with Spain, linking the Atlantic Ocean through the Bay of Biscay to the Mediterranean Sea. It was one of the great works of French engineering commissioned by Louis XIV, 150 miles long, second in cost only to Versailles, intended to ship French wheat and wine to the Mediterranean without enduring the long voyage around Spain and risking the piratical Strait of Gibraltar.[2]

Bridgewater brought the concept home. He had collieries in Lancashire, but hauling coal on packhorses on the execrable English roads and paying exorbitant fees to ferry it across three rivers to Manchester made delivery uncompetitively expensive. In 1757, a year before he removed from London to Worsley, the duke hired an agent to guide his planning. John Gilbert began surveying the canal route; Bridgewater went to work on Parliament, which authorized a version of his canal in March 1759.[3]

Gilbert introduced the duke to James Brindley, a celebrated mill and canal builder. Brindley spent six days in discussions with Bridgewater at Worsley Old Hall in July 1759. By the end of that time, he had convinced the duke to chart a different canal route, one that would connect the Manchester navigation to a larger network Brindley was planning that would link the region's canals to Liverpool, London, Birmingham, and the world.

Young though he was, the duke was an astute businessman, with good judgment of risks. The most controversial structure in Brindley's revision was a three-arched stone aqueduct to carry the canal *over* the River Irwell. In London in January 1760, Brindley demonstrated this radical innovation to a parliamentary committee with a model he had carved from a wheel of English cheddar.[4] A prominent consulting engineer brought in to assess the new plan dismissed it contemptuously: "I have often heard of castles in the air," he testified, "but never before saw where one was to be erected."[5] Nevertheless, the duke prevailed. By March, he had won royal assent as well. A year and a half later, the first segment of the canal, with its aerial aqueduct, was finished, to the astonishment of travelers. One wrote to a Manchester newspaper of the "pleasure" the novel aqueduct had given him:

> At Barton-bridge [Mr. Brindley] has erected a navigable canal in the air; for it is as high as the tree-tops. Whilst I was surveying it with a mixture of wonder and delight, four barges passed me in the space of about three minutes, two of them being chained together, and dragged by two horses, who went on the terras of the canal, whereon, I must own, I durst hardly venture to walk, as I almost trembled to behold the large river Irwell underneath me.[6]

The Bridgewater Canal began *inside* the duke's Worsley coal mine, where an enlarged, brick-lined adit allowed the miners to load boxes of coal into barges for transfer onto canal boats, the first such system in Europe. (As the mine deepened, the adit system lengthened, eventually extending more than fifty-two miles underground.[7]) Unlike previous canals, which merely cut off river meanders or paralleled rapids and shallows, the Bridgewater was an arterial system, crossing valleys between watersheds and requiring aqueducts or high embankments. It reached Castlefield Wharf in Manchester in 1765, fulfilling its purpose even as it refilled the duke's purse. A Swedish traveler and industrial

spy, Eric Svedenstierna, noted on his visit to Manchester in 1802, "With such a large demand for coal, it is no small advantage that at even the present high prices, Manchester can have coal at about 50 per cent cheaper than the coal cost a little over 40 years ago, before the Duke of Bridgewater's Canal was finished, from whose coal mines practically the whole of Manchester is supplied."[8]

The British population that used coal for heating and cooking was increasing, from 5.2 million in 1700 to 7.8 million in 1800, and on up to 12 million by 1831. Industry used coal for Newcomen engines pumping out coal mines and pumping water, although much of that coal was essentially mine waste. But iron smelting with coked coal began a major expansion after 1750, radiating outward from the Darby enterprise at Coalbrookdale and rapidly replacing smelting with charcoal made from wood.

This transformative conversion from charcoal to coke just as Britain was beginning to industrialize has customarily been attributed to improvements in the quality of coke-smelted iron. Recent scholarship finds a more decisive factor to be the rising cost of charcoal. Demand, pushing up prices, was one reason for the increase. Another, once again, was the increasing scarcity of wood available for charcoaling. Even though iron smelted with coke cost the ironmaster more to finish than iron smelted with charcoal, the higher price of charcoal after 1750 justified the expense.[9]

Coal, dug from point sources (coal mines) rather than wood gathered at large—*punctiform* rather than *areal*—concentrated the transportation of fuel efficiently along a small number of routes. That concentration justified major investment in the routes' improvement. And the most efficient routes in that era were canals. (Of the 165 parliamentary acts passed between 1758 and 1802 that authorized canals, 90 listed coal as their preponderant anticipated freight.) [10]

If coal was moved more efficiently by canal boat, it still had to be transported first from the pithead to the waterway. Wooden rails wore down quickly under loads of coal or iron ore. Richard Reynolds, who managed the Darbys' Ketley ironworks, near Coalbrookdale, in the 1760s, introduced cast-iron plates and then cast-iron rails to protect the wooden rails from wear and tear or to replace them. He had another reason as well for using iron: as an ingenious storage system. A depression following the end of the Seven Years' War in 1764 reduced demand for iron products. Prices fell. Reynolds wanted to keep his furnaces going and his employees at work. Rather than warehouse the excess production, Reynolds used it for rails. Then, if iron prices went back up, he could have the rails removed and sold. Reynolds "tried it at first with great caution," his granddaughter recalled, "but found it to answer so well, that very soon all their railways were made with iron."[11]

With iron wheels on iron rails, a single horse could haul thirty tons of coal or ore. Mineral transport by rail no longer had to depend on gravity for its energy supply. At the same time that the old wagonways were being shod, new horse-drawn mineral railways began to be laid as direct feeders to canals, like multiplying capillaries draining into veins. Parliament authorized the first such railway on 13 May 1776, from Caldon Low Quarries in Staffordshire, below Liverpool, to Froghall Wharf on the Trent & Mersey Canal, a distance of 3.1 miles.[12] From that beginning, the British network of feeder railways grew with the canal network.

Three other developments, two of them fortuitous, then stimulated innovative change. James Watt's steam engine patent expired in June 1800. Its patent protection had discouraged other inventors from exploring improvements.

Casting iron pigs, so-called because of the resemblance of the casting molds to rows of suckling pigs.

Now, as both Watt and Boulton retired, the field of steam engine design opened up to new ideas.

In 1803, another fortuity put the cost of railway horses and fodder almost out of reach: England declared war on the France of Emperor Napoleon Bonaparte, beginning twelve years of European military conflict. This war required the services not only of millions of men but also of several million horses. The British cavalry of that era, unlike Britain's allies and enemies in Europe, had no formal remount system. Rather than maintaining a stable of studs and breeding warhorses, it bought horses from civilian breeders and agents on the open market. Officially, the cavalry would pay only £30 for a light cavalry horse and £40 for a heavy cavalry or artillery draft horse. But civilian horse dealers charged whatever the traffic would bear. On the war front, one historian notes, "men would prove easier to replace than horses."[13] Good horses went for £80 to £100 (today £6,400, or $9,000). The cost of hay and fodder doubled.[14] Just when coal and iron ore were becoming essential commodities in an industrializing nation, for civilian goods as well as military ordnance, the cost of their transport increased significantly. That opened a way to potentially lucrative improvement.

Already in the 1790s, a young engineer named Richard Trevithick Jr., born in Cornwall in 1771, had begun to develop a different kind of steam engine. Trevithick, whose father was a mine captain, agent, and engineer, was a giant of a man by the standards of the day. At six foot two, rawboned and muscular, he was strong enough to draw crowds when he demonstrated lifting a blacksmith's mandrel, a thousand-pound cone of cast iron. Once, at a public dinner, he picked up a sturdy six-foot colleague, rotated him upside down and stamped his boot prints onto the ceiling. Such exploits earned him the nickname "the Cornish Giant." Besides an imposing physical presence, Trevithick was gifted at building steam engines, usually introducing innovations that increased their efficiency. In 1795, when Boulton & Watt enjoined an atmospheric engine Trevithick built at Ding Dong mine in Cornwall for interfering with its patents, the young

engineer responded by rebuilding the machine, inverting its big cylinder, removing the cylinder cover, and foregoing the benefits of Watt's separate condenser.

Trevithick's father died in 1797 when Richard was twenty-six, and such was the son's gathering reputation that he was elected to replace his father as Cornish mining's leading engineer, taking charge of engine operations at a colorful list of mines: Ding Dong, Wheal* Bog, Wheal Druid, Hallamanin, Wheal Prosper, Wheal Hope, Wheal Abraham, Dolcoath, Rosewall, Polgrane, Trenethick Wood, Baldue, Trevenen, Wheal Rose, Wheal Malkin, East Pool, Wheal Seal-hole, Cook's Kitchen, Camborne Vean. He married that year as well, to Jane Harvey, the tall, fair daughter of a foundry owner.[15] Richard invented the plunger pump, which thrust a loose wooden plunger down a cast-iron pipe to force water up into a drain, replacing the easily fouled chain-and-bucket mechanism of previous mine drainage systems.[16] (The following year, he inverted the design to make an engine driven by water pressure, useful when a flow of water was available on higher ground: the force of falling water drove the pump rod up and down, motion that could then be harnessed for work.[17])

Primed by challenges from Boulton & Watt and already practiced at invention, Trevithick looked beyond the familiar precincts of the low-pressure atmospheric engine to the unexplored territory of high-pressure steam—"strong steam," as engineers then called it. The atmospheric engine condensed low-pressure steam with cold water to make a vacuum under a piston, which the weight of the atmosphere then drove to perform work. High-pressure steam could drive a piston directly, with greater force and without the cumbersome addition of a separate condenser. To increase the capacity of an atmospheric engine required increasing the size of the cylinder: the largest such engine Boulton & Watt ever built, in 1792, had a cylinder five feet nine inches in diameter, with a nine-foot stroke.[18] In contrast, the cylinder of a high-pressure steam engine could be kept small and made even smaller by increasing the pressure of the steam—provided that its boiler could be constructed sturdily enough to withstand the additional pressure without exploding.

A sturdy boiler was one necessity, Trevithick understood. There was room

*Wheal is Cornish for "mine."

for significant improvement in the boilers of the day, some of which actually used hemp packing stuffed between the small iron plates that were riveted together to make the walls. At low pressure, the packing sealed the plates against steam loss, but they leaked badly under the pressure of strong steam. Using high-pressure steam directly, Trevithick no longer needed to bleed off the steam into a separate condenser. It could be vented into the air. But he needed to know what his engine would lose and what it might gain if it did so. Who could tell him?

He traveled to London in late 1796 to testify in the Watt patent lawsuits, the first time he had ever been out of Cornwall.[19] In London, he met and befriended twenty-nine-year-old Davies Giddy,* a mathematician and former high sheriff of Cornwall and a friend of Trevithick's father, who had also been called to testify. Back in Cornwall, Giddy remembers, "On one occasion, Trevithick came to me and inquired with great eagerness as to what I apprehended would be the loss of power in working an engine by the force of steam, raised to the pressure of several atmospheres, but instead of condensing [the steam,] to let the steam escape. I of course answered at once that the loss of power would be one atmosphere." That is, whatever the pressure of the steam in Trevithick's engine, the only loss from his design compared with an atmospheric engine would be the loss of the vacuum: his engine would have to work not against a vacuum but against atmospheric pressure, 14.7 pounds per square inch. And, added Giddy, such loss would be partly offset in Trevithick's simpler direct-steam design by having eliminated some of the other inefficiencies of an atmospheric engine—no air pump with its friction, no friction or work raising the condensing water from its reservoir. "I never saw a man more delighted," Giddy concludes.[20]

Within a month, Trevithick had built his first working model: a tabletop engine executed in bright brass. Everyone in West Cornwall seems to have been involved in the long battle with Boulton & Watt, including Francis Basset, a member of Parliament whom the diarist and biographer James Boswell calls "a genteel, smart little man, well-informed and lively." Basset had recently been

*Davies Giddy changed his name to Gilbert in 1817 to inherit the substantial estates of his wife's uncle, who wanted the Gilbert family name perpetuated. He served as president of the Royal Society from 1827 to 1830.

made Baron de Dunstanville.[21] One day in 1797, Giddy and Lord and Lady de Dunstanville arrived at the Trevithicks' house in Camborne to see the model at work. "A boiler," writes Trevithick's Victorian biographer, "something like a strong iron kettle, was placed on the fire; Davies [Giddy] was stoker, and blew the bellows; Lady De Dunstanville was engine-man and turned the cock for the admission of steam to the first high-pressure steam engine." Trevithick built a second model that combined the boiler and the piston in one cylinder. Set on wheels, it "ran round the table, or the room." A third model, with a spirit lamp for heating, went off to London as an exhibit in the lawsuit.[22]

Odd as it sounds today, there was general doubt at the end of the eighteenth century if engine-driven iron wheels would move a load or simply spin in place. Wagons were hauled by horses, of course, and rolled downhill in response to gravity, but under those conditions, they merely freewheeled: the horses' hooves did the work. In the extreme case of a railway, the actual contact between an iron wheel and an iron rail is minimal—for a modern railroad wheel, an area about the size of a dime—which is why rail transport is so efficient and why a team of horses in Trevithick's day could easily pull a train of tramway wagons loaded with tens of tons of coal. But whether or not a steam engine could drive iron wheels on a dirt road, much less on a railway, was still an open question.

Trevithick answered the question to his own satisfaction in the summer of 1801 by enlisting Davies Giddy to help him move a carriage by hand. "Trevithick and myself tried an experiment on a one-horse chaise," Giddy recalls, "as to the hold of the wheels on the ground for moving it up an ascent." Having unharnessed the horse, they turned the wheels of the chaise by hand to force it forward. "We discovered that none of the acclivities were sufficient to make the wheels slide in any perceptible degree," Giddy concludes.[23]

The question would come up again, repeatedly, and railway innovators would have to prove to would-be investors that iron-to-iron wheel systems in particular actually worked. It influenced British railroad construction across the next decades: grades on British railroads would be shallower, and curves wider, than US railroads built later for the more rugged American terrain.[24] Even on modern diesel-electric railroad engines, a funnel of sharp sand in the nose of the cab (the "sander") allows the engineer to strew grit onto the rails ahead of the drive wheels when traction needs improving.

With models built and tested and traction confirmed, Trevithick proceeded to build a common road locomotive powered by strong steam. The engine was about the size of a modern automobile but cylindrical, with a single vertical piston entering from the top at the end opposite the smokestack and the wrought iron firebox and boiler flues fitted inside. Steam at 60 pounds per square inch (psi) drove the vertical piston, vented around the feed-water pipe to preheat the water and then exited midway up the wrought iron chimney to function as a steam blast, drawing air through the fire to make it burn hotter and to burn up the smoke. (Black smoke from a steam engine stack is coal smoke; white smoke is steam.)

Trevithick—"Captain Dick" to Cornwall locals—tested his road engine on Christmas Eve 1801, one participant remembered, "out on the high-road. . . .

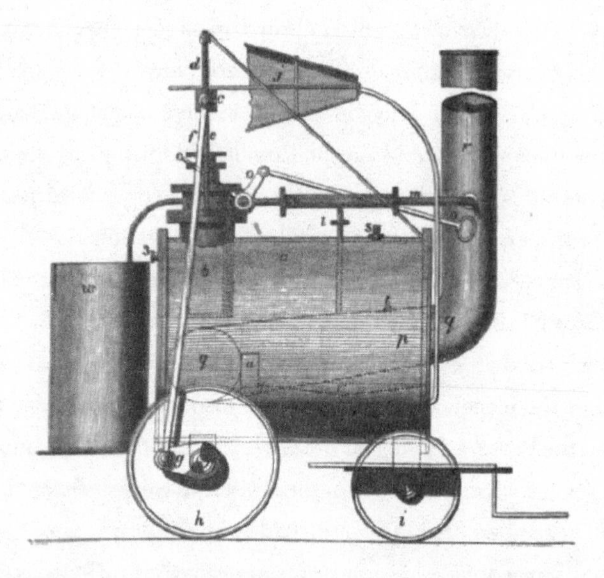

A reconstruction of Trevithick's first common road locomotive, 1801. Feed-water tank at left. Flanged steam piston, set vertically above engine body at left, drives large wheels through connecting rods. Dashed lines on engine body indicate U-shaped firebox inside, opening to right beside smokestack. Pipe above and paralleling engine body from piston to stack feeds waste steam to draw air through firebox. Fireman stood on platform at right to shovel coal from a tender wagon, not shown, into the open firebox, and steered the smaller wheels with a steering handle—the black bar projecting slightly past the platform edge, right. (Oval handle silhouetted against chimney is throttle for controlling engine speed.) The actual engine was wood framed, however, with a larger platform.

When we see'd that Captain Dick was a-going to turn on steam, we jumped up as many as could; maybe seven or eight of us. 'Twas a stiffish hill going from the Weith up to Camborne Beacon, but she went off like a little bird."[25] She went off like a little bird heading for Lord de Dunstanville's house, but, rounding a wall, the engine hit a gully and turned over. No one was reported hurt. Giddy recalls the sequel:

> The carriage was forced under some shelter, and the parties adjourned to the hotel, & comforted their hearts with a roast goose & proper drinks, when, forgetful of the engine, its water boiled away, the iron became red hot, and nothing that was combustible remained either of the engine or the [shelter].[26]

Undiscouraged, Trevithick and his cousin Andrew Vivian headed for London two weeks later, in mid-January 1802, to file for a joint patent, armed with letters of introduction from Lord de Dunstanville and from Giddy. The letters won them meetings with the chemist Humphry Davy, the American-born physicist Benjamin Thompson—Count Rumford—and other notables. Someone, probably their patent attorney, recommended that they have a steam carriage built and exhibit it in London. The patent, for "Methods for improving the construction of Steam engines and the application thereof for driving Carriages and for other purposes," was granted on 24 March 1802.[27]

At Coalbrookdale that summer, Trevithick designed and built an engine intended to run under no less than 145 psi, a boiler pressure far above anything anyone had dared attempt before. The cast-iron boiler was only four feet in diameter, Trevithick wrote Giddy in August 1802, but one and a half inches thick. Its cylinder measured a mere seven inches in diameter. "The engineers about this place," he scoffed, "all said that it was impossible for such a small cylinder to lift water to the top of the pumps. . . . They are constantly calling on me, for they all say they would never believe it unless they see it, and no person here will take his neighbor's word even if he swears to it." They came, they saw, and "after a short time, they set off with a solid countenance and a silent tongue."[28]

By February 1803, a foundry was casting a wrought iron cylinder and boiler for the planned three-wheeled London carriage that Trevithick and Vivian had been advised to exhibit. Its cylinder would be only five and a half inches in diameter and fixed horizontally rather than vertically, making for a much smoother ride. With a boiler no larger than a kettle drum and with wrought-iron parts rather than cast-iron, the whole engine would weigh only six hundred pounds. In London, coach maker William Felton of Leather Lane was building the elegant coach, set on leaf springs between the big eight-foot driving wheels and designed to carry eight to ten people. The engine and boiler shipped by sea from Falmouth to London in August to be assembled at Felton's coach works.[29]

Sometime in autumn 1803, Richard Trevithick's steam-powered horseless carriage was finished and tested.[30] It cost in total £207 (today £17,000, or $25,000).[31] For the next six months, it chugged around the city streets, delighting ordinary Londoners and scaring the horses.[32] Humphry Davy had already named it "Trevithick's Dragon."[33]

Few stories remain of the adventures of the first steam carriage to make the rounds of London. Strangely, the newspapers of the day reported none of its activities, leaving the field to later reminiscences. Felton, the coach maker, and his sons occupied the coach on its first day out of the shop, a further reward for their work building it. Vivian recalls of that first outing that the Dragon wheeled along "from Leather Lane . . . through Liquorpond Street into Gray's Inn Road by Lord's Cricket Ground to Paddington and Islington and back to

Leather Lane."[34] The ride was jarring despite the coach springs, since the big drive wheels couldn't be sprung; doing so would dislodge their gearing. The streets were, at best, cobblestone.[35]

A Mrs. Humblestone remembered seeing the Dragon pass through Oxford Street, evidently an organized public trial. The shops had been closed for the occasion and the streets cleared of carriages. People along the route cheered and waved handkerchiefs.[36] London's busmen and cabdrivers didn't cheer, however. They recognized the danger of the new technology to their employment. They "pelted the engine with cabbage stalks and rotten eggs,"[37] Mrs. Humblestone recalled.

Trevithick was a master of engines, but he had not yet found a reliable method of steering. Nineteen-year-old John Vivian, Andrew Vivian's nephew, recalled a day starting at four in the morning when he drove the Dragon along Tottenham Court Road and then the New Road, where a canal paralleled the street, thinking how deep the canal was if they should veer into it:

> I was steering, and Captain Trevithick and someone else were attending to the engine. Captain Dick came alongside of me and said, "She is going all right." "Yes," I said, "I think we had better go on to Cornwall." She was going along five or six miles an hour, and Captain Dick called out, "Put the helm down, John!" and before I could tell what was up, Captain Dick's foot was upon the steering-wheel handle, and we were tearing down six or seven yards of railing from a garden wall. A person put his head from a window, and called out, "What the devil are you doing there? What the devil is that thing!"[38]

Going on to Cornwall was a joke, not that they could have made the journey. The roads were too bad.

The extended demonstration of Trevithick's common road steam locomotive turned up no investors interested in supporting its further development. The Cornish Giant concluded that the future of the steam engine wasn't transporting people but working machinery and hauling coal. He set his engines to work boring brass cannon, pumping water, and blowing furnaces.

Along the Thames in Greenwich, one of Trevithick's engines had been installed to drain a building foundation. One day in September 1803, the boiler exploded, strewing great chunks of cast iron as far as 125 yards away. Trevithick wrote Giddy that the boy assigned to attend the engine had tied down the throttle, left it in the care of one of the laborers, and gone off to catch eels in the foundation pool. The engine began racing, the laborer stopped it without releasing the throttle, and it blew up. "It killed three on the spot," Trevithick concluded, "and one other is since dead of his wounds."[39] Although human error caused the accident, Boulton and Watt seized on it as proof of the dangers of high-pressure steam. "They have done their uttermost to report the explosion both in the newspapers and private letters very different to what it really is," Trevithick complained to Giddy. In the future, he told his friend, he would build his engines with two safety valves and a mercury steam gauge, "so that the quicksilver shall blow out in case the two valves should stick. . . . A small hole will discharge a great quantity of steam at that pressure."[40] Boiler explosions would plague steam engines well into the late nineteenth century, until steel replaced iron in their construction.

Even before the London demonstration, Trevithick had run short of money. A wealthy businessman, Samuel Homfray, ironmaster of the Penydarren Ironworks at Merthyr Tydfil in South Wales, rescued him, buying a quarter of the high-pressure patent for £10,000 (today £800,000, or $1.1 million).[41] If Boulton and Watt saw no future in high-pressure steam, Homfray did.

Trevithick built several stationary steam engines at Penydarren, after which Homfray agreed to support his developing a high-pressure steam locomotive. Homfray was interested in particular in steam transport on a new tramway that extended from Merthyr to Abercynon, where a wharf on the River Cynon allowed canal boats access to the River Taff. The tramway had been laid down by the Welsh ironmasters who had built a canal along the course from Merthyr on to Cardiff. "The canal was not just successful," writes the historian Anthony Burton, "it was too successful. With forty-nine locks in just 24 miles of waterway, congestion soon reached chronic proportions."[42]

The solution the ironmasters found to this early canal-boat gridlock marked another step in the evolution of the railroad. Parliament had authorized "collateral cuts or railways" within four miles of the canal. Its intention had been to allow direct access to the canal from any mines or factories along the way. Taking advantage of the ambiguous language, the ironmasters built a nine-and-a-half mile railed tramway, the Merthyr Tramroad, opened in 1802. It kept within four miles of the canal but paralleled it from Merthyr to Abercynon, thus bypassing the canal's congestion.[43] Unintentionally, the ironmasters set up a classic experiment testing the efficiency of two different transport technologies: gravity-powered water transport versus land transport powered by steam.

Several secondary technologies came together at this point as well. Most tramways were now railed with cast-iron rails set on stone blocks. They were not yet linked with what Americans call "crossties" and English call "sleepers," because the space between the rails had to be left open as a path where the horses could walk to pull the wagons. The rails themselves were now flanged, with a raised rim on the inside edge to hold the iron wheels on the track. Flanging, of either the rails or the wheels, is the reason why locomotives don't have steering wheels: the flanging does the steering, guiding the wheels along.

Richard Crawshay, another Welsh ironmaster, owned a controlling interest in the canal. Not surprisingly, he disdained steam. He believed a steam locomotive under load would spin its wheels and go nowhere. Trevithick's sponsor Samuel Homfray, besides being a competitor of Crawshay's in the iron trade, was a gambling man. He bet Crawshay 500 guineas even money—a total purse of 1,000 guineas (a guinea was then pegged at 21 shillings, so £88,000 today, or $126,000)—that steam could do at least what one horse could do: haul ten tons of iron down the tramway from Penydarren to the Abercynon wharf and return the empty wagons to Penydarren, thus traveling a total distance of about nineteen miles.[44]

Trevithick was up to the challenge. He built an engine with one 8¼-inch cylinder with a 54-inch stroke, the cylinder set into the boiler to keep it hot. Steam drove the piston and then exhausted up the smokestack, drawing air through the firebox for a hotter fire. As with Trevithick's earlier engines, a large flywheel mounted on one side carried the piston across the dead center of its stroke, a point where at lower speeds it might sometimes hang. The engine was odd-looking, with a crosshead connection from the piston rod on the right side to the flywheel on the left side and to gears that drove the wheels. The crosshead "crashed to and fro like some overgrown trombone slide," writes Burton. He wonders "how anyone manage[d] to fire the engine without having his head knocked off."[45]

Early in February 1804, Trevithick wrote Giddy to share his excitement. "Last Saturday we lighted the fire in the Tram Waggon and worked it without the wheels to try the engine," he told his friend, "and Monday we put it on the tram road. It worked very well and ran up hill and down with great ease, and very manageable. . . . The bet will not be decided until the middle of next week." Five days later, Trevithick wrote again. He had operated the "Tram Waggon," his steam locomotive, several times, he wrote, and it worked "exceedingly well, and is much more manageable than horses. We have not tried to draw but ten tons [of iron] at a time yet, but I doubt not but we could draw forty tons at a time very well, for ten tons stands no chance at all with it."[46]

Finally, both Crawshay and Homfray arrived to witness the contest, as did a large crowd of spectators, many of whom wanted to ride along in the wagons with the pig iron. Confident as always, Trevithick let them. "We carried ten tons of iron, five wagons, and 70 men riding on them the whole of the journey," he wrote Giddy the next day. One nine-mile stretch when they had to cut down trees and remove large rocks that blocked the road took four hours. Once they ran clear, they managed almost five miles per hour. Trevithick was happy to win the bet but even happier to confound his skeptics: "The gentleman that bet five hundred guineas against it, rid the whole of the journey with us and is satisfied that he have lost the bet. The public until now called me a scheming fellow, but now their tone is much altered."[47]

The date of the trial, Tuesday, 21 February 1804, marked the first time a steam locomotive running on rails hauled a loaded train of freight cars—in this case, about twenty-five tons of engine, iron, wagons, and men.

Yet only Trevithick and Homfray seem to have judged the trial a success. Although the engine performed well enough, its weight was more than cast-iron rails and tram plates could bear. "She worked very well," one of the men involved in the contest recalled, "but frequently from her weight broke the tram plates."[48] The continuing concern that iron wheels would lack traction on iron rails had led them to concentrate the weight of the engine onto only four wheels. "In consequence of this great pressure," Giddy assessed the problem, "a large number of rails broke, & on the whole, the experiment was considered as a failure."[49]

For the next several years, Trevithick and Homfray were too busy manufacturing and selling high-pressure steam engines to worry about tramways or railroads. English mines continued to be deepened to expose new veins of coal or iron ore; expensive horses were less than adequate at lifting water or minerals from deeper levels. Trevithick steam engines—smaller, more efficient, and less expensive than the big atmospheric engines of Boulton & Watt—filled a need. "Whim engines," they were called: *whim* a contraction of *whimsy,* a name borrowed from the merry-go-rounds that the winding drum once walked by horses whimsically resembled. By 1803, Trevithick had erected twelve winding engines. Others—in Coalbrookdale, Manchester, Liverpool, Bridgenorth, Newcastle, London, and beyond—followed up to 1808.[50] He built engines for dredging the Thames as well and also tried his hand at tunneling under that wide river. Trevithick managed to dig his way a thousand feet along and within two hundred feet of the farther shore before what he calls "a quick sand" broke through and flooded out his crew. His investors consulted the Royal Engineers, who had no experience with tunneling. They concluded that no one could dig a tunnel under the Thames, and no one did until Marc Isambard Brunel succeeded in 1843.

Settled in London with his family in 1808, Trevithick tried once again to launch a railroad engine. This time he used a small dredging engine of his own design: a superbly simple locomotive with one vertical cylinder driving the rear wheels through connecting rods. Giddy's sister, Philippa, is supposed to have supplied a name for the machine in the form of a witticism that Trevithick borrowed: she said its purpose was "Catch me who can." From promoting his steam engines for work, Trevithick had decided to try promoting one for sport, advertising in the *London Times* of 19 July 1808 that he would begin exhibiting *Catch me who can* that day at eleven o'clock; the advertisement was headed "Racing Steam Engine."[51] Trevithick had offered a bet, as another London newspaper announced, that *Catch me who can* "was matched to run twenty-four hours against any horse in the kingdom."[52]

He had built a circular racetrack about a hundred feet across on open land at what is now North Gower Street and Euston Road, near the Wellcome Trust; had enclosed it with a high wooden fence; and was preparing to sell tickets for rides in an open carriage the engine would haul. His exhibition didn't open on 11 July, however. "The ground was very soft," Trevithick wrote Giddy several

A ticket to the exhibition.

weeks later, "and the engine, about 8 tons, sunk the timber under the rails and broke a great number of them, since which I have taken up the whole of the timber and iron, and have laid balk [wooden blocks] of from 12 to 14 inches square, down in the ground, and have nearly all the road laid again, which now appear to be very firm, as we prove every part as we lay [it] down, by running the engine over it by hand."[53]

Once the roadbed had been reinforced, Trevithick began selling one-shilling (today £4, or $6) rides on his train. A reliable observer, an engineer named John Hawkins, recalled timing his ride at twelve miles per hour—about the speed of a trotting horse—and hearing Trevithick claim *Catch me who can* would be good for twenty mph or more on a straight track. Hawkins speaks of "a ride for the few who were not too timid," suggesting that many were. Riding in an open car pulled by a strange, wheezing, smoke-belching black-iron contraption with a fire in its belly, feeling an unfamiliar centrifugal force around the continuously curving circular track, fearing an accident or an explosion, would make most people timid. "It ran for some weeks," Hawkins concludes to the point, "when a rail broke and occasioned the engine to fly off in a tangent and overturn, the ground being very soft at the time." No record survives of any twenty-four-hour mileage contest against a racehorse.

After London and *Catch me who can,* Richard Trevithick's struggles and setbacks continued until, in 1816, he left behind his long-suffering wife and children and set off for South America to try his luck clearing drowned silver mines in Peru. He was gone eleven years, won and lost fortunes, returned to England penniless, lived long enough to see railroads established in England, and died a pauper in 1833.

Once again, infrastructure had set a limit on the pace of development of a new technology. But by the second decade of the nineteenth century, malleable-iron rails were beginning to replace brittle cast iron on the wagonways of England, and stone or wooden sleepers strengthened the roadbeds for heavy locomotives. With such improvements, Trevithick's circular track on vacant land in London would open out into a network of fast, reliable transportation—but not before its inventors and engineers endured a final long haul of challenges and trials.

SIX

UNCONQUERED STEAM!

The year 1800, the turn of a new century, hinged in Britain between the old organic economy and the new economy of industry powered by fossil fuel. In America, with a population of 5.3 million, only half that of Britain, a new generation began advancing westward in horse-drawn wagons. The new nation's rivers were its highways: steam there would first drive steamboats. Britain, in contrast, eruptive with steam and braided with canals, looked beyond wagon and water carriage to the railway.

"The steam engine meant that coal could be exploited to supply mechanical energy as readily as heat energy," writes the economic historian E. A. Wrigley, "thus overcoming the last remaining barrier to the application of fossil-fuel energy to all the main productive processes."[1] For William Blake, that newfound mechanical energy, turning drive belts and working looms, might culminate in "dark satanic mills." The continuing increase of the British population, from 6 million in 1700 to 10.5 million in 1801, suggests a more optimistic outcome. Coal provided about half of British domestic energy in 1700, but such was the growth of industry across the eighteenth century that the share of coal used for domestic heating and cooking had dropped to less than one-third by 1800 even as total consumption had greatly increased.

"When coal could be substituted for other energy sources," Wrigley elaborates, "expansion could occur without simultaneously creating a matching rise in the pressure on the land. Access to the store of the products of past photosynthesis [such as coal] could relieve pressure on the current supply [i.e., wood and falling water]."[2] Among unexpected consequences, "the average number of hours worked per year in London rose by 27 percent" between about 1760 and 1800. For Wrigley, a desire "to gain access to goods or services" drove this increased labor.[3] The emergence of the fossil-fuel economy fueled the opening of the age of consumption.

Yet Britain was still green in 1800, as America was still largely primeval. William Wordsworth, a poet with a private income, would have much to say in other poems about the stifling effect of industrial labor on the body and the soul, but in July 1802 he stood on London's Westminster Bridge and found

A sight . . . touching in its majesty:
This City now doth, like a garment, wear
The beauty of the morning; silent, bare,
Ships, towers, domes, theatres, and temples lie
Open unto the fields, and to the sky;
All bright and glittering in the smokeless air.[4]

(If the air was smokeless that morning, it wasn't usually so, Dorothy Wordsworth noted in her journal, the source of the poem. "It was a beautiful morning" in London, she wrote, because "the houses were not overhung by their cloud of smoke."[5])

Wordsworth's enthusiasm for an urban world comparable in its majesty to the world of nature—"Earth has not anything to show more fair," the sonnet begins—embodied an optimism common to the day. Finding that turnpike mileage tripled in England between 1750 and 1770, Sidney and Beatrice Webb quote "an able and quite trustworthy writer" in 1767 declaring the development "an astonishing revolution. . . . The carriage of grain, coals, merchandise, etc., is in general conducted with little more than half the number of horses with which it formerly was. Journeys of business are performed with more than double expedition, . . . *Everything wears the face of dispatch* . . . and the hinge which has guided all these movements and upon which they turn is the reformation which has been made in our public roads."[6]

The turnpikes may have offered firmer footage to those who could afford the tolls. The writer Thomas De Quincey encountered more familiar conditions on the English highroads on "a most heavenly day in May" 1800: "vast droves of cattle . . . upon the great north roads, all with their heads directed to London, and expounding the size of the attracting body, together with the force of its attractive power, by the never-ending succession of these droves, and the remoteness from the capital of the lines upon which they were moving."[7]

American cotton was unknown in Britain in 1784, when an American ship arrived at Liverpool with eight bales among her cargo. These, two observers report, "were seized by the officers of Customs, under the conviction they could not be the growth of America!" By 1806, American cotton commanded a 53 percent share of the British market.[8] Leather, cotton, wool, and building construction in roughly equal shares totaled 68 percent of value added in British industry in 1801.[9] From 1788 onward, the quantity of iron England produced doubled every eight or ten years, an early industrial version of Moore's law.[10] What major product did England manufacture from all that iron? Nails, says Samuel Smiles, the Victorian chronicler, "nails of iron made with pit coal."[11] It was still a wooden world, the craftsman's essential tool a hammer.

Erasmus Darwin, Charles Darwin's grandfather, a physician, poet, and naturalist, predicted a steam-propelled future in his 1791 poem *The Botanic Garden,* extending even to steampunk aircraft:

Soon shall thy arm, unconquered steam! Afar
Drag the slow barge, or drive the rapid car;
Or on wide-waving wings expanded bear
The flying-chariot through the fields of air.[12]

Yet not many of those alive in 1800 recognized the degree to which their world was changing. "I am astonished," the American steamboat inventor James Rumsey wrote George Washington in 1785, "that it is so hard to force an advantage on the public."[13] The future is a hard sell. "The man in the street in the 1790s,"

Wrigley argues, "would be in no doubt about the occurrence of a revolution across the Channel in France, but he would have been astonished to learn that he was living in the middle of what future generations would also term a revolution." Nor was the man in the street the only person in denial, Wrigley adds. "The three greatest of the framers of classical economics, Adam Smith, Thomas Malthus, and David Ricardo, not only were equally unaware of it, but were unanimous in dismissing the possibility of what later generations came to term an industrial revolution."*[14]

Wagonways and railways extending to and from canals were numerous by 1800. A few railways hauling coal, like the Merthyr Tramroad, bypassed canals where traffic was heavy. But a colliery engineer, William Thomas, only explored the project of a railroad as we know it today, carrying passengers and freight between cities, on the record for the first time at a meeting of the Literary and Philosophical Society of Newcastle on 11 February 1805.[15] Thomas proposed what he called "a middle line," a transportation system with many of the advantages and few of the drawbacks of a canal or a public road.[16] Rather than "the common wood rail used in collieries," Thomas's railway would run on cast-iron "plates" four and a half feet long and five inches wide, wide enough to accommodate the wheels of ordinary horse-drawn carriages, with a flanged edge "to prevent the carriage slipping off the road."[17] The cars could carry grain to market and "return . . . manure from the town," Thomas explained, and as the railway passed through more densely populated areas, people would "avail themselves of so cheap and expeditious a conveyance" as well.[18] Riding in open cars with loads of animal and human waste must seem unappealing today, but exposure to the sight and smell of manure was an everyday occurrence in a world of animal transportation.

Speed would be a benefit, Thomas calculated: "It is expected that the present coach which passes daily between Newcastle and Hexham with four horses, and takes four hours and a half, may with two horses travel the same distance in one hour less time" on his railway.[19] Two tracks side by side would allow trains

*Two centuries later, the greatest physicists of the early twentieth century—Ernest Rutherford, Albert Einstein, and Niels Bohr—would similarly dismiss the possibility of splitting the atom to release nuclear energy as "moonshine."

traveling in opposite directions to pass each other. And, merging the old with the new, as technologies in transition often do, the bed between the rails could serve as a path for riding horseback for people who chose to continue to do so, "as, no doubt," Thomas says parenthetically, "many on horseback will prefer this level road to the present uneven one." The possibility of improving "the present uneven one"—the public highroad with all its cattle and fowl herding along—seems not to have occurred to Thomas or anyone else in Britain in this era.[20]

Jolly Jack Phosphorus—John Anderson, the Scottish professor of natural philosophy who was James Watt's friend—responded to Thomas's proposal with such enthusiasm that many of his friends thought he had gone mad. Anderson's vision of a world made generous and peaceful with technology parallels Wordsworth's vision of manmade beauty as nature's equal. Both visions celebrate the new, energy-rich world just then emerging and discount the associated smoke pollution and debilitating mining and factory labor. But Anderson was too humane an educator not to assess those consequences. He seems to have thought them a reasonable price to pay for the technological paradise he envisioned.

"If we can diminish only one single farthing in the cost of transportation and personal intercommunication." Anderson proposed, "you form, as it were, a new creation—not only of stone and earth, of trees and plants, but of men also; and, what is of far greater consequence, you promote industry, happiness, and joy." The benefits Anderson expected from more efficient transportation included reducing the cost of living, improving agriculture, and connecting town and country. "Time and distance would be almost annihilated," Anderson imagined grandly; "the number of horses to carry on traffic would be diminished; mines and manufactories would appear in neighborhoods hitherto considered almost isolated by distance; villages, towns, and even cities would spring up all through the country; and spots now silent as the grave would be enlivened with the busy hum of human voices, the sound of the hammer, and the clatter of machinery"—as if silence were a burden and noise a virtue. And, concluding, "the whole country would be, as it were, revolutionized with life and activity, and a general prosperity would be the result of this mighty auxiliary to trade and commerce throughout the land."[21] The railroad, when it came, would meet high expectations.

It came quickly enough, but before the necessary technologies converged into a successful system, variety flourished. Passengers were first carried on 25 March 1807 on the Oystermouth Tramroad on the Gower Peninsula in Swansea, northwest of Cardiff in Wales. The cars were horse-drawn, and the operator paid tolls to the company that owned the road.[22] On the opposite coast of England, south of the Tyne at Bewick Main Colliery, a great crowd assembled in May 1809 to witness the inauguration of a succession of inclined planes: railed inclines up and down which stationary steam engines winding strong ropes would haul coal wagons to and from the river. "Four wagons of small coals were brought up the first plane by the steam engine amid discharges of artillery," Robert Galloway reports, "to the great admiration of the spectators."[23]

How to move a train on iron rails was still unsettled in 1812, when John Blenkinsop, the twenty-nine-year-old manager of Middleton Colliery in Leeds, commissioned an engine with a geared drive wheel he had patented that engaged a rack rail paralleling one side of the track. Blenkinsop was, once again, concerned about adhesion, especially of heavy coal wagons. His rack-rail engine hauled coal successfully at Leeds, he wrote a newspaper in 1814, "even during the great falls of snow in January last; and more waggons of coals were

An inclined plane with fixed steam engine at the top hauling a coal wagon.

conveyed to Leeds in that severe month, by the locomotive engine, than in any preceding one by horses."[24]

Other inventors worked other combinations, including an engine that pulled itself forward along a chain, later adapted for canal and river ferries. The most ingenious, and unsuccessful, pushed itself forward on iron hind legs, which tore up the roadbed and stuck in the mud.[25] Finally, in 1813, Christopher Blackett, a newspaper publisher who owned the Northumberland Colliery in Wylam, on the Tyne ten miles west of Newcastle, commissioned an experiment to determine if iron wheels would adhere to iron rails without gearing. He had a large four-wheeled handcar built, worked by a windlass. With six men aboard for added weight, the handcar rolled along without slippage.[26] Richard Trevithick had proven the same point as early as 1803, but Blackett could publicize the fact in his newspaper. He acted on it as well, commissioning a locomotive, *Puffing Billy,* a two-cylinder, four-wheeled engine that hauled coal on the Wylam Wagonway from Wylam to the Tyne.

George Stephenson may have been the most remarkable of all the self-taught engineers of the early years of steam locomotion. As with many people of exceptional skill, he acquired his intimate knowledge of his specialty in childhood. His father, Robert, was a foreman of the Wylam Colliery pumping

Blenkinsop's rack-rail engine; note center geared drive wheel engaging rail rack.

engine when George was born in Wylam, in a cottage two yards back from a wooden wagonway, on 9 June 1781.[27] The boy first worked herding milk cows and hoeing turnips for a widowed farm woman for twopence a day; then led plow horses for fourpence a day; then, a little older, sorted stones—"bats and brasses" (shale and pyrites)—from good coal for sixpence.[28] At twelve, he drove the gin horse at Black Callerton Colliery, west of Newcastle; moved up to working under his father as an assistant fireman at Dewley Burn Colliery; then worked as a fireman at two other pits. (A fireman *made* fires rather than put them out, shoveling coal into steam engine fireboxes at the right pace to keep up the steam.) By seventeen, George was given charge of a new pumping engine at Water Row, on the banks of the Tyne a few miles west of Newcastle.

Stephenson acquired the skills necessary to this succession of jobs despite being illiterate. He learned to read and write only in his eighteenth year, 1799, and his writing would always be labored. In his scant spare time, he made and repaired boots and shoes and repaired clocks and watches to supplement his wages. By 1801, the twenty-year-old was earning a pound a week as brakeman in charge of the winding engine at the Dolly Pit, Black Callerton, and had put aside enough money to marry. His first and second courtships failed, but he succeeded on the third try, marrying Fanny Henderson, a housemaid twelve years his senior, in November 1802. Their son Robert was born a year later.

Sadly, Fanny was consumptive in an era when about half of young adults with tuberculosis died of the disease. In 1805 she lost a daughter at three weeks and died herself the following spring. Stephenson, stricken, hired a housekeeper to care for three-year-old Robert. Then he walked north two hundred miles into Scotland to Montrose, on the North Sea northeast of the city of Dundee, and took up work there operating a Boulton & Watt steam engine at a spinning mill. He worked at the mill for more than a year and saved £28 (today £2,000, or $2,900) before returning to the Newcastle upon Tyne area, only to encounter more misfortune: his father had been scalded and blinded in a steam engine accident. Stephenson used his savings to pay his father's debts and helped support him for the rest of his life.

Besides working with steam engines from the coal up, as it were, George also, on Saturdays when the engines were idle, took them apart and put them back together, examining each component as he did so to understand its function. Between feeding them, operating them, dismantling and repairing them, he learned their mechanism at a level of tacit knowledge unlikely even for a formally educated professional engineer. Those who worked with him, including professionals, found his ability to diagnose engine troubles uncanny. It was bred in his bones.

An opportunity to advance presented itself in 1811, when an atmospheric engine installed the previous year on a new pit at Killingworth Colliery failed to clear the deepening pit of water. The manager hired several mechanics, one after another, to improve it, without success: the mine remained flooded. "A rumor went forth," reports Stephenson's lifelong friend Thomas Summerside, "that the entire undertaking would be a failure."[29] Stephenson studied the great beam engine and concluded that its cold-water injection system was inadequate, which meant it couldn't develop an adequate vacuum and was thus underpowered.[30] He was only a brakeman, but everyone else had failed; the manager invited his help. Stephenson raised the level of the injection water in its cistern by ten feet, enlarged the injection valve, and, most controversially, doubled the steam pressure from 5 psi to 10 psi. The doubters scoffed, especially when the engine cranked the beam so powerfully that it banged down on its stops and shook the engine house—"came bounce into the house," Stephenson said. That panicked the viewer, but as the pit shaft began to clear of water, the engine settled down to running smoothly.[31] Two days later, it had pumped the shaft dry, earning Stephenson a £10 reward. The following year, the Grand Alliance, the powerful cartel of mine owners, promoted Stephenson to chief enginewright in charge of all the machinery in their collieries at an annual salary of £100 (today £6,300, or $9,000). That salary soon became a retainer, and Stephenson an industrywide consultant.

Besides his tacit knowledge of steam engines, Stephenson's other secret weapon was his son, Robert, the price of whose formal education was teaching his father in turn. Once Robert had completed his training as an engineer, he partnered with his father throughout George's life.

Stephenson built his first steam locomotive in 1814. He modeled it on Blenkinsop's patented rack-rail locomotive but avoided the cost of licensing the patent by using smooth wheels—with the additional innovation of flanging. That engine, the *Blucher,* rolled past Stephenson's birth cottage in Wylam on

25 July 1814.[32] "Two days later," Galloway writes, "it was tried upon a piece of [rail]road . . . ascending about one [foot] in 450; and was found to drag after it, exclusive of its own weight, eight loaded carriages, weighing altogether about thirty tons, at the rate of four miles an hour; and after that time continued regularly at work."[33] Another, similar engine soon followed. Late in life, Stephenson recalled them happily and conflated the two: "The first locomotive that I made was at Killingworth Colliery, and with Lord Ravensworth's money. Yes! Lord Ravensworth and partners [the Grand Alliance] were the first to entrust me with money to make a locomotive engine. . . . We called it *My Lord*. I said to my friends, there was no limit to the speed of such an engine, if the works could be made to stand it."[34]

The story of Stephenson's invention of a safety lamp for collieries contaminated with flammable gas falls outside the range of this history, but his breakthrough occurred prior to chemist Humphry Davy's better-known invention and was equally effective. Davy's lamp used a gauze screen, Stephenson's, a metal tube punched with a grid of fine holes. But both lamps worked on the same principle: the flame of the oil lamp ignited any small quantity of flammable gas that seeped through the mesh into the lamp's interior, with the mesh cooling the ignited gas enough to prevent it from propagating beyond the mantle into the mine. Davy, the discoverer of five elements—barium, calcium, boron, strontium, and magnesium, all in the same year, 1808—a Royal Society lecturer, and, later, its president, received a prize of £2,000 (today £132,300, or $188,000) for his invention. The award incensed the Grand Alliance and other Newcastle colliery owners, who knew that Stephenson's invention had priority. In response, they raised a purse of £1,000 for their man. He had previously received a consolation prize of £100 as well.

Although Stephenson began to manufacture steam locomotives in steady numbers after *Blucher* and *My Lord* had demonstrated their utility, railway infrastructure continued to limit development. Early-nineteenth-century cast iron was far more impure and brittle than cast iron is today and often broke under the weight of heavy steam engines. Consequently, rail sections had to be short, about three feet, which in turn introduced numerous unstable joints. Allowing for a horse path between rails—as late as 1828, Stephenson's first major British railway still hauled 43 percent of its tonnage with horses—meant that rails had to be supported on stone blocks rather than connected with crossties, making it difficult to keep them aligned.[35]

Cast-iron rails, despite their limitations, met a characteristic requirement of new technology: lower cost. Haulage by rail cost less than by packhorse or horse cart. "In an account of Dunfermline, published in 1815," Galloway writes, "we are informed that 'within these five years, coals have been sent to the [River] Forth, for exportation, on cast-iron railways'; and this mode of conveyance now saves the labor of not fewer than one hundred horses."[36] Thirteen million tons of coal were consumed within Britain that year as well.[37]

Wrought iron began to replace cast iron before 1820, when a Northumberland railway engineer named John Birkinshaw patented a method of rolling wrought iron rails in various shapes in fifteen-foot lengths that could withstand the weight of steam locomotives pounding and running over them. The Scottish engineer Robert Stevenson, the writer Robert Louis Stevenson's grandfather and a former student of Jolly Jack Phosphorus, praised the new material in a report on a proposed new railroad: "Three miles and a half of [malleable iron rails] have been in use for about eight years on Lord Carlisle's works, at Tindal Fell, in Cumberland, where there are also two miles of cast-iron rail; but the malleable iron road is found to answer better in every respect." It was, Stevenson added, "not only considerably cheaper in the first cost than the cast-iron railway, but is also much less liable to accident."[38]

George Stephenson recognized the new material's long-term value, writing

Stone-block sleepers supporting cast-iron rails.

the Scottish engineer in 1821, "Those rails are so much liked in this neighborhood, that I think in a short time they will do away [with] the cast-iron railways. They make a fine line for our engines, as there are so few joints compared with the other."[39] Stephenson liked the new rails so much that he recommended them for the first public railway he was engineering that used steam locomotives—the Stockton & Darlington—even though the recommendation cost his own company, which still made the older cast-iron rails, a major sale.

The Stockton & Darlington opened to fanfare on 27 September 1825. Its coal cars loaded that day with officials and spectators, it ran from Stockton, on the River Tees thirty miles below Newcastle upon Tyne, twenty-five miles inland to Witton Park Colliery. Since it was primarily a low-speed coal-hauling venture, with passengers an afterthought, half its length had been laid with cast-iron rails, the other half with wrought iron.[40]

By then, Stephenson was working on the next important railway under development in England: the Liverpool & Manchester. Unlike the Stockton & Darlington, which had won through the parliamentary authorization process with little difficulty, the Liverpool & Manchester encountered fierce resistance from canal owners, stagecoach operators, turnpike trusts, and innkeepers who had come to understand that railway competition was likely to be fatal to their businesses and investments. Nor did the landed gentry whose wayleave the new railway needed to acquire want any part of so noisy and smoky a fire hazard. (The Duke of Cleveland had resisted the Stockton & Darlington because it would interfere with his fox cover—the area of his property where foxes denned—and had demanded that the plan of the line be amended to avoid it, which it was.[41])

George Stephenson faced further harassment when he surveyed the course

Opening day on the Stockton & Darlington Railway, 27 September 1825.

for the Liverpool & Manchester. He and his men had "sad work with Lord Derby, Lord Sefton, and Bradshaw the great Canal Proprietor," he wrote his business manager. They had blockaded their grounds on every side, he complained, to bar the surveying from going forward. Worse, "Bradshaw fires guns through his ground in the course of the night to prevent the Surveyors coming on in the dark." The railway was determined to force a survey through, he added, even though Lord Sefton warned them he'd have a hundred men out on his grounds to prevent them.[42] Stephenson would testify in Parliament that he was "threatened to be ducked in the pond if I proceeded." He managed only a hasty survey, "by stealth, at the time when the persons were at dinner."[43]

The prospectus for the Liverpool & Manchester line claimed that it would deliver goods much faster than shipment by canal. The unreliability of early steam engines justified questioning that claim, but travel we would not consider rapid today also seemed impossible in a world where no one traveled faster than a horse could gallop. "What can be more palpably absurd and ridiculous," asked a reviewer for London's *Quarterly Review* who *favored* a plan for a railway to Woolwich, "than the prospect held out of locomotives traveling *twice as fast* as stagecoaches! We should as soon expect the people of Woolwich to suffer themselves to be fired off upon one of Congreve's . . . rockets,* as trust themselves to the mercy of such a machine going at such a rate. . . . We trust that Parliament will, in all railways it may sanction, limit the speed to *eight or nine miles an hour,* which . . . is as great as can be ventured on with safety."[44]

George Stephenson faced a far more brutal challenge than a reviewer's skepticism about locomotive speeds in three days of testimony before the parliamentary committee investigating the plans for the Liverpool & Manchester. Beginning on 25 April 1825, the engineer recalled, he suffered cross-examination from a group of hostile barristers. "Some member of the Committee asked if I was a foreigner, and another hinted that I was mad," he recalled.[45] The "foreigner" question ridiculed what Smiles calls Stephenson's "strong Northumbrian accent," and certainly he was attacked in part because his origins were

*Sir William Congreve developed the first British military rockets from Indian models in 1805. It was their "red glare" over Fort McHenry during the War of 1812 that Francis Scott Key evoked in his "The Star-Spangled Banner."

working class and he lacked formal education. He may have, but Stephenson understood the physics of motion better than the barrister leading the pack:

> You *say that the machine can go at the rate of twelve miles an hour; suppose there is a turn in the road, what will become of the machine?*
> It would go round.
>
> Would *it not go straight forward?*
> No.[46]

Stephenson's real humiliation came when his interrogator, a shrewd barrister named Edward Alderson, challenged his survey for the new line, found numerous faults in its measurements, and extracted from Stephenson an admission that he had not taken the surveying levels himself on which he had based his cost estimates:

> You *do not believe that you are out in your levels?*
> I have made my estimate from the levels which I believe are correct.
>
> Do *you believe, aye or no, that your levels are correct?*
> I have heard it reported that they are not.
>
> Did *you take the levels yourself?*
> They were taken for me.
>
> Other *people have taken them for you, and upon their estimate you have made your estimate?*
> Yes.[47]

Stephenson knew that he could lay out the railway properly even if the formal survey was inaccurate, but having that inaccuracy exposed made him look like a fool or a fraud. It left him shaken. The bill failed that year, but with adjustments in the route to bypass several estates and a gift of a thousand shares of railway stock to the Marquis of Stafford, a canal owner, for withdrawing his objection, it passed the House of Commons the following year and cleared the House of Lords on 27 April 1826.

Stephenson planned the railway to cross a singular feature called a moss—in this case, Chat Moss, an area of about twelve square miles lying west of Manchester. A moss is a peat bog, a wet marsh, a shallow basin formed by glacial scouring that fills with dead and waterlogged vegetation to form a domed, quicksand-like mass.[48] "This huge fungus," as a contemporary writer called it, ranged in depth from ten to forty feet. "Who but Mr. Stephenson would have thought of entering into Chat Moss," the solicitor who summed up for the railway opposition asked, "carrying it out almost like wet dung? It is ignorance almost inconceivable. It is perfect madness, in a person called upon to speak on a scientific subject, to propose such a plan."[49]

To lawyers and politicians, it may have seemed perfect madness, but Stephenson set his work crews building a railway across Chat Moss in June 1826, after the Liverpool & Manchester bill received royal endorsement. He knew that boats floated on the ocean, and wet though Chat Moss was, it was much more dense with its freight of living and dead vegetation than seawater was. If he could build a floating platform sufficiently stable to support the rails, his railroad should go through.

Stephenson ordered deep longitudinal drains cut along both sides of the planned roadbed. When the moss surface between the drains dried and firmed, he judged, his workmen could lay the roadbed with single or double layers of hurdles—four-by-nine-foot mats woven of hazel branches and covered with heather. Sand and earth ballast would then be spread over the hurdles and crossties and rails laid on the ballast.

At first, the moss defeated Stephenson's attempt at drainage. "[W]hen the longitudinal drains were first cut along either side of the intended railway," Smiles writes, "the oozy fluid of the bog poured in, threatening in many places to fill it up entirely."[50] Stephenson thought over the problem and decided to fill the drains with a makeshift sewer that would carry away the water and prevent it from overflowing the roadbed. In Liverpool and Manchester, he bought up every old tallow barrel his men could find, knocked out the ends, had the drains recut, and laid down the barrels with their ends shoved together loosely to allow water to flow in.

The drainage was only superficial, however. Below was ten or twenty feet of watery bog: the wooden barrels bobbed up like corks and drifted apart. Stephenson had them weighted with clay to hold them down. That worked. His men spread tons of hurdles, sand and earth over the roadbed between the drains, says Smiles, "but it was soon apparent that this weight was squeezing down the moss and making it rise up on either side of the line, so that the railway lay as it were in a valley, and formed one huge drain running across the bog."[51] Stephenson had the hurdles and earth extended thirty feet out on each side of the railway, forcing the bog down and raising the rails again.

On a lower-lying area of Chat Moss, Stephenson needed a twelve-foot embankment to level his railway. His men excavated the moss itself and piled it up along with sand and gravel to make the embankment. "After working for weeks and weeks in filling in materials to form the road," Stephenson recalled, "there did not yet appear to be the least sign of our being able to raise the solid embankment one single inch; in short, we went on filling in without the slightest apparent effect." The directors began to worry that the expensive effort would fail. They consulted other engineers, who found against Stephenson's plan. Seriously alarmed, they called a board meeting on Chat Moss to decide whether or not Stephenson should end the work. But it was too far along, with too much invested. "So the directors were *compelled* to allow me to go on with my plans," Stephenson writes, "of the ultimate success of which I myself never for one moment doubted."[52]

The filling operation continued. Stephenson hired hundreds of workers to cut the moss into blocks for a half mile around and skin it off with sharp spades like turf. Once the turf blocks dried, Stephenson had them laid to form the embankment. At first, the turf sank to the bottom. Eventually the growing pile rose above the wet surface of the Moss, then settled slowly to merge with the floating road. The finished road, says Smiles, "looked like a long ridge of tightly pressed tobacco-leaf."[53]

The work went on for six months. Stephenson estimated his workmen moved, with pick and shovel, some 520,000 cubic yards of material.* They completed the four-mile Chat Moss crossing in December 1829. The

*About half the volume of the Empire State Building.

View of the Railway Across Chat Moss, by T. T. Bury, 1833.

opponents of the line had claimed that mastering Chat Moss would cost "upwards of £200,000."[54] It cost £27,719.[55]

The Liverpool & Manchester Railway hauled its first test load of freight and passengers between its two eponymous cities on Saturday, 1 December 1830.[56] In eighteen wagons, it moved 135 bags and bales of American cotton, 200 barrels of flour, 63 sacks of oatmeal, 34 sacks of malt, and 15 passengers. Engine, train, and contents weighed 86 tons and maintained an average speed of twelve and a half miles per hour. The completed line opened to general traffic nine months later, on 15 September 1831.

Before then, however, in 1829, the Stephensons, father and son, had to prove to the Liverpool & Manchester directors that a steam locomotive was better fitted to moving freight and passengers than rope haulage with a stationary engine, odd as that choice sounds today. "[I]t is not too much to say that the whole future of the steam locomotive was now in the balance," writes Stephenson biographer W. O. Skeat. ". . . To be fair to the stationary-engine supporters, they could point to a higher degree of reliability in their favored motive power, contrasting with the steam locomotive, still in its infancy and largely an unknown quantity; at that stage in its development, dependability had hardly yet been achieved."[57]

In late 1828 the Liverpool & Manchester directors sent several of their members on a tour of northern England to find out what other railway operators used for locomotion. The delegation returned "decidedly adverse to

Horse Power," a participant wrote, "and rather in favor of fixed Engines."[58] Next Stephenson visited the region, compared fixed and locomotive power, and reported himself in favor of locomotives. The directors then hired two expert engineers, who repeated the northern tour for the third time but returned undecided, although they thought fixed engines would operate at lower cost.[59]

The directors had faith in George Stephenson. If he believed in locomotives, they were inclined to follow his lead. But which locomotive? How could they decide which of the five types of locomotive then in operation—Blenkinsop's rack-rail, Hedley's *Puffing Billy*, Robert Stephenson's *Lancashire Witch*, Timothy Hackworth's *Royal George*, John Urpeth Rastrick's *Agenoria*—or some new design yet to emerge would be best suited to reliable, daily, scheduled operation hauling both passengers and freight? Rural England routinely compared animal breeds in cattle shows, horse shows, sheep shows, and more: head-to-head comparison should work as well for steam engines. So the directors decided to stage a contest: a substantial prize of £500 (today £40,000, or $57,000) for the locomotive engine hauling freight and passengers that delivered the best time on a prescribed course. One of the expert engineers advising them, James Walker, suggested that the course might be at Rainhill, a village about ten miles east of Liverpool on the new Liverpool & Manchester line. There were stationary engines at Rainhill, Walker advised, that would "enable you to judge of the comparative advantages of the two systems."[60] The directors agreed.

On 25 April 1829 they published a list of "Stipulations & Conditions" for the Rainhill trials. First on the list was a requirement that the "engine must effectually consume its own smoke."[61] Black, sulfurous coal smoke blowing across their lands had offended the rural gentry in particular. This requirement had already been written into law in the 1825 Railway Act of George IV. To comply, locomotives had been designed to burn coke, a much cleaner fuel than coal, and to route waste steam up the chimney to increase the draft and fan the fire.

If the engine weighed six tons, the rules next stipulated, it "must be capable of drawing after it, day by day, on a well-constructed railway on a level plane, a Train of Carriages of the gross weight of twenty tons . . . at the rate of ten miles per hour with a pressure of steam in the boiler not exceeding 50 pounds per square inch."[62] Lighter engines were assigned lighter loads.

The next stipulation called for two safety valves, neither of which could be fastened down and one of which "must be completely out of the reach of contact of the engine-man." Boilers blew up. The directors wanted no disasters marring their contest, one purpose of which was to advertise the railway as a new and better means of common transportation. They stipulated that the boiler and its related machinery should be able to withstand test filling with water under a pressure of 150 psi.[63] Further stipulations limited the engine's total weight to six tons supported on springs and standing no more than fifteen feet high "to the top of the chimney." It could cost no more than £550 delivered.[64]

Proposals poured in for engines powered by steam, by compressed air, by perpetual motion. Five contestants had passed muster when the crowd, estimated at ten to fifteen thousand, gathered at Rainhill on a cool autumn morning at the beginning of October 1829. Two horses harnessed together on a treadmill powered one of the entries, the *Cycloped*, hardly a serious contender.[65] The remaining four were *Novelty*, a new engine which two London engineers, John Braithwaite and John Ericsson, built especially for the trials; the Stephensons' *Rocket*; Darlington engineer Timothy Hackworth's *Sans Pareil*; and Leith engineer Timothy Burstall's *Perseverance*.[66]

In *Rocket*, the Stephensons built what was essentially a racing locomotive. Light, fast, and powerful, it was made to appear even lighter by its black and bright-yellow color scheme, patterned on the customary color scheme of a fast stagecoach, with a tall smokestack painted white. Instead of the usual large single or double pipe carrying fire through the boiler barrel to make steam, *Rocket* carried the hot combustion gases from a separate firebox through twenty-five small-diameter, thin-walled copper tubes that passed through the water in the boiler barrel before exiting to the chimney. Even though they were lighter, the multiple tubes exposed almost three times the surface area that a larger single or double tube would offer for heat transfer.[67]

No one had ever tried attaching multiple copper tubes to the end plates of an iron boiler. They were wedged into place with a hollow, cone-shaped iron fitting called a ferrule. When Stephenson tested the results with pressurized water, however, the boiler barrel bellied out enough to loosen the tubes. Improvising, he put in what he called "stays": iron rods—long, narrow bolts—that bolted together the two ends of the boiler barrel and kept it rigid.

Rocket's innovative design proved its worth. More reliable than the other

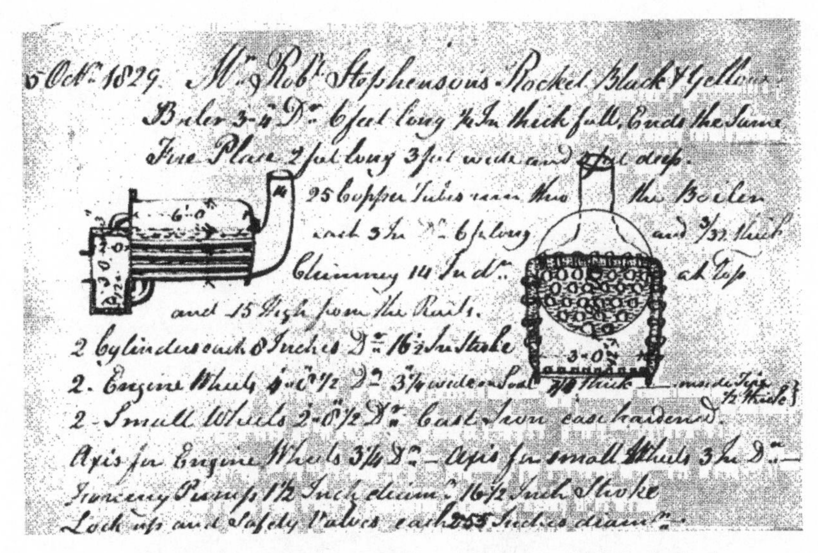
5 Oct. 1829. Mr. Robt. Stephenson's Rocket Black & Yellow
Boiler 3-4 Dr. 6 feet long ½ In thick full, Ends the Same
Fire Place 2 feet long 3 feet wide and 3 feet deep.
25 Copper Tubes run thro the Boiler
and 3 In Dr. 6 feet long and 3/32 thick
Chimney 14 Inch Dr. at Top
and 15 High from the Rails.
2 Cylinders each 8 Inches Dr. 16½ In Stroke
2. Engine Wheels 4-8½ Dr. 3¾ wide on Sole ... thick ... made Tyre ½ thick
2 - Small Wheels 2-8½ Dr. Cast Iron case hardened.
Axis for Engine Wheels 3¼ Dr. – Axis for small Wheels 3½ Dr.
Forcing Pump 1½ Inch diam. 16½ Inch Stroke
Lock up and Safety Valves each 2.5 Inches diam.

Judge's notes on Rainhill trials showing Rocket's multitube boiler.

contestants' engines, it was the only locomotive still running on the last day of the contest, which it won. On 8 October, the first to be timed, it had easily hauled the thirteen-ton load of stone and passengers, averaging about sixteen miles per hour in fifty passes back and forth along the one-and-a-half-mile test track—a challenge that simulated traveling the seventy-mile round trip from Liverpool to Manchester and back.[68]

When one of the other locomotives broke down midway through the week of trials, Stephenson had uncoupled *Rocket*'s tender and raced back and forth seven times on the test track to demonstrate its speed, inspiring a reporter for the *London Times* to exclaim in print that "the engine alone shot along the road at the incredible rate of 32 miles in the hour. So astonishing was the celerity with which the engine, without its apparatus, darted past the spectators, that it could be compared to nothing but the rapidity with which the swallow darts through the air. . . . The power of steam is unlimited!"[69] On another such exhibition run, *Novelty* hauled a carriage load of about forty-five people at thirty miles per hour—so fast, according to a reporter, "that we could scarcely distinguish objects as we passed them by."[70] Apparently our ancestors had yet to learn to pan their heads for rapid viewing.

A shortage of wood had driven the English to take up burning coal. Digging

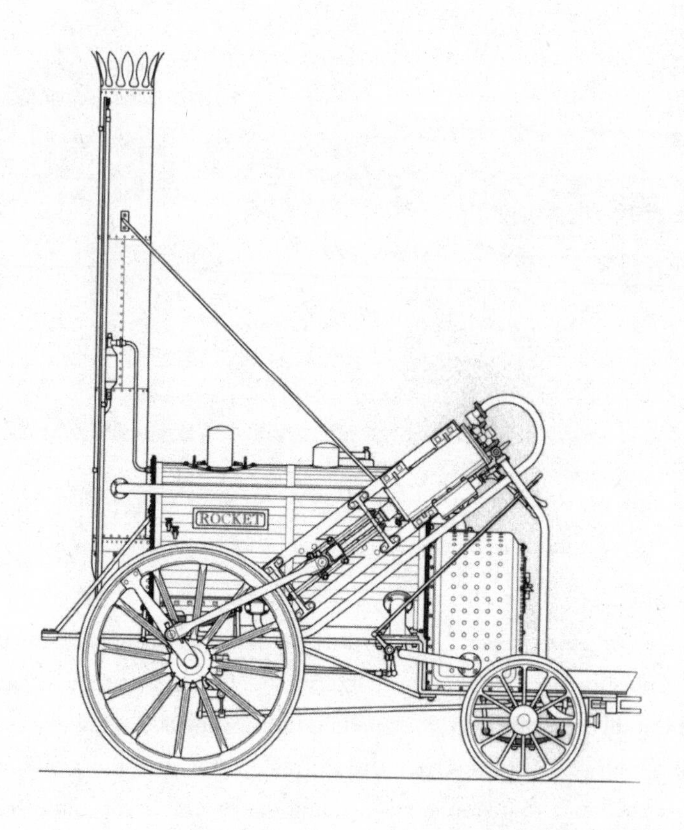

Stephenson's *Rocket* configured for the Rainhill trials.

ever deeper for coal, they found their mines flooding, driving them to invent engines to pump out the water. Raising water with fire, as they said—they liked the phrase—demonstrated that heat energy could be converted to mechanical energy. And if heat energy could pump water, could it not also turn wheels? It could, in mills, in factories, on the open road clumsily, on the railed road with unimaginable power and speed. And that changed almost everything, first in England, later in America and throughout the world.

James Walker, the Georgian engineer, saw with remarkable foresight, as early as 1831, the revolution this abrupt transition from organic to fossil-fuel energy had begun. "Perhaps the most striking result produced by the completion of this Railway," he wrote of the new Liverpool & Manchester line, "is the sudden and marvelous change which has been effected in our ideas of time and space. . . . Speed—dispatch—distance—are still relative terms, but their meaning has been totally changed within a few months: what was quick is now slow; what was distant is now near; and this change in our ideas will not be limited to the environs of Liverpool and Manchester—it will pervade society at large. Our notions of expedition, though at first having reference to locomotion, will influence, more or less, the whole tenor and business of life." As we traveled faster, Walker meant, so would we live faster, leaving the slow, vegetative world behind, blurring past animals and one another, seeing more but also seeing less—seeing, at least, differently.

A corollary was that living faster meant embracing the mechanical; the machines that augmented our rates of movement and of change. "From west to east," Walker concludes, "and from north to south, the mechanical principle, the philosophy of the nineteenth century, will spread and extend itself. The world has received a new impulse."[71]

It had, and the transformation would be profound. But the human world still largely lingered in the dark for half the earth's each turning. There were remedies for that condition as well: oils, rushes, tallow, the fat of pigs, coal gas, whales. All would serve in their time.

PART TWO

LIGHT

SEVEN

RUSHLIGHT TO GASLIGHT

The common soft rush grows bright green along stream margins, around ponds, and in marshes throughout the warmer world. Two to three feet tall, perennial, it makes a cheap substitute for a candle. "The largest and longest are best," writes Gilbert White, the eighteenth-century naturalist whose 1789 *The Natural History and Antiquities of Selborne* was the first book of its kind in England. Gather them at the height of summer, White advises, but they will still serve the purpose even into autumn.[1]

Peeling them is a nice trick: stripping off the outer rind "so as to leave one regular, narrow, even rib from top to bottom that may support the pith." Children learned it quickly enough, White writes, "and we have seen an old woman, stone-blind, performing this business with great dispatch, and seldom failing to strip them with the nicest regularity." Then the peeled piths are spread on the grass "to bleach and take the dew for some nights, and afterwards be dried in the sun."[2]

The pith is the secret. It's a cylinder of hollow tubes. In the living plant, the tubes carry sap; in a rushlight, they hold fuel. For that, they're dipped into hot fat—common kitchen grease of almost any kind. "The scummings of a bacon-pot" will do, White says, and cost nothing. "If men that keep bees will mix a

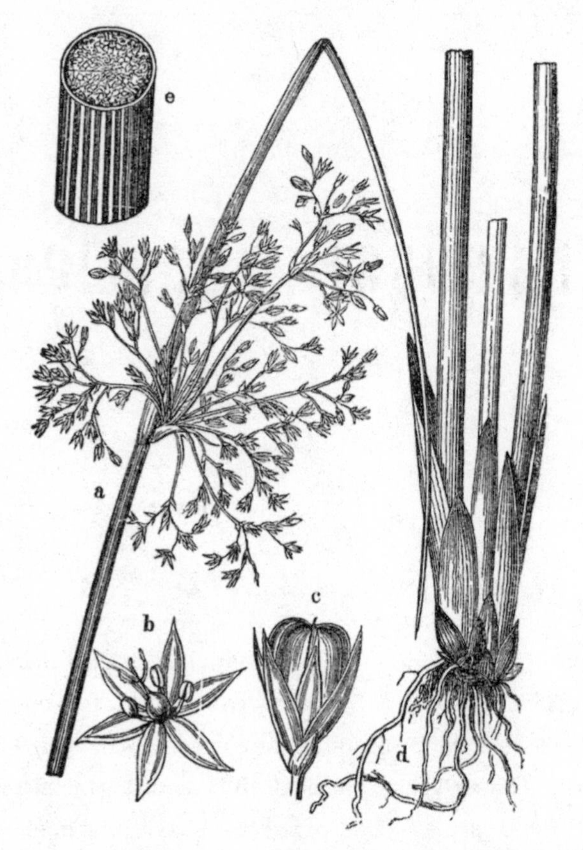

Common soft rush (*Juncus effusus*). Cross section, *e*, shows stem with pith.

little wax with the grease, it will give it a consistency, and render it more cleanly, and make the rushes burn longer; mutton-suet would have the same effect."[3]

White bought a pound of dry rushes for one shilling and counted 1,600 stems. Enough grease to prepare them—six pounds—cost two shillings more. White timed one of these rushlights: it burned for fifty-seven minutes. If his sixteen hundred rushes averaged only a half hour each, he calculated, "then a poor man will purchase eight hundred hours of light, a time exceeding thirty-three entire days, for three shillings. . . . An experienced old housekeeper assures me that one pound and a half of rushes completely supplies his family the year round, since working people burn no candle in the long days, because they rise and go to bed by daylight." Small farmers, White adds, burned rushlights morning and evening in winter in the dairy and kitchen.[4]

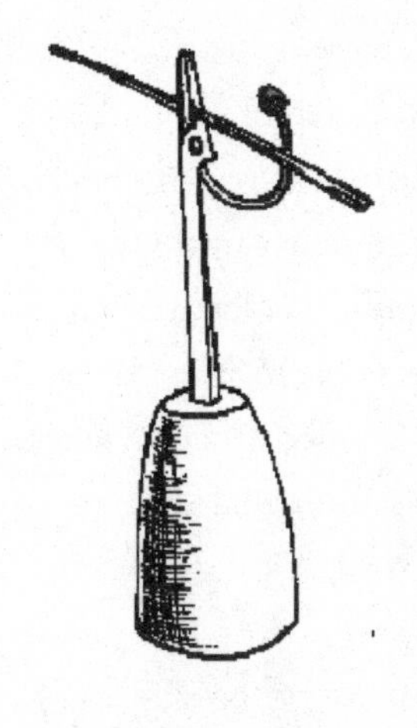

Rushlight in lamp stand.

Eight hundred hours of light, but no more than a candle flame's worth at a time. Oil lamps, like miniature gravy boats, burned even more feebly with their wicks of twisted rag. The oil might be flax, rape, walnut, or fish liver, and, around the Mediterranean, the industrious olive. On St. Kilda, in the Hebrides west of Scotland, the stomach oil of the fulmar, an oily, all-purpose seabird, made lamplight. "The Shetland Islanders," writes a folklore historian, "as recently as the end of the nineteenth century, threaded wicks through stormy petrels [killed and dried for the purpose], birds so fat and oily that they eject oil through the digestive tract when caught."[5] For the poor, rushes and hearth fire served for light; for the yeoman and squire, smoky tallow candles; for the rich, candelabra of beeswax backed by mirrors.

Without adequate lighting, the country night was dark, though lustered by starlight or full moon. Eighteenth-century Birmingham's Lunar Society—country neighbors Erasmus Darwin, Matthew Boulton, James Watt, Josiah Wedgwood, and chemist Joseph Priestley—convened when the moon was full, bright enough to cast shadows, so they could walk to their meetings.[6] But night in the city was dark and threatening. In ancient Rome, a historian warns, "night fell over the city like the shadow of a great danger. . . . Everyone fled to his home, shut himself in, and barricaded the entrance."[7] John Stow, the Elizabethan chronicler, says that in the eleventh century, King William I—William the Conqueror—"commanded that in every town and village, a bell should be nightly rung at eight o'clock, and that all people should then put out their fire and candle, and take their rest." We call such a prohibition a curfew, a word derived from Norman French *covre le feu:* "cover the fire!" Henry I lifted his father's curfew, Stow adds, but "by reason of wars within the realm, many men . . . also gave themselves to robbery and murder in the night."[8]

Because rotting fish phosphoresce faintly, dried fish skins were tried for explosion-proof lighting in coal mines before Trevithick and Davy invented their safety lamps. Skin light was too dim for mining but bright enough that Erasmus Darwin, returning home at night from medical school in Edinburgh in 1754, picked up a discarded fish head for a light by which to read his pocket watch.[9] A Victorian chronicler of eighteenth-century London thought that oil-burning streetlamps at best "served to shed a faint glimmer of light, or rather to make the darkness visible at street corners and crossings from sundown to midnight."[10] Oil lamps were uncommon in early and frontier America, where reading and handwork depended on hearth light (as in sentimental illustrations of the young student Abraham Lincoln reading by the fire), and resinous splinters of fatwood pine replaced candles.

A threat to Britain's national security—a consequence of the shortage of wood—catalyzed British development of a superior form of light from coal in the transition decades between the eighteenth and nineteenth centuries. This time shipworm, a worm-shaped mollusk that gnaws burrows that can perforate a hull in months, threatened the Royal Navy's wooden warships. Coatings of wood-derived tar or pitch deterred the animals, but a country short of timber lacked the necessary raw material. Coal tar was a potential substitute.

Archibald Cochrane, born in 1748, a Scotsman and the ninth Earl of Dundonald, earned the nickname "daft Dundonald" for his many original inventions, including the economical extraction of coal tar from cannel coal.*[11] Dundonald inherited an estate largely signed away, his son writes ironically, from "support[ing] one generation of the Stuarts [and] rebellion against another." His problem, his son believed, was "too many irons in the fire": a nice cliché for an industrialist and inventor who worked with kilns. "One by one, his inventions fell into other hands, some by fair sale, but most of them by piracy, when it became known that he had nothing left wherewith to maintain his rights. In short, with seven children to provide for, he found himself a ruined man."[12]

Lesser British warships in Dundonald's day hobnailed their hulls against shipworm with large-headed iron nails. Dundonald, who went to sea at twenty, first thought of using a coating of coal tar instead of hobnails to prevent

*Cannel coal is an older name for what is today called oil shale, a hard, shiny, bituminous mineral with a high oil content.

shipworm damage while still a midshipman. The idea stayed with him through two years of naval service. His family estate, Culross, two square miles along the Firth of Forth inland from Edinburgh, held resources of timber, coal, salt, iron ore, and fire clay. Dundonald came home to supervise his coal mines and struggle with his family's and his own increasing debt. He assumed the family title when the eighth earl, his father, died in 1778.

The coal-tar extraction method Dundonald invented in 1780 and patented in 1781 involved smoldering burning coal in a kiln by controlling the air intake rather than heating the kiln externally to roast the coal inside, thus saving the cost of fuel. (This was the process that Lord Sheffield had included in his list of British inventions that would replace the loss of the thirteen American colonies: "Lord Dundonald's discovery of making coke [from coal] at half the present price.")

Coal tar came out of Dundonald's kilns, but so did coal gas, a mixture of hydrogen, carbon monoxide, methane, volatile hydrocarbons, and traces of carbon dioxide and nitrogen. His son says Dundonald discovered the illuminating qualities of coal gas by accident: "Having noticed the inflammable nature of a vapour arising during the distillation of tar, the Earl, by way of experiment, fitted a gun barrel to the eduction pipe leading from the condenser. On applying fire to the muzzle, a vivid light blazed forth across the waters of the Firth, becoming, as was afterwards ascertained, distinctly visible on the opposite shore." If so, Dundonald dismissed the product as "merely a curious natural phenomenon."[13]

Certainly the earl was preoccupied with the practical matter of recouping his family fortunes—"He has beggared himself," his mother wrote bitterly around this time, "and forgets it was his own doing"—but missing the value of an inflammable by-product of his coal-tar and coke manufacturing process seems at least unimaginative.[14] An elderly employee, a blacksmith who had assisted Dundonald with his experiments, told a eulogist, "His Lordship . . . was in the habit of burning the gas in [Culross] Abbey [House] as a curiosity, and for this purpose he had a vessel constructed resembling a large tea urn; this he frequently caused to be filled and carried up to the Abbey to light the hall, especially when he had any company with him."[15] The eulogist speculates that Dundonald had coal tar on his mind, that the illuminating quality of the raw gas was poor, and that the high cost of cast-iron water pipes for distributing the gas made their use unusual in the Scotland of the day.

Whatever the reason, Dundonald missed developing a practical product of great value. The British Board of Admiralty rejected Dundonald's proposed coal-tar shield against shipworm, preferring instead to clad its ships with copper. The earl had to sell off what was left of his estates in 1798 to pay his debts and died in poverty in Paris in 1831.[16]

Yet Dundonald almost certainly helped alert Boulton & Watt to the promise of coal gas as a lighting fuel. Thomas Cochrane "vividly" remembered a trip to London with his father when he was seven, in 1782, when they stopped off to visit James Watt. "Amongst other scientific subjects discussed during our stay," he writes, "were the various products of coal, including the gas-light phenomenon of the Culross Abbey tar-kiln."[17] Similarly, Watt's industrialist partner, Matthew Boulton, discussed manufacturing projects at dinner with Lady Dundonald during a tour of Scotland and Ireland in 1783.[18]

Many others experimented with coal gas during the late eighteenth century, particularly Frenchmen looking for a source of inexpensive gas for sustained ballooning after the first hot-air balloon flights of the Montgolfier brothers in 1783. Jan-Pieter Minckelers, a Dutch chemist at the University of Louvain, made gas from coal as well as other raw materials as various as straw, wood, bones, and nuts.[19] Minckelers reported his findings in a 1784 memoir. He concluded that coal gas was the easiest to produce in quantity and in 1785 used coal gas to light his lecture hall. He abandoned further exploitation, however, when he escaped Louvain in 1790 during the Brabant Revolution.

Alessandro Volta, before he invented the first battery, in 1799, invented a lighter fueled with coal gas and ignited with a spark that became popular among science cognoscenti throughout Europe and in England. Ironically, the purpose of Volta's elaborate apparatus was merely to light candles (the friction match was not invented until 1828), although it was also used as a lamp.

Charles Diller, a Dutch instrument maker, used distilled gases to produce what he called "philosophical fireworks" of colored flames for public displays. Sir Gilbert Elliot, a Whig member of Parliament (MP), attended one of Diller's theater exhibitions in London in 1788 with a group of friends. He wrote an acquaintance afterward that it was "most beautiful and most ingenious. . . . It

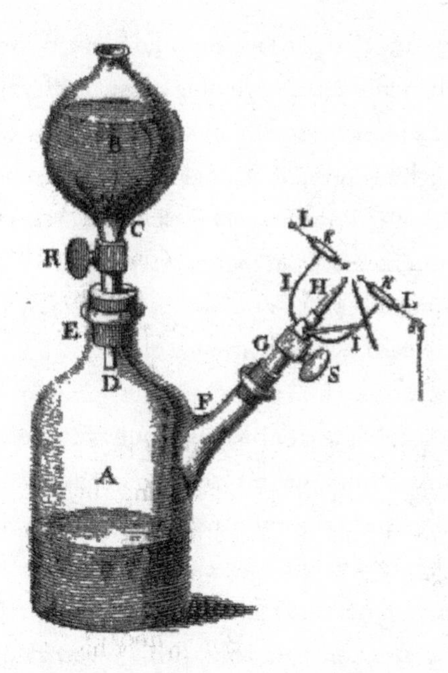

Volta's lighter. Water in upper chamber is released into lower chamber *A* to force gas out through nozzle *H*, which is then ignited with a spark fired between the two poles *L-L* (spark-generating apparatus not shown).

is an imitation of fireworks, but without any noise or any smoke." By pumping gases that burned in various colors through outlet pipes, Diller made flames seem to grow from stems to plants and then to flowers. "He represents different insects and animals," Elliot continues, "and has a most curious chase of a viper after its prey, and of a little flying dragon after a butterfly." One of the gases Diller used in quantity was ether, Elliot says, calling (and misspelling) it "Hoffman" after Hoffmann's drops, a well-known medical compound of ether and alcohol: "The room smelt so strongly of Hoffman as to add very much to my pleasure and to that of Mrs. Johnston, who has the same affection for Hoffman that I have. Everybody else was loudly complaining of the stench, while we were whiffing it up and agreeing that it was a nosegay, and that it smelt of a *good night.*"[20]

An Italian fireworks company, M. Amboise, presented a display like Diller's in Philadelphia in 1796, as did a sideshow operator, Benjamin Healy, at Haymarket Gardens in Richmond, Virginia, in 1802.[21]

All this activity made gaslight known to both engineers and inventors and to at least the British public at the beginning of the nineteenth century. But producing laboratory- and theater-scale demonstrations was easy; generating and distributing gas for lighting on an industrial scale was far more difficult. The difficulty had less to do with the technology itself than with the large financial commitment necessary. Boulton & Watt commanded such resources; the firm made the first major investment in gaslight, somewhat grudgingly supporting the work of one of its most talented engineers, a young Scot named William Murdoch.

Murdoch's father worked as a miller and millwright for Alexander Boswell, the father of the biographer James Boswell. Young Murdoch, born in 1754, attended a good Scottish school, and his father taught him a range of skills. Father and son raised a fine stone bridge over the River Nith, near Dumfries. Together they conceived and built a wooden precursor of the bicycle. At twenty-three, in 1777, young Murdoch traveled to Boulton & Watt's Birmingham factory and presented himself to Boulton, probably with a recommendation from James Boswell in his pocket—Boswell had visited the factory the previous year.[22] Legend has it that Murdoch, powerfully built and more than six feet tall, presented Boulton with a wooden hat that he had designed himself and turned out on a lathe. These qualities and skills, plus an endorsement from a wealthy and well-known Scottish laird, impressed Boulton to hire him.

In less than a year, Boulton & Watt had promoted Murdoch to principal pattern maker. In autumn 1779 the firm sent him to Cornwall to erect steam engines for pumping out the copper and tin mines there; Cornwall had little coal, making the old Newcomen engines prohibitively expensive to operate.

Settled in the town of Redruth, Murdoch not only erected Boulton & Watt engines in Cornwall but also often improved them. Boulton wrote appreciatively to Watt in 1782 that Murdoch was "indefatigable," describing a week when he "slav[ed] night and day" keeping steam engines running in three different collieries. In 1784 Boulton called him "the most active man and best engine erector I ever saw."[23] But Murdoch didn't always deploy his energy to the partners' approval. "I wish William could be brought to do as we do," Watt complained to Boulton in 1786, when their young protégé was preoccupied with building a steam-powered road carriage, "to mind the business in hand, and let [others] throw away their time and money, hunting shadows."[24]

The most important shadow Murdoch hunted during his Cornwall years was gaslight. Like Dundonald and others, Murdoch initially investigated distilling various materials to produce wood preservatives. In 1791 he patented "a method of making . . . copperas [a dye fixative], vitriol, and different sorts of dye or dying stuff, paints, and colors; and also a composition for preserving the bottoms of all kinds of vessels, and all wood required to be immersed in water." The method involved roasting not coal but pyrites (iron disulfide) "or other minerals or ores."[25] Sometime during that patent year, Murdoch moved on from pyrites to coal.[26] In a paper read before the Royal Society sixteen years later, he recalled experimenting with distilling gases from coal "as well as from peat, wood, and other inflammable substances . . . and being struck with the great quantities of gas which they afforded, as well as with the brilliancy of the light, and the facility of its production."[27]

Historians have not agreed when Murdoch first moved from experimenting with gaslight to installing a system in a building. Though Victorian sources claim he illuminated his house in Redruth with gas in 1792, the best evidence points to the late 1790s, after he had returned to Birmingham from Cornwall.[28] James Boswell visited Murdoch at his Redruth home in September 1792, admired some Cornish minerals in Murdoch's collection, and commented snobbishly in his diary afterward that it was "a curious sensation . . . to find a tenant's son in so good a state." But he made no mention of any gas lighting, which he would have done had any been installed.[29]

Others who worked with Murdoch in Redruth remember him experimenting instead at a nearby foundry, using a metal case set over a fire to distill gas, piping it through an old gun barrel, allowing the flame to jet out several feet into the air or attaching a thimble to the muzzle punched with holes and spraying out the flame in multiple smaller jets. "Bags of leather and of varnished silk," one of Murdoch's colleagues recalled, "bladders, and vessels of tinned iron were filled with the gas, which was set fire to and carried about from room to room, with a view of ascertaining how far it could be made to answer the purpose of a movable or transferable light. Trials were likewise made of the different quantities and qualities of gas produced by coals of various descriptions."[30] The bladder investigations taught Murdoch how to use gas as a lantern on his dark walks home from the colliery: he would fill a bladder fitted with a pipestem with gas, light the pipestem, and work the bladder under his arm like a bagpipe. When he got home, he would blow out the flame, dump the remaining gas, fold up the bladder, and tuck it into his pocket.[31]

Urgent with the prospect of an entirely new lighting system that Boulton & Watt could pioneer, Murdoch presented the results of his research to James Watt Jr. in Birmingham in 1794, proposing that the firm should apply for a patent. James Jr. and Boulton's son, Robinson, were moving into leadership at Boulton & Watt as their fathers neared retirement, but the cautious James Jr. warned off Murdoch. "I told him I was not quite certain if it were a proper object for a patent," he testified several years later, "and I was induced to be rather nice [i.e., wary] upon the subject of patents, from being at that time engaged in carrying on the defense of a patent which my father had obtained for improvements in the steam engine." James Jr. pointed as well to the previous researches of Dundonald and others, concluding, "I advised Mr. Murdoch not to prosecute his experiments for the present, until the question respecting the steam engine had been decided, and until we had an opportunity of considering the subject more maturely. Murdoch acquiesced, and nothing was done until 1801."[32]

Which was not quite true. In 1794, independent of Murdoch, James Watt Sr. began work under painful circumstances on a device for generating medicinal gases, a device that would influence the development of gaslight technology. The new field of treatment was called pneumatic medicine, the invention of Thomas Beddoes, an Oxford- and Edinburgh-educated physician (and another student of John "Jolly Jack Phosphorus" Anderson).

Beddoes was based in Bristol beginning in 1793 after a stint as a chemical reader at Oxford: despite his gifts, he had been denied a prestigious regius chair in chemistry because he was politically active in support of the French Revolution. His prospective father-in-law described him around this time as "a little fat Democrat of considerable abilities, of great name in the Scientific world as a naturalist and Chemist—good humored good natured—a man of honour and Virtue, enthusiastic & candid. . . . If he will put off his political projects till he has accomplish'd his medical establishment, he will succeed and make a fortune."[33]

Beddoes was preparing to open a combined research laboratory and clinic in Bristol Hotwells, a spa center for patients with pulmonary consumption, and needed prominent endorsement and investment. He had already won over Erasmus Darwin, who in turn recommended him to Watt. Beddoes first wrote to Watt on 4 March 1794, emphasizing his commitment to experiment, an attitude that Watt would have approved of as the scientific approach both men had learned from Anderson. In weekly letters that followed, Beddoes's biographers report, he described "his experiences with patients breathing 'airs'—oxygen, hydrogen, fixed air (carbon dioxide), and hydrocarbonate (water gas; i.e., carbon monoxide and hydrogen)—and his problems with the breathing apparatus."[34] It was those problems he hoped Watt might help him solve.

Watt had more grievous trouble. His beloved fifteen-year-old daughter, Jessy, a late child of his second marriage, was mortally ill with consumption. When her condition worsened that spring, Watt called in Darwin. He "gave little hopes," Watt told his friend Joseph Black, "but prescribed for the fever and other urgent symptoms." Darwin proposed they try Beddoes's gases. Beddoes attended Jessy daily for a week, but he too could offer Watt little hope. He arranged for Jessy to breathe effervescent carbon dioxide. Her seizures—Watt called them "hystericks"—and weakness prevented him from trying other gases or medicines.

Jessy died on 6 June 1794, Watt wrote Black on the Monday following: "My Amiable and lovely daughter expired on Friday morning after long suffering, the fever she had when I wrote you last proved a hectic* of the most violent kind, which perhaps we might have seen sooner if we had not been misled by her violent hystericks."[35]

To distract himself from grief, and in hope of saving others—consumption was a terrible scourge in that era, killing one in four of its victims—Watt agreed to help Beddoes by designing and manufacturing a device for generating medicinal gases. On 30 June 1794, when he wrote Darwin to thank the physician for his condolences, he explained why he had signed on with Beddoes. "I have long found," he wrote, "that when an evil is irreparable, the best consolation is to turn the mind to any other subject that can occupy it for the moment."[36]

Remarkably, by then, only twenty-four days after his daughter's death, Watt had already designed and built his device: "I have made an apparatus,"

*The type of flushing fever then associated with consumption.

he continued in his letter to Darwin, "for extracting, washing, and collecting of poisonous and medicinal airs." He would send Darwin one, he added, "with which you may try the whole round of poisonous and salutiferous airs; and I hope, in your hands, not without success."[37]

Two weeks later, Watt wrote Beddoes, agreeing to offer the apparatus to the public at a price yet to be determined but "as moderate as we can make it."[38] Beddoes included the letter along with Watt's description of the device in a book he published later that year under both their names, *Considerations on the Medicinal Use of Factitious* [Artificial] *Airs, And on the Manner of Obtaining Them in Large Quantities.*[39]

Watt's apparatus consisted of a pot to hold materials to be gasified—Watt called it an "alembic"—with a tight lid, set over a fire, and connected by a pipe to a water tank where the gas generated in the pot could be washed and cooled. Another pipe then led the cleaned and cooled gas to a gasometer—a bellows—where the gas could be accumulated and transferred into oiled silk bags. Patients would breathe the gas from one of the bags.[40]

Watt's factitious-airs generator was essentially identical to a gaslight generator except for the additional types of gases it produced and the fact that the gases were intended to be inhaled rather than ignited. Watt introduced it at a time when gases and their production were under investigation in Britain and throughout Europe.

With support from Darwin, Josiah Wedgwood's invalid youngest son, Thomas, the Duchess of Devonshire, and others, Beddoes opened his laboratory and clinic in Bristol Hotwells on 21 March 1799. It drew large numbers of patients, whom Beddoes treated at first with conventional remedies rather than gases, to establish his competence as a physician. For a laboratory superintendent, he had hired twenty-year-old Humphry Davy, fresh from a chemistry apprenticeship in his native Cornwall. Together they began experimenting with nitrous oxide, experiments that would eventually involve James Watt and the poets Samuel Taylor Coleridge and Robert Southey, and would tarnish Beddoes's reputation.

They manufactured the gas in the Watt generator by heating crystals of ammonium nitrate. The first time Davy inhaled the gas experimentally, he erupted "shouting, leaping, running" around the laboratory, his actions

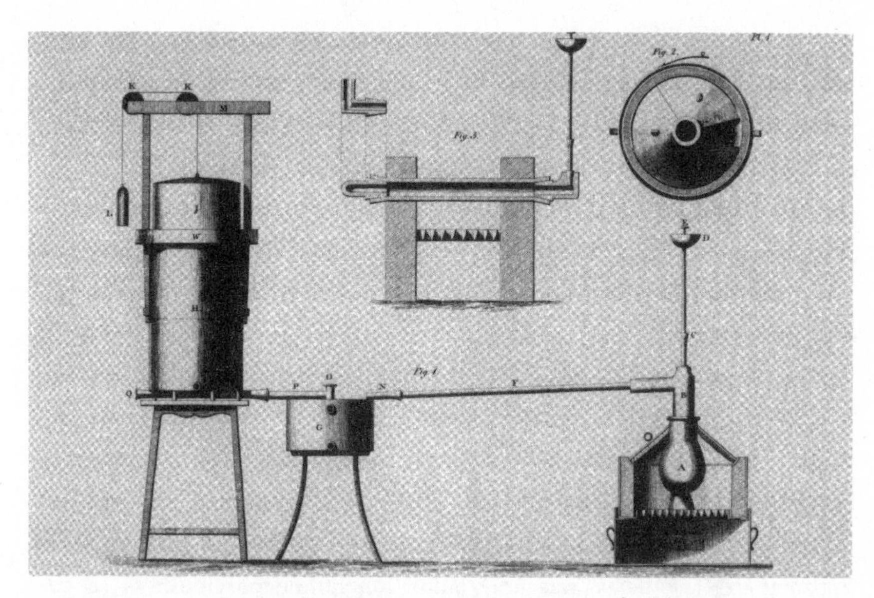

Watt's "factitious airs" machine: alembic on right over fire grate, washer-cooler tank in middle, gasometer at left. Funnel and pipe above alembic admit liquids into alembic. Cleaned and cooled gas was collected from gasometer into oiled silk bags. Second figure above first depicts smaller "fire-tube" version of device; cylinder upper right is end-on view of funnel.

becoming "various and violent," with a "highly pleasurable thrilling in the chest and extremities."[41] He was intoxicated, but neither he nor Beddoes interpreted his reaction as mere intoxication. For them, it was indication as well of a new medical treatment of powerful effect—perhaps something that might restore motion in the limbs of paralytics or energize invalids such as Tom Wedgwood.

Soon they were sharing the discovery with their friends for recreation as well as treating patients, many of whom found relief and at least temporary animation. Davy became addicted, hauling the green silk bag that stored the gas to his room to breathe it alone or on solitary walks along the Avon River at night. He recorded in his notebook his intoxicated vision that "I seemed to be a sublime being, newly created and superior to other mortals."[42] Southey speculated that "the atmosphere of the highest of all possible heavens must be composed of this gas."[43] More than one spoke of entering a state, in Coleridge's words, "of more unmingled pleasure than I had ever before experienced."[44]

Davy also noticed that the gas blocked the sensation of pain. The historian Mike Jay comments that one sentence in Davy's 1800 volume *Researches, Chemical and Philosophical, Chiefly Concerning Nitrous Oxide, or Dephlogisticated Air, and Its Respiration*, has "attracted more subsequent attention than the rest of the book put together."[45] That sentence is: "As nitrous oxide in its extensive operation appears capable of destroying physical pain, it may probably be used with great advantage during surgical operations in which no great effusion of blood takes place."[46]

If a gas was available in 1800 that could eliminate the terrible pain of surgical cutting or tooth extraction, why did humanity have to wait until 1842 for its first use, in dentistry? Jay answers this question brilliantly: not because Beddoes and Davy somehow "missed" nitrous oxide's anesthetic properties—Davy's *Researches* observation makes that clear—but because, first, the most apparent effect of the gas was giddy animation, not unconsciousness; second, the dominant medical and religious opinion of the era held that pain was " 'the voice of nature,' a necessary condition of life," even "a stimulus that kept traumatized patients alive"; and third, anesthesia was perceived to be unnecessary and even insulting to the surgeon, alien to "the crucial elements in an operation," which were "the surgeon's skill and the patient's bravery." Anesthesia, Jay concludes, "when it emerged in the 1840s, was as much a response to surgeons' needs as to patients': technical advances had led to more sophisticated operations, and the ability of the patient to endure them had become a limiting factor that needed to be addressed."[47]

The era of Beddoes's and Davy's nitrous oxide researches was also when the English novelist Fanny Burney endured a total mastectomy without anesthesia, in 1811, and described the experience in a famous letter to her sister Esther. "When the dreadful steel was plunged into the breast," Burney's description begins, "cutting through veins—arteries—flesh—nerves—I needed no injunctions not to restrain my cries. I began a scream that lasted unintermittingly during the whole time of the incision . . . so excruciating was the agony." The operation lasted twenty minutes.

The same Watt generator that led off in one direction to pneumatic medicine, not to say frolic, led in another to the development of large-scale gas lighting. When William Murdoch moved back to Birmingham in 1798, he continued his earlier work on gas lighting at the new, advanced Soho Foundry which Boulton & Watt, now reorganized as Boulton, Watt & Co., had built just outside the city. James Jr. testified that Murdoch's coal-gas machinery "was applied during many successive nights, to the lighting of the [foundry]. . . . Experiments on different apertures were repeated and extended upon a large scale. Various methods were also practiced of washing and purifying the [gas], to get rid of the smoke and smell."[48]

James Jr. resolved any remaining doubts about entering the business of manufacturing gaslight apparatus after his half brother Gregory visited Paris in late 1801. "Tell Murdoch," Gregory wrote home, "that a man here has not merely made a lamp with the gaz [*sic*] procured by heat from wood or coal but that he has lighted up his house and garden with it and has it in contemplation to light up Paris."[49]

The French consider that man, Philippe Lebon, an engineer with the Department of Bridges and Roads, to be the inventor of illuminating gas. Lebon joined the department when he was twenty-five, in 1792, and that same year received a National Reward of 2,000 francs (today £24,000, or $34,000) for inventions related to the steam engine.[50] Beginning around the same time and for most of the decade, Lebon experimented with illuminating gas made from sawdust before patenting a gas generator he called a Thermolamp in 1799. The Thermolamp, like Watt's system, consisted of a retort set over a fire connected to a water tank for washing the gas, connected to a gasometer, with a pipe leading from the gasometer into the space to be illuminated that terminated in gas outlets.

Lebon's patent specified not only wood fuel but also "coal, oil, resins, tallow, and other combustible materials."[51] ("Oil" here doesn't mean liquid petroleum, which had not been identified yet except as a flammable seepage in a few limited locations. Lebon probably meant vegetable oils such as flaxseed or rape.) In a prospectus distributed in 1801 for the lighting exhibition that Gregory Watt had visited, Lebon pitched his invention as a labor-saving device while anthropomorphizing it as a creature more obedient than any servant:

1802 German version of Lebon's thermolamp.

> This ethereal substance can travel, in a cold state, along a pipe only an inch square within the thickness of walls and ceilings.* . . . In the twinkling of an eye the light can be transferred from one room to another in a way that you cannot do with ordinary fires, no more sparks, cinders or soot; no more heavy buckets of fuel to lug upstairs. By day and by night, light and heat are there under your hand without servants. The heat can take the form of hot air, fluorescence, or flame. It is suited to every form. It can if you wish cook your meat and even reheat it at your table, dry your linen, and heat the water for your baths. You can control it, order it to appear or disappear, and it will obey as even the most obedient servant will not.[52]

The news from Paris aroused James Jr. to ask Murdoch to revive gaslight development at Soho. "Murdoch is going to Cornwall upon his own affairs," he

*Just as some had doubted the possibility of a light without a wick, so did others, including members of Parliament and the Royal Society, assume that the pipes carrying gas would be so hot they would be a fire hazard, especially if confined within walls.

wrote Gregory at the end of 1801. "Upon his return here, some decisive experiments are to be made which will determine whether we shall proceed upon his plan or not."[53] They did proceed, deciding to add gaslights to a celebration of the Treaty of Amiens planned for their Soho Foundry. Britain, France, Spain, and the Batavian Republic (the Netherlands) were scheduled to sign that treaty in the French city in March 1802. The treaty was supposed to end the war with Napoleon I and relieve the British of an onerous income tax.[54] Boulton, Watt & Co had already ordered several hundred colored lamps and candles and fourteen gallons of oil. To that array, they decided to add two large gaslights set in copper vases, one at each end of the main factory building. "Bengal lights," they called them: gaslit versions of blue signal flares used at sea that threw off sparks like enlarged fireworks of the type Americans call sparklers.

The historian William Matthews attended the Soho event. "This remarkable illumination was the first public display of its kind in this country," he recalled it. Besides oil lamps and gaslights, fireworks lit the sky, and "three very splendid Montgolfier Balloons ascended in succession from the courtyard within the Manufactory at proper intervals, on a signal from the discharge of cannon."[55]

A Manchester manufacturer, George Augustus Lee, a technology enthusiast, visited Soho in 1800 specifically to view Murdoch's gaslight technology.[56] He liked what he saw, and proposed to install it in a new cotton mill he and his partners George and John Philips were then constructing in Manchester. It would be only the second iron-framed mill to be built in Britain and one of the two largest, steam heated, with steam engines powering belt-driven looms.[57]

Boulton, Watt & Co. invested more than £4,000 (today £329,000, or $478,000) in gaslight manufacturing equipment in 1803.[58] After four years of development work, the firm installed gas lighting in Lee's home in 1804 and lit the Philips & Lee mill in 1805, replacing oil lamps and candlelight. Murdoch was happy that the gas didn't release what he called "the Soho stink." Lee's wife and several daughters had visited the factory that night, "and their delicate noses have not been offended." Lee was trying different lamps (gas outlets) to identify the most effective flame arrangement. Gaslight—cheaper, brighter, and safer—improved mill working conditions but also allowed owners to extend their hours unmercifully. A strikers' pamphlet published in Manchester about another, similar mill describes its conditions in terms familiar today from third world sweatshops:

> At Tyldesley they work fourteen hours per day, including the nominal hour for dinner; the door is locked in working hours, except half an hour at tea time; the workpeople are not allowed to send for water to drink, in the hot factory; and even the rain water is locked up, by the master's orders, otherwise they would be happy to drink even that.[59]

Lebon's invention proved a dead end in France. Thousands toured his exhibition and were impressed, even astonished, at the quality and brilliance of the light compared with oil or candles, but the French government refused to support building a distribution system. Few private individuals were prepared to pay 1,000 livres (today £2,600, or $3,800) or more for a Thermolamp. Lebon himself allowed his lighting work to lapse. He was more interested in manufacturing tar for ships, like his British contemporaries, than in developing gaslight.

The French government agreed. In 1803 it granted the engineer a concession of a pine forest near Le Havre with the proviso that he use it to distill a daily production of five hundred pounds of pine tar. To that end, he built a tar manufactory, but a fire partly destroyed it, and a storm unroofed his house. Called to Paris to assist in organizing the city for the coronation of Napoleon as emperor, Philippe Lebon was stabbed to death by an unknown assailant in the Champs-Elysees on the night of 2 December 1804.[60] As late as 1837, Paris remained lighted only with oil lamps.

Lebon's exhibition in Paris inspired a flamboyant German entrepreneur named Frederick Albert Winsor to carry the word of distributed gas lighting to London in 1803. Devoid of technological gifts, Winsor was a master salesman. By 1807, he had arranged for gas lighting in fashionable Pall Mall, the first gas street lighting in the world, and formed a company, the New Patriotic Imperial and National Light and Heat Company, to pipe gas to public and private establishments in London.

Friedrich Accum, a German chemist testifying before Parliament in support of New Patriotic's charter application, had to explain to the incredulous members that a gas flame "may be increased or diminished by admitting more or less of the gas, by turning a stopper, it is contained in the pipe, just as water or beer is in a barrel, and by turning the cock you let it out."[61] Humphry Davy scoffed, "You would have to fill St. Peter's dome to get as much gas as you need, and then it would explode."[62] English investors eventually ousted Winsor from his company, which was renamed the London and Westminster Chartered Gas Light and Coke Company when Parliament approved its charter in 1810. By 1814, Westminster Bridge glowed with gaslight, and by 1815, the company had laid thirty miles of gas mains in Britain's capital, to be followed in the years ahead by miles more in cities throughout the country.

The district gas factories that fed those mains washed the gas of its Soho stink with lime cream, a mixture of slaked lime (calcium hydroxide) and water. Untreated coal gas is foul with hydrogen sulfide, which smells like rotten eggs and is highly toxic. Ammonia is another contaminant, as is carbon monoxide, odorless but deadly. Gasified solids such as coal tar and soot can be scrubbed from coal gas by bubbling it through water. Lime cream removes the toxic gases. As it does so, it becomes increasingly contaminated and eventually turns blue, a state that earned it the nickname "blue billy." A ferro-ferricyanide compound, blue billy releases cyanide gas, which makes it smell of bitter almond or marzipan. Carted through the streets for disposal, it created a serious nuisance. Gas water and tar dumped into sewers, streams, and rivers foully polluted them.

Gaslight had too many advantages to resist, however. Pollution is seldom the first concern when new technologies are introduced. Glasgow, Liverpool, and Dublin lit up next, in 1817–18, and a continuing progression of British cities thereafter. By then, gas lighting had crossed the Atlantic to the United States. David Melville, a hardware merchant in Newport, Rhode Island, began investigating gas lighting sometime before 1810, when he received a patent for a gas lamp. A second patent in 1813 covered a gas manufacturing system largely identical to those being used in Britain, with a retort, a water tank gasometer, and burners. The gasometer was counterbalanced with weights that could be removed to lower the upper tank, creating pressure to force the stored gas through a pipe to the burners. Melville first lighted his house and the street in front of his house, possibly as early as 1806. Between 1813 and 1817, he installed gasworks in cotton mills in Watertown, Massachusetts, and Providence, and also in a Rhode Island lighthouse.[63]

Advertisement for Peale's Baltimore Museum, 1816.

Philadelphia's notable Peale family, artists and museum proprietors, installed gas lighting in two of their museums: the Peale Museum in Independence Hall in Philadelphia in 1814, and an offshoot museum of the same name established by Rembrandt Peale, one of patriarch Charles Willson Peale's sons, in Baltimore in 1816. Both probably built their gasworks after Melville's design, but the Philadelphia gasworks was fueled with pine tar rather than coal after Rubens Peale tried coal and suffered so many complaints from visitors of noxious fumes—his gasworks was installed under a stairwell, with inadequate ventilation—that he had to limit operation.[64] In any case, coal was expensive in Philadelphia until railroads crossed the Allegheny Mountains in the late 1840s.

Rembrandt Peale fared better in Baltimore. He built his gasworks in a building behind his museum and washed his coal gas with both water and lime cream. He put some showmanship into his 13 June 1816 opening night, advertising widely and debuting gaslight in what he called a "magic ring" of one hundred burners. A control valve allowed him to turn the ring of flame up and down, something impossible with candlelight and hardly possible with oil.[65] Spectators crowded the museum throughout the exhibition.

Four days after the magic-ring debut, Baltimore acted on an ordinance permitting Rembrandt Peale and four Baltimore businessmen to form the first commercial gaslight company in the United States, the Gas Light Company of

Baltimore.[66] Less than a year later, the GLCB lit its first street light and, after that, its first commercial building and private home.

Gas lighting expanded slowly in America, however. "By 1850," writes historian Christopher Castaneda, "about 50 urban areas in the United States had a manufactured gasworks. Generally, gas lighting was available only in medium-sized or larger cities, and it was used for lighting streets, commercial establishments, and some residences. . . . Other than gas, whale oil and tallow candles continued to be the most popular fuels for lighting."[67]

Tallow is the rendered fat of cattle. Whale oil comes from whales. Castaneda is wrong about whale oil; not only the world's largest mammal rendered up its fats and oils for lighting in the first half of the nineteenth century, or whales would have been extirpated. They very nearly were.

EIGHT

PURSUING LEVIATHAN

Though the Nantucket Quaker Francis Rotch owned one of the ships that hosted the Boston Tea Party, the Quakers of Nantucket Island wanted nothing to do with the American Revolution. They were pacifists; their narrow, undefended island, out in the Atlantic thirty miles south of Cape Cod, harbored a wealth of whaling ships the British would capture or burn. Forty years later, when he was eighty years old, Francis's brother William Rotch chronicled how both sides had misused Nantucket's Quakers during the war and afterward, so much so that they sought to move their whaling enterprise to England and considered moving it to France.

Nantucket Island is a terminal moraine of glacial sand and gravel left behind by the melting of the great ice sheet that covered the upper half of North America until about thirteen thousand years ago. The island's poor soil (Thomas Jefferson would call it "a sand bar"[1]) and isolation make it almost totally dependent on supplies shipped from the mainland. That was why the island had early founded its economy on the sea.

Oppression as well as opportunity had marked Nantucket's settling. In 1657 Thomas Macy, a Massachusetts Bay Colony Baptist sawmill owner, sheltered four Quakers during a downpour for part of an hour, breaking a recent edict against harboring Quakers and other "cursed sects of heretics."[2] The Massachusetts General Court fined Macy thirty shillings for his effrontery (today £208, or $300) and ordered the governor to admonish him. The offense was serious: two of the Quakers Macy sheltered were subsequently hanged. Obed Macy, one of Thomas's direct descendants and the author of the earliest history of Nantucket, writes that after this incident, his ancestor "could no longer live in peace and in the enjoyment of religious freedom among his own nation" and "chose therefore to remove his family to a place unsettled by the whites, to take up his abode among savages." In autumn 1659 Macy sailed for Nantucket with his family in an open boat.

Nantucket had a previous European owner: an English merchant named Thomas Mayhew. King Charles I had "owned" Nantucket by right of English discovery. He had given it to two English noblemen, who in turn sold it to Mayhew. On 2 July 1659 Mayhew sold the island to nine men, including Thomas Macy, for £30 "and also two Beaver Hatts one for myself and one for my wife."[3] Before that sale, Mayhew had purchased the rights to part of the island from two Nantucket Wampanoag chiefs, or sachems, for £12.[4] In 1662 the new Nantucketers would purchase further rights from the chief sachem, Wanackmamak, for £5 "to be paid to me in English goods." Another £40 in 1671 bought the remainder of the island from the Wampanoags.[5] Fifty-seven pounds, the total paid to the original residents, whose oral tradition recalls them walking out to the island on the firm surface of an Ice Age glacier, is about £7,700 today, or $11,200.[6]

Nine proprietors and their families settled among a Nantucket Island population of Wampanoag Indians in the years after 1659 and built a town they named Sherburne after their English place of origin, renamed Nantucket in 1795. European diseases had reduced the Wampanoags to little more than a remnant since the island's discovery in 1602, when an estimated three thousand Wampanoags had lived there. A hundred years later, by 1763, only 358 Wampanoags survived on Nantucket. In the following year, another 222 died of an unidentified epidemic that one medical expert speculates might have been louse-borne relapsing fever, a spirochetal infection similar to Lyme disease or syphilis little known today outside Southeast Asia.[7]

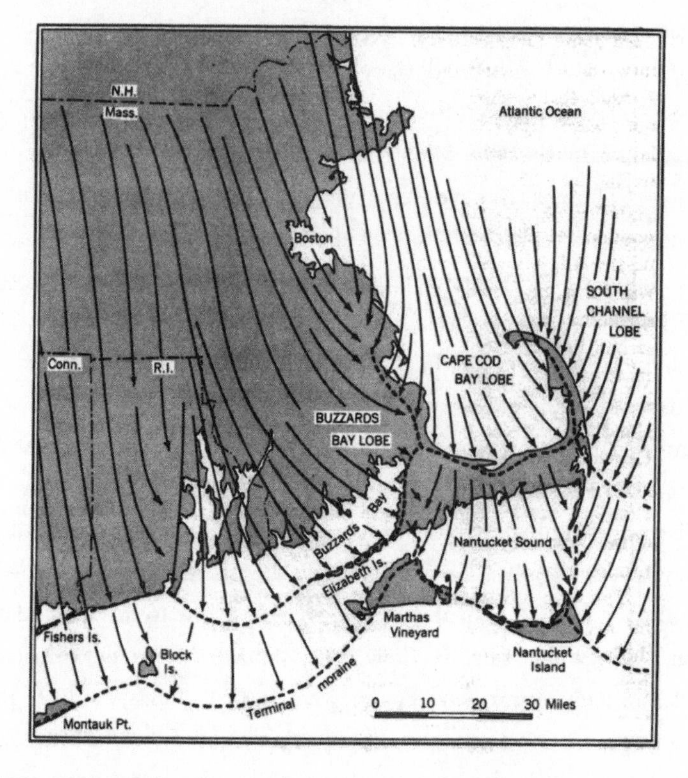

Glacial lobes dropped their sand and gravel to form Cape Cod, Nantucket, and Martha's Vineyard when they melted about thirteen thousand years ago.

The new Nantucketers learned whaling in stages. Drift whaling—harvesting dead or dying whales washed up on shore after storms—is a self-evident technology, requiring only the necessary tools to cut up so large an animal and a strong stomach against the stink. Inshore whaling—rowing out in open boats to attack coasting whales spotted from beach lookout towers—may have been a Wampanoag practice. It was common among the European settlers on Cape Cod, Martha's Vineyard, and the New England coast. The historian Alexander Starbuck reports that in 1690 the Nantucketers, "finding that the people of Cape Cod had made greater proficiency in the art of whale-catching than themselves," hired an experienced Cape Cod sailor named Ichabod Paddock to teach them how to hunt whales.[8]

Europeans hunted whales for their lighting oils, not their meat. By the

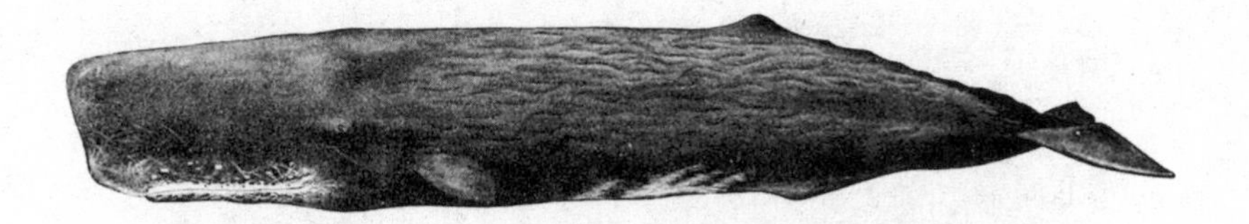

Sperm whale. Length, fifty to sixty feet (the length of a railroad boxcar); weight, thirty-five to forty-five tons. It could cool or warm its prize oil, spermaceti, stored in a cavity in its head, increasing or decreasing its density thereby to help maintain neutral buoyancy.

second decade of the eighteenth century, the Nantucketers had identified the sperm whale as the source of the highest-quality oil. They had advanced from shore whaling to whaling in ships within a few day's sail of shore, returning to shore towing a whale carcass each time they caught one to process it.[9] But sperm whales are pelagic, living in the open sea, and to hunt them, the Nantucketers had to build larger ships and prepare for extended voyages. So long as they hunted in the North Atlantic, they could store unrendered strips of blubber in barrels on shipboard for processing when they returned home.

When they began hunting in hotter climates, however—the Cape Verde Islands off the west coast of Africa, the Caribbean, and the Brazil Banks—the heat spoiled the unrendered fat. It became necessary to render the blubber in transit, for which purpose they mounted on deck an iron and brick tryworks, set with two or more large cast-iron pots, its fires fed initially with wood and then with the oily cracklings left over from the rendering itself. With ships capable of carrying several thousand barrels of oil and with a tryworks on deck for rendering at sea, the Nantucket fleet was prepared to range throughout the world's oceans, wherever sperm whales congregated.

In 1774 Nantucket's whaling fleet consisted of 150 vessels with an average "burden"—carrying capacity—of 100 tons. The fleet returned about 26,000 barrels of spermaceti oil that year, the product of some 3,000 sperm whales. Another 210 ships from other US ports also hunted sperm whales in 1774, bringing the total number of ships to 360 and the total production to at least 45,000 barrels.[10] Milky spermaceti, scooped from the sperm whale's head chamber (the "case") hardens into white wax when exposed to the air. Blubber renders an oil

Tryworks rendering blubber.

of lower quality and lower value—so-called train oil*—but spermaceti for illumination was the prize.

After 1775, with the coming of war, this industry of great brutality and rich harvest collapsed. Early that year, the British House of Commons voted punitively to prohibit the rebellious New England colonies from commercial fishing "on the Banks of Newfoundland, or any other part of the North American coast."[11] The Quakers of Nantucket, through their English kin, managed to win exemption for whaling, an important reason why they remained neutral during the war. The British navy captured 134 of their ships and impressed 1,200 of their seamen anyway, as William Rotch had feared, and another 15 ships were lost to shipwreck. The islanders sustained themselves with local fishing and subsistence farming. In the hard winter of 1780, with the harbor frozen and the peat swamps and fields deep in snow, many suffered cold and near starvation.

*"Train" derives from Dutch and Germanic roots identifying secreted substances such as tears and tree resins.

"The announcement of peace" in 1784, says Starbuck, "came to a people whose commerce was sadly devastated. . . . The business of whaling was practically ruined and required rebuilding."[12] The economic depression that followed the war added to the island's burden, as did lessened demand from Americans grown accustomed to the tallow candles and vegetable oils they had reverted to in wartime. Worse, the Nantucketers were now US citizens and were thus subject to the punishing alien duty the British imposed to support their own whaling industry. The duty, which reversed the return on Nantucket sperm oil from a profit to a loss of nearly £8 per ton, effectively closed off a lighting market in London alone of some four thousand tons of sperm oil per year, worth about £300,000 (today £34.5 million, or $49 million).[13] That was when William Rotch, whose losses from the war, whale ships in particular, came to "about $60,000" (today £715,000, or $1.02 million), traveled to England to see if he could negotiate the resettlement there of the Nantucket whale fishery.[14]

Rotch sailed for England with his twenty-year-old son, Benjamin, on 4 July 1785, arriving at the end of the month. The British government advised Rotch that it was preoccupied with domestic issues and suggested he wait several months before pressing his cause.[15] In any case, he had reconnoitering to do. He and Benjamin set out for the west of England with a Quaker friend to tour the seacoast "in search of a good situation for the Whale Fishery." Of the several ports Rotch found suitable, he favored Falmouth, on the south coast of Cornwall, for its large harbor, which sheltered several smaller harbors as well. Other locations welcomed him and made "very favorable offers," but he was only inspecting until he learned what support the British government might propose. Back to London then, stopping off in Bristol to visit the grave of one of his brothers, buried there eighteen years before.[16]

By now, it was November, but with an introduction from a member of Parliament, Rotch won an audience with William Pitt the Younger, whom he identifies in his memoir as the chancellor of the exchequer.[17] Rotch neglects to mention that Pitt was also prime minister, the youngest ever, elected at twenty-four and then only twenty-seven years old. Nor does the Nantucket Quaker seem to have known of Pitt's late-August discussions of the American whale fishery with the inaugural United States ambassador to Great Britain, John Adams.

Adams had described his meeting with Pitt in a 24 August letter to John Jay, the American secretary of state. "He . . . led me into a long, rambling conversation about our whale fishery and the English," Adams wrote. The conversation wasn't worth repeating, Adams added, but it had served Pitt's purpose, preparing the way to ask Adams "a sudden question, whether we had taken any measures to find a market for our oil anywhere but in France. . . . I answered that I believed we had, and . . . there could not be a doubt that spermaceti oil might find a market in most of the great cities in Europe."[18]

At which point, Adams presumed to educate the young prime minister on the security benefits of good street lighting, informing him, "The fat of the spermaceti whale gives the clearest and most beautiful flame of any substance that is known in nature, and we are all surprised that you prefer darkness, and consequent robberies, burglaries, and murders in your streets, to the receiving, as a remittance, our spermaceti oil."[19]

Adams spoke so bluntly from concern that Britain intended to punish his upstart new nation. "There are many ways in which they may hurt us," he wrote Jay further on 30 August.[20] Yet William Rotch was nearing London by then, expecting to move an entire whale fishery to England, or at least the considerable portion of it he and his partners controlled. When Rotch met with Pitt in late November, he informed the prime minister of Nantucket's "ruinous situation," reminding him that Nantucketers had wanted no part of the war and had remained neutral throughout. "Nevertheless, you have taken from us about Two Hundred sail of Vessels," Rotch accused, "valued at 200,000 pounds Sterling, unjustly and illegally." His key argument, which he repeated several times, was that Nantucket had remained part of England's dominions "until separated by the peace"—and therefore deserved redress of its losses.

Pitt paused to think over Rotch's claim. "Most undoubtedly, you are right, Sir," he responded. "Now, what can be done for you?"[21]

Without relief, Rotch told the prime minister, the majority of Nantucket's population would have to leave the island. "Some would go into the Country," he said, meaning the United States, while a part "wish to continue the Whale Fishery, wherever it can be pursued to advantage." He had traveled to England to inform Pitt of Nantucket's distress and to see if the English government would encourage the fishery to resettle there.[22]

The question raised, Rotch returned to his London lodgings to await an answer. A few days later, a note arrived confirming that Pitt had presented Rotch's proposal to the Privy Council, which advised the king on matters of state. Rotch then languished in London for more than four months until, losing patience, he asked the government to appoint *someone* to confer with him.

That someone would be dour Lord Hawkesbury, George III's champion and close adviser since the beginning of his political career and secretary of war through the latter half of the American Revolution. He bore no love for the United States. Rotch proposed to him that England encourage Nantucket whaling families to move by paying them to do so: "100 pounds-Sterling for transportation for a family of five persons, and 100 pounds settlement."[23] (In today's currency, about £23,000, or $32,000, per family, or a total of about £2.3 million, or $3.2 million.)

"Oh!" Hawkesbury complained. "This is a great sum . . . at this time when we are endeavoring to economize in our expenditures. And what do you propose to give us in return for this outlay of money?"

Rotch replied proudly, "I will give you some of the best blood of the island of Nantucket."[24] They worked over their differences in a long conversation, after which Hawkesbury invited Rotch to call again.

Rotch did so a few days later, upping the ante: in addition to his previous request for compensated resettlement, he wanted to move thirty American ships to England to establish the proposed fishery. Impossible, Hawkesbury responded: England required English ships and English seamen as a reserve for the Royal Navy, a long-standing British national security policy. He proposed shaving Rotch's £100 per family to £87.

Hawkesbury's quibbling over a few pounds exasperated Rotch. The Nantucket whale fishery would add nearly £150,000 per year to the English economy. He sailed for France to offer his fishery there, with Hawkesbury all but chasing him across the Channel. The French gave him a warmer reception. "The Government of France could not be inattentive to these proceedings," Thomas Jefferson would write of the affair; "they saw the danger of letting 4[000] or 5000 seamen, of the best in the world, be transferred to the marine strength of another nation, and carry over with them an art which they possessed almost exclusively."[25]

The French government offered the Nantucket Quakers liberal terms, including freedom of religion and freedom from military requisitions, if they chose to settle in Dunkirk on the Channel coast of France, but its terms were not liberal enough to entice the Nantucketers from their small Atlantic island. Only nine families, thirty-three people, actually moved to Dunkirk.[26]

Jefferson painted a discouraging picture of the economics of whaling in his government communications. Because common whale oil competed with cheaper vegetable oils, he wrote in 1789, "the whale fishery is the poorest business into which a merchant or sailor can enter."[27] He judged the distinctive products of the sperm whale to be exceptions, however. Sperm whale oil was "luminous," Jefferson explained, "resists coagulation by cold" to 41°F, "and yields no smell at all. It is used therefore within doors to lighten shops, and even in the richest houses for antechambers, stairs, galleries, &c. It sells at the London market for treble the price of common whale oil. This enables the adventurer [the supplier] to pay the [British] duty of £18 5*s* Sterling the ton, and still to have a living profit." Besides the oil, Jefferson added, the sperm whale's head "yields 3. or 4. Barrels of what is called head-matter, from which is made the solid Spermaceti used for medicine and candles. This sells by the pound at double the price of the oil."[28]

Here was the future of American whaling, though Jefferson was pessimistic about its profitability once competitors and government fees and subsidies reemerged. In the years after the American Revolution, the wealthy returned from tallow candles to spermaceti, and the poor to whale oil, which was cheaper, in any case, than the vegetable oils. New England whalers extended the hunt all the way to the Falkland Islands and Patagonia, at the southernmost tip of South America.[29] The largest territorial expansion of American whaling followed in 1791, when a Nantucket whaler, the *Beaver,* rounded Cape Horn and hunted for seventeen months in Pacific waters, the first of the American fleet to do so.[30] (A British whaler, the *Amelia,* had passed Cape Horn into the Pacific Ocean in 1789.) A ship from New Bedford, Massachusetts, the *Rebecca,* followed the *Beaver* in 1791.

Recovery continued into the nineteenth century. By 1807, Nantucket's fleet numbered forty-six whalers; New Bedford sailed another forty.[31] Having rounded the Horn, American ships now worked a populous sperm whale confluence off the coast of Chile: the onshore grounds. But Britain wasn't through with its former colony yet. Having reignited war with Napoleonic France in May 1803, and with a much smaller population than France, it needed manpower for its navy. It began acquiring men by boarding American ships and impressing sailors, some ten thousand in all between 1800 and 1815, whom it claimed were British deserters. In fact, only about a thousand had been.[32]

President Thomas Jefferson had responded to these coercive recruitments by proposing an embargo on the exportation of American whaling products, which passed Congress on 21 December 1807. Whaling could continue—and, to some degree, did—but the domestic market for oil and other whale products was limited. The embargo was such a disaster that Jefferson asked Congress to modify it to a nonintercourse act, removing all countries except Britain and France from the embargo and banning those two nations' ships from US waters. He signed the modified act on 1 March 1809, three days before the end of his presidency.

The War of 1812, returning the United States to armed conflict with Britain, found most of the Nantucket fleet at sea in the Pacific; those ships that learned of the war rushed back to Nantucket and sailed from there to defend harbors at Boston, Newport, or New Bedford. The British managed to capture fourteen Nantucket whalers. Peruvian privateers, claiming that they were allies of the British, attacked the Pacific fleet as well. By the end of that fruitless war in 1815, Nantucket had lost half its shipping.[33]

It quickly recovered. "Immediately all was hurry and bustle," Starbuck reports. "The wharves, lately so deserted, teemed with life; the ships, lately dismantled, put on their new dress; the faces of the people, lately so disconsolate, were radiant with hope." In May, two ships sailed; in June, seven more. As 1816 opened, more than thirty Nantucket whalers once again worked the North and South Atlantic, the Indian and Pacific Oceans.[34]

Five years later, Nantucket counted seventy-two whaling ships. The American fleet continued to increase across the decade. Whale numbers declined correspondingly on heavily hunted grounds, forcing the hunters to search farther afield. In 1818 the whaler *Globe* of Nantucket had discovered a new cruising ground in the wide ocean west of South America: a 3,600-mile region extending along the equator from Peru well out into the central Pacific.

By the mid-1820s, more than fifty ships could be found hunting this offshore ground in season.[35]

If coal gas lit shops in the largest American and British cities and the aisles of factories, candles and oil still illuminated the private homes, commercial and public buildings, lighthouses, and farms where Americans lived and worked. John James Audubon, the French-American artist who sailed to England from New Orleans in 1826 to supervise the engraving of his great four-volume work *The Birds of America*, was dazzled by the gaslight he encountered in Liverpool in the first weeks after his arrival. The new market there, he wrote in his journal, "a large, high, and long building divided into five spacious avenues, each containing their specific commodities," was so brightly lighted with gas "that at 10 o'clock this evening I could plainly see the colors of the eyes of living pigeons in cages." Though he had visited and worked in US cities from Boston to New Orleans, Audubon had not yet observed such extensive gas lighting in his adopted country.[36]

But the American population was increasing rapidly, from 5.3 million in 1800 to 12.9 million in 1830, and from sixteen states in 1800 to twenty-four in 1830, most of the increase across the mountains in the trans-Appalachian west. The river steamboat from 1807, the Erie Canal between Albany, New York, and the Great Lakes from 1825, railroads from 1829, penetrated the American wilderness and fostered its settlement. These new places and people needed lighting.

The whaling era that opened after the War of 1812, from about 1817 to the mid-1850s, has come to be called whaling's golden age. Its primary anchorage shifted from Nantucket Island to New Bedford, below Cape Cod at the mouth of the Acushnet River. A shoal at the entrance to Nantucket Harbor—the Nantucket Bar—impeded the passage of the golden age's larger and more heavily loaded ships; the thousands of barrels of oil in their holds had to be laboriously off-loaded into lighters and rowed over the bar into the harbor at burdensome additional cost.

Eventually a Nantucket businessman, Peter Ewer, met the challenge.[37] Drawing on the experience of the Dutch moving ships over shallows into the Zuider Zee, Ewer designed and had built what he called "Camels": double floating drydocks 135 feet long, which could be flooded and positioned port and starboard alongside the whaler with chains slipped under the hull cradling the ship between them. Steam engines then pumped out the two drydocks, which floated them and their cradled ship high enough to clear the bar when a steam tug hauled them into harbor.

Conservative Nantucketers resisted Ewer's invention when the Camels first launched in early September 1842, especially when the improvised chains snapped one after another and damaged the copper sheathing of a ship being hauled, the *Phebe.* Ewer had already ordered stronger chains. On 21 September his Camels lifted the whale ship *Constitution* over the bar and out to sea. More dramatically, on 15 October they brought in a loaded whale ship.

Other troubles plagued Nantucket in the 1840s and 1850s: a disastrous fire in 1846, which burned down the waterfront and the center of town; the California gold rush of 1848 to 1855, which drained the island of some eight hundred vigorous young sailors eager to pan for gold. Many crews from America's whaling ports jumped ship on the West Coast in those years, officers among them. Many in the gold rush years signed on in the first place for a free ticket to California.

By the 1850s, whales retreating farther and farther away from their relentless Lilliputian hunters, who impaled them with harpoons and cut them to death with sharpened spades, had withdrawn to the Japan grounds off the northeastern coast of that archipelago or up into the Arctic Ocean. As had been the case of firewood in Elizabethan England, increasing the distance from collection point to market increased the cost, and whale oils, in any case, were costlier than the alternatives, most of them little known today, developed to replace them. What were those alternatives? And what became of them?

NINE

BURNING FLUIDS

Settlers on the Minnesota frontier in the 1850s could fill their lamps with camphene or burning fluids at the local general store. The store sold whale oil as well, less volatile and therefore popular for lanterns carried outdoors, with glass fronts and tin sides and backs perforated with star- and diamond-shaped holes. A memoirist recalls early settlers speculating about what might be "a possible source of supply for [lamp] oil when all the whales had been killed."[1] It was general knowledge, then, that whales were overhunted in whaling's so-called golden age, their populations declining.

The US whaling fleet had reached its maximum extent in 1846, with 736 ships totaling more than 233,000 tons burden.[2] Whale oils were a depleting asset, inherently limited by their limited source: far more gallons of camphene and burning fluids than of whale oils were produced for lighting in the United States in the first half of the nineteenth century. Castor, rape, and peanut oils, tallow and lard were widely used as well, as were wood and grain alcohol. But camphene, at 50 cents a gallon, was cheaper than whale oil at $1.30 to $2.50 a gallon, and cheaper even than lard oil at 90 cents a gallon.

Burning fluids included naphtha and benzene, both distilled from coal. Camphene was distilled turpentine. The most common burning fluid was a

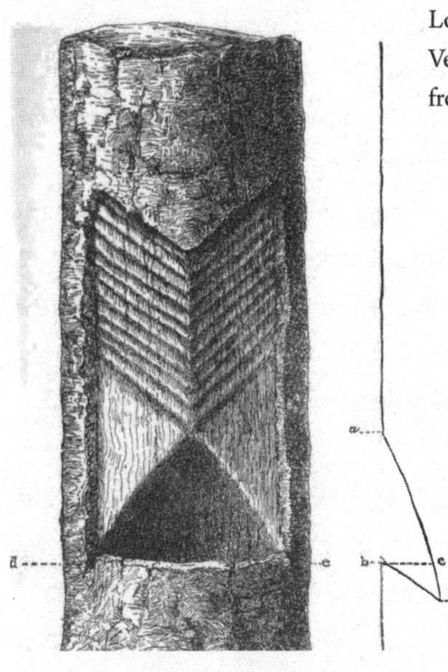

Longleaf pine boxed for collecting raw turpentine. Vertical cross section at right shows catch-basin from which turpentine was scooped.

mixture of high-proof grain alcohol blended with 20 percent to 50 percent camphene to color the flame and deodorized with a few drops of camphor oil.[3]

The primary turpentine tree in America, the longleaf pine, *Pinus palustris*, grew in great abundance in the American South, including some 400,000 acres (625 square miles) of sandy pine forests in eastern North Carolina.[4] Workers collected crude turpentine, the liquid resin of the longleaf pine, by "boxing": cutting a large, chevron-shaped drainage into a tree with a deep notch below it that collected the liquid, which was then scooped out and barreled.[5] The process was similar to maple-sugaring, although boxing was far more damaging to the tree. Nor is resin the same as sap: rather than carry nutrients, resin is exuded as a natural bandage when a tree is wounded, the plant version of a scab.

To early American settlers, the primordial longleaf pine forests of the southeastern United States seemed an inexhaustible resource. Mature trees grew to heights of one hundred feet or more and diameters of as much as two feet, clear of branches for two-thirds of their height. Their combined canopy shaded the forest floor from undergrowth, covering it instead with a thick mat of golden twelve-inch pine needles. "A squirrel could travel through longleaf pine treetops," the saying went, "from Virginia all the way to Texas and never touch the ground."[6]

Before the American Revolution, producers barreled the raw turpentine and shipped it to England for distilling. After the Revolution, distilleries established in Philadelphia and New York processed the turpentine into camphene. Distilleries moved closer to the pine forests with the coming of the railroad in the 1830s. By then, turpentine distilled for lighting had displaced pitch and tar as the primary product of the southern forests. The value of turpentine production in the United States approached $7.5 million in 1860 ($210 million today), of which North Carolina accounted for more than $5 million.[7]

Boxing had to be repeated from year to year, progressively damaging the trees. Damage extended up as high as fourteen feet above ground. Double boxing cut away two faces of a tree, leaving a limited strip of sapwood to feed the canopy. "Chipping"—scarifying the tree every two or three days to restart the flow of turpentine after the box face had sealed over with hardened resin—further challenged the tree's resources.

First-season turpentine was called "virgin dip" and earned the largest profit. Less valuable second- and after-year turpentine was called "yellow dip." A tree might survive seven seasons despite the progressive damage of boxing and chipping to its structure. Loggers then harvested the dead and damaged trees for their dense heartwood.[8]

By the 1850s, turpentine production was declining in America. A new competitor surged into the lamp oil market: coal oil, distilled from cannel coal (oil shale) or asphalt/bitumen, a heavy hydrocarbon found naturally in semisolid pools such as Pitch Lake on the Caribbean island of Trinidad and the La Brea Tar Pits in Los Angeles. (*Tar* and *pitch* are traditional terms for asphalt/bitumen, but are also used, confusingly, for tree exudates.) The world's most extensive reserve of bitumen occurs in the Canadian province of Alberta, where the mineral in the form of oil sands—sand mixed with asphalt/bitumen—covers some fifty-five thousand square miles, an area larger than England.

A Canadian physician and entrepreneur named Abraham Gesner pioneered the development of coal oil, initially as a source of coal gas for lighting. One of twelve children, Gesner was born in Nova Scotia in 1797. His early life is unknown, but at nineteen, in 1816, he demonstrated his gumption by attempting to sell a string of starving horses. The horses were starving in that "year without a summer" because the ash from an erupting Indonesian volcano, Mount Tambora, had drifted around the world and blocked enough sunlight to cause hard freezes in July in the more northern latitudes of North America and Europe. Crops failed; two hundred thousand people died of starvation in Europe that year. Paying for his own passage by working as a deckhand, Gesner shipped his horses to the West Indies, sold them there, and used the small profit to tour the islands, collecting rocks, shells, and minerals. He investigated tar springs on Barbados and inspected Trinidad's Pitch Lake. Christopher Columbus had encountered Trinidad in 1498, after which the Spanish had used Trinidad pitch for caulking their ships. Gesner hauled home a mass of the lake's bitumen. He tried horse trading twice more without success: both voyages were shipwrecked.[9]

After these ventures, the enterprising young Canadian had the shrewdness or good fortune to marry a physician's daughter. His father-in-law probably supported Gesner's three-year course in medicine and surgery at London hospitals. He returned to Nova Scotia in 1827 with a medical degree and took up practice as a country doctor.[10]

Medicine at that time was one of the few avenues to learning and doing science. Gesner's real interest was geology. By 1836, he had investigated the subject thoroughly enough to write a book of his own, *Remarks on the Geology and Mineralogy of Nova Scotia,* which included a discussion of the region's coal mines and potential iron resources. Two years later, he was appointed provincial geologist of the province of New Brunswick, northeast of Maine, and undertook a five-year geological survey of the region with an eye to deposits with commercial value.

Thereafter Gesner concentrated on developing a fuel for lighting. He used

Pitch Lake, Trinidad.

the pitch he had collected in Trinidad as feedstock, conducting some two thousand separate experiments. By 1846, he had successfully distilled coal oil, as it was commonly called, from this bitumen.[11] He demonstrated it at public lectures he delivered that year on Prince Edward Island and subsequently in Halifax, Nova Scotia.[12] Newspaper stories about his lectures caught the attention of none other than the tenth Earl of Dundonald, Thomas Cochrane, admiral of the British North American and West Indian fleet, based in Halifax. The earl happened to be negotiating for the mineral rights to Trinidad's Pitch Lake. He enlisted Gesner as a consultant, and in 1851 won full control of the lake's pitch deposits by acquiring all the surrounding land.[13] The relationship formalized when Gesner moved to Halifax in 1849.

In a matter of months, the Canadian physician developed a distinctive process for making illuminating gas from bitumen with coal oil as an intermediary. When he applied for a Nova Scotia patent on his process in June 1849, he used the patent to protect his products' brand names as well, calling them kerosene and kerosene gas (from *keros*, Greek for "wax," and *-ene* to associate the new products with familiar camphene).[14]

A complex series of lawsuits over the next four years barred Gesner from developing the Nova Scotia and New Brunswick bitumen mining claims he believed he controlled. He lost a patron and partner in 1851 when Dundonald, then seventy-six, completed his naval service in North America and returned to England to spend the rest of his life—he died in 1860—promoting ventures in Trinidad. In 1853 Gesner packed up his wife and five sons and moved to New York. He had already applied for US patents for his kerosene products. He soon found a promising new market for lighting gas on Long Island.

That year, with US partners, Gesner formed a new venture, the North American Kerosene and Gas-Light Company, with a manufactory in Brooklyn along Newtown Creek. But the lack of intercity piping limited gas sales, which led Gesner and his company to concentrate on producing lamp fuels for the extended East Coast market. The Kerosene Oil Company, as people called it, announced in its prospectus that it would produce "Mineral Naphtha, Hydraulic Concrete, Burning Fluids, Mineral Tar and Pitch, and Railway Grease" from "Asphalte Rock [i.e., bitumen] . . . found in inexhaustible quantities in the Province of New Brunswick."[15] It emphasized that kerosene was ideal for the production of burning fluids less expensive than those currently in use. Gesner's US patents were issued in June 1854. For feedstock, his company would initially use cannel coal from New Brunswick.

In the meantime, Gesner kept busy perfecting kerosene. The crude coal oil that emerged from his stills smoked badly when it burned and smelled worse. After treating the oil with acids and processing it with lime, he succeeded in creating a kerosene that burned, he reported, "with a brilliant white light [and] without smoke or the naphthalous odor so offensive in many hydrocarbons having some resemblance to this but possessing very different properties."[16] Independent chemists, according to a company brochure, attested that kerosene burned "13 times brighter than Sylvic,* Lard, and Whale Oils; 6 times as bright as sperm oil; 2½ times as bright as Camphene or rapeseed oil; 26 times as bright as 'burning fluid'; and 4 times as bright as even that paragon of the age, the gaslight."[17] Even more impressively, it was cheap at a dollar a gallon. Adjusting for its more intense light output, it was seven times cheaper than burning fluid, six times cheaper than sperm oil, four times cheaper than lard oil, and half the price of gaslight. "[It] must soon be used in every house in the country," the sales brochure concludes.[18]

It nearly was. After fighting its way into the market over the determined resistance of the turpentine and alcohol interests, Gesner's company by 1859 was producing five thousand gallons of kerosene per day, three hundred days a year, for a total of more than 1.5 million gallons annually.[19] It now distilled its product from Boghead coal, a kind of cannel coal imported from a Scottish mine in Bathgate, eighteen miles west of Edinburgh. Ironically, the Kerosene Works flared away the incidental gas it generated, since it lacked a pipeline system for distribution.[20]

Imitators sprang up on every side. By 1860, between sixty and seventy-five coal-oil plants operated up and down the Eastern Seaboard, from Maine to Philadelphia and along the Ohio River system from Pittsburgh to western Kentucky, compared with some two hundred plants producing coal gas in US cities, a mature forty-year-old industry.[21] Coal-oil production in early 1860 totaled some 20,000 to 30,000 gallons per day, or about 7 million to 9 million gallons

*A patented oil distilled from wood pulp.

per year.[22] By comparison, the whale-oil harvest had peaked in 1854 at about 10.3 million gallons and begun a sharp decline.[23]

One of the expert witnesses against Abraham Gesner in the trial challenging his Canadian bitumen mining claims had been Benjamin Silliman Jr., an 1837 Yale College graduate and subsequently a professor of chemistry there. Silliman's father, a pioneering chemist born during the American Revolution, was a principal founder of graduate science education in America. While teaching at Yale and elsewhere, father and son both supplemented their incomes by serving as expert witnesses and as mine evaluators. In 1847 young Silliman had joined other investors in establishing the New Haven Gas Light Company—his house was the first in the city to be piped and lighted—and continued to serve as a director. New Haven, Connecticut, at that time was the smallest city in the United States to upgrade to coal gas, with four miles of mains connecting the gas plant with the town center. Silliman also served on the city council.

With these exceptional credentials, the Yale chemist was an obvious choice to evaluate a potential new feedstock for kerosene production when two businessmen approached him with that project in autumn 1854. It was generally known that petroleum—rock oil, as it was commonly called, translating its Latin compound—could be refined into kerosene using much the same distilling process that Gesner and others used for solid bitumens. The two men judged that a domestic liquid source would allow them to produce kerosene less expensively for the large and expanding market for lamp fuel than mined sources did. They had acquired a farm in western Pennsylvania, one hundred miles north of Pittsburgh in Venango County, near the town of Titusville, where seeps of brown, greenish-tinged oil gave their name to a stream called Oil Creek.

The oil had been seeping since its first notice by early Spanish and French explorers. In the 1760s, a Moravian missionary to the Seneca Indians, David Leisberger, described seeing several kinds of oil springs in the area. The Senecas, Leisberger noted, purified the oil by boiling it and then "used [it] medicinally as an ointment for toothache, headache, swellings, rheumatism, and sprains. Sometimes it is taken internally. It is of a brown color, and can also be used in lamps. It burns well."[24] The steady seepage into Oil Creek argued for a substantial underground reservoir. To find that reservoir and tap it would require investment. Silliman, the two men hoped, would certify the value of the petroleum to potential investors.

They hired the right man. Silliman was well aware of the Oil Creek flows. His father had mentioned them in a report on a similar seep in Seneca country near Cuba, New York, published in the *American Journal of Science and Art* in 1833. Silliman Sr. noted that "the large quantities of petroleum used in the eastern states, under the name of Seneca oil" came not from the Cuba seep but from a source "about one hundred miles from Pittsburgh, on the Oil Creek . . . in the township and county of Venango. . . . By dams, enclosing certain parts of the river or creek, it is prevented from flowing away, and it is absorbed by blankets, from which it is wrung."[25]

So the younger Silliman, who had worked for years as his father's laboratory assistant and lecture demonstrator, would have been familiar with the petroleum from Oil Creek, Pennsylvania, which the two businessmen had brought him. He had almost certainly distilled it for his father. The senior Silliman was more interested in the possibility that the two oil springs, in southwestern New York State and in northwestern Pennsylvania, might contain coal beds as well. If so, he recommended with blind foresight in his 1833 report, "it would not be wise, without more evidence, to proceed to sink shafts; for they would be very expensive and might be fruitless. It would be much wiser," he emphasized, "to *bore*." For coal, he meant, not for oil; that idea had not yet taken hold. Like many mineralogists of his day, the elder Silliman believed that the two hydrocarbons were related and that an abundant show of petroleum at the surface "affords a strong presumption in favor of the existence of coal beneath."[26]

The two men who consulted with Silliman Jr. were George H. Bissell and Jonathan G. Eveleth, partners in the Wall Street law firm of Eveleth & Bissell. Eveleth was an 1854 graduate of Harvard Law School, but Bissell was a man of previous careers. On his own since he was twelve, he had taught school, worked his way through Dartmouth College, written for newspapers, and served as superintendent of schools for the city of New Orleans before returning to his native New England in ill health in 1853 and taking up investing.[27]

Bissell and Eveleth were following out a series of events so fortuitous it seems almost accidental. In 1849 a Titusville sawmill owner, Ebenezer Brewer, had sent to his son Francis Brewer, a young physician practicing in Vermont, five gallons of Seneca oil from the creek that ran below his sawmill, "with the assurance," his son said later, "that it possessed great medicinal and curative properties."[28] The younger Brewer had used the oil to treat patients and had shared some with Dixi Crosby, a former professor of his who taught surgery at the medical school of Dartmouth College. He also shared some with a Dartmouth chemistry professor, O. P. Hubbard, who thought the oil would never be common.[29] Discouraged, the young physician did nothing further about it until 1853.

By then, Brewer had given up his medical practice to join his father and his father's partners in operating the Titusville sawmill and selling the lumber. Settled in Pennsylvania, Brewer recalled, "for the first time, [I] examined the oil spring, in the vicinity of one of our saw mills." What he found contradicted Hubbard's hasty dismissal of its prospects: "I became satisfied there was oil in abundance." He discussed with his partners how to exploit it. They decided to try to increase its flow and hired a local man, Jacob D. Angier, to do the work. On 4 July 1853, the sawmill owners signed a lease with Angier, the first oil lease known to have been executed in the United States.

Even with trenching and other improvements, however, the material increase amounted to only a few gallons a day, not enough to justify marketing. "It was used for lighting the mill and lubricating the machinery," Brewer said, hardly the valuable uses the sawmill owners had envisioned.[30]

Dixi Crosby's son Albert, a twenty-six-year-old lawyer, was Francis Brewer's cousin.[31] When Dixi Crosby showed his visiting former student, George Bissell, a sample of Oil Creek oil and praised its valuable properties, young Crosby proposed traveling to Pennsylvania to inspect the oil spring on Bissell's behalf. "From the first," Brewer recalled, his young cousin "was enthusiastic in his estimate of its value and its abundance." Albert and Francis worked their way along the creek, visiting the various places where oil welled up into the water, "as far down as the McClintock farm. Here was a well in the middle of the creek. Crosby became excited, and recommended the purchase of the entire oil district then known." Brewer recalls:

> As we stood on the circle of rough logs surrounding the spring and saw the oil bubbling up, and spreading its bright and golden colors over the surface,

> it seemed like a golden vision, and Crosby at once proposed to purchase the whole farm, which we could have done for $7,000, but as our pecuniary ability was limited to a much smaller sum, I was obliged to decline the tempting opportunity. . . . When I told him we did not wish to take capital from our lumber business to put into oil, he said, "Damn lumber, I would rather have McClintock's farm than all the timber in western Pennsylvania."[32]

Seven thousand dollars in 1854 would be about $198,000 today—a great deal to invest when no one had determined how much oil could be recovered or what its real value might be. In lieu of buying John McClintock's farm, Crosby proposed to the farmer a thirty-day option dividing his farm into three parts: the land on one side of Oil Creek for $1,000, the land on the other side for $5,000, and the oil spring for $2,000—or the entire farm for $7,000. Evidently disbelieving the promise of oil riches under his property, McClintock agreed.

Back in Titusville, Albert Crosby negotiated an agreement with Brewer and his sawmill partners, who owned the farm where the original oil spring was located. (Bissell had preauthorized Crosby to do so if he liked what he saw.) The young attorney proposed to organize a joint-stock company capitalized at $250,000 ($7 million today), dividing among the parties the shares to be sold to investors. With that capital, the enterprise, to be called the Pennsylvania Rock Oil Company, would buy the hundred-acre Hibbard farm outright for $5,000. The company's public offering would include the oil rights to several thousand more acres that the sawmill partners owned in the area.

A period of confusion ensued as the various parties sorted out what they thought their agreements meant, but by autumn 1854, Bissell and Eveleth had formed the Pennsylvania Rock Oil Company of New York and were preparing to sell stock. They were also attempting to recoup expenses by selling some of their newly acquired Oil Creek oil rights to a new body of potential buyers in New Haven, Connecticut. Why they chose New Haven is obscure, but the choice proved significant when the New Haven group required Bissell and Eveleth to commission Benjamin Silliman Jr. to analyze the Oil Creek petroleum and report on its merits.[33]

They did so in October 1854. Within a month, they had an early response from Silliman: the petroleum was more likely to be valuable as a paint solvent "than as a medicinal, burning, or lubricating fluid." The finding was premature and must have been discouraging, but the two men judged that the expense of Silliman's forthcoming chemical analysis would still be money well spent.[34]

Yale was not, in those days, the munificently endowed university it has become. When Silliman was appointed professor of chemistry in the new School of Applied Chemistry, established in 1847, the Yale Corporation made no provision for paying him or anyone else on the school's faculty. Silliman proceeded to rent the substantial campus house allotted to Yale's president, who preferred to live elsewhere, convert it into laboratories, and collect fees from his students.[35] In 1849 the University of Louisville lured him away by offering him a salary to teach in its medical school. When Silliman returned to Yale in 1854 upon his father's retirement to take up the college's chemistry professorship, he complained to a friend that he was "making a very large pecuniary sacrifice" in leaving Louisville.[36] Yale would pay him $1,000 for his first year as a chemistry professor ($28,000 today) plus $300 ($8,500) for laboratory expenses.[37] He had no intention of making a further pecuniary sacrifice consulting on Pennsylvania oil. He had a college laboratory to furnish.

After his preliminary report to Bissell and Eveleth, Silliman continued analyzing the Oil Creek petroleum sample to establish its potential commercial value. By late December, he could write the partners more optimistically: "I am very much interested in this research, & think I can promise you that the result will meet your expectations of the value of this material for many useful purposes."[38] The two men were happy to hear it.

Suspecting that the crude petroleum consisted of several different liquids with different boiling points, Silliman had begun separating it into its component parts by fractional distillation. The process involved heating an oil-filled retort in a liquid bath—water until the water boiled, and for higher temperatures, linseed oil—and slowly raising the temperature until distillation began, then holding the temperature there until the fraction that boiled at that temperature completely distilled away. He then increased the temperature by stages of 10 degrees. Each component would boil at a certain temperature, just as water does, and would distill out as a gas, after which increasing the temperature would allow the next fraction, with a higher boiling point, to do the same.[39]

Silliman called fractional distillation "tedious," and, in fact, it took him more than three weeks to complete the experiment. But he identified seven different component oils mixed together in the crude petroleum, ranging in color, viscosity, and odor from product no. 2, "perfectly colorless . . . thin and limpid . . . [with] an exceedingly persistent odor," to product no. 8, "the color and consistency of honey, and the odor . . . less penetrating than that of the preceding oils." (Product no. 1, the first to boil off, was merely water.) Left behind in the retort was "a dark, thick, resinous-looking varnish, which was so stiff, when cold, that it could be inverted without spilling."[40]

The Yale professor discovered something else as well: the real breakthrough for which Bissell and Eveleth hoped. Several of the oils he'd distilled from the crude sample had what seemed to be varying boiling points. That anomaly implied that they were themselves mixtures of yet other oils, he told his sponsors, "produced by the heat and chemical change in the process of distillation" and "evolving new bodies not before existing in the original substance."[41] Distilling these mixtures in turn came to be called "cracking" them—breaking them open, as it were.

Silliman had struck oil—oil suitable for lighting. "Having met unexpected success in the use of the distilled product of *Rock Oil* as an illuminator," he began a new request for funds on 1 March 1855, he proposed to test the distilled product "in various lamps and also in comparison with *various oils.*" He listed the various "oils": "Sperm, Colza & Sperm Candles." To compare them, he needed a supply of standard candles and an expensive, custom-made photometer (which he also planned to use to experiment with gaslight for the New Haven Gas-Light Company).[42] In for a dime, in for a dollar; what else could Bissell and Eveleth do?

On 16 April 1855, when Silliman finally finished his *Report on the Rock Oil, or Petroleum, From Venango Co., Pennsylvania, With Special Reference to its Use for Illumination and Other Purposes,* the two investors judged the analysis to be everything they had hoped for, even if it had cost them, as Eveleth complained, "a good deal of money." It cost somewhat more than $600 (about $17,000 today). But it would "make a stir," Eveleth added. "Stock will then sell. Things look well."[43]

Yet four more years would pass before rock oil began to flow in volume in Pennsylvania. Once the Silliman report established that petroleum had value as an illuminant, the problem remained how to extract enough from the Oil Creek site to make the venture profitable.

In the meantime, the company went through several more reorganizations as the New Haven investors contrived to cut the New York men out of the deal. In 1858 a new entity, the Seneca Oil Company of New Haven, Connecticut, swallowed up the Pennsylvania Rock Oil Company.[44] Banker James Townsend and his fellow New Haven investors took control, brought in a new associate, a local man named Edwin L. Drake, and elected him president. It was Drake, improbably, who would find a way to release petroleum from its stone detention underground.

Townsend had befriended Drake in the Tontine, the New Haven hotel where they both lived. Drake had moved there with his young son in 1855 after the death of his wife in childbirth the previous year. In 1858 he was thirty-eight years old, tall, slim, black-bearded, and engaging, a religious man who rarely swore but whose fund of stories filled hotel evenings with laughter. Born in Greenville, New York, near Albany, Drake had grown up on farms in New York and Vermont and worked at clerk- and transportation-related jobs from Massachusetts to Michigan. His primary qualification for overseeing oil production, besides his respectable presence, seems to have been his possession of a railroad pass good for free travel, earned from working as a train conductor for the New York & New Haven Railroad from its inception in 1849 until he retired because of ill health in 1857. The condition that led to his retirement was malaria, a disease endemic in those days throughout the lower Middle West and as far northeast as Pittsburgh.[45]

Between 1855 and 1858, besides negotiating to acquire land and oil leases along Oil Creek, the investors in what came to be the Seneca Oil Company had sent various people out to Titusville to try to increase production. The trench along the creek where seeping oil accumulated had been enlarged until it was some eighteen feet deep and four feet wide. Even so, the seep seldom delivered as much as six gallons on a good day. The best year's production so far, in 1857, had totaled not more than twelve large barrels holding about a thousand gallons of crude oil, at a time when coal-oil production was approaching seven *million* gallons annually.[46] When Townsend replaced Drake as company president, early in 1858, Drake was appointed superintendent, at an annual salary of $1,000 ($29,000 today). He had remarried in 1857. Late that year, he had traveled to Titusville on his own to inspect the Seneca properties. Now, in May 1858, he packed up his wife, son, and five-month-old infant and prepared to move to Titusville.[47]

Along the way to Titusville in December 1857, Drake had stopped in Syracuse, New York, to inspect a number of salt wells there. No one had drilled for oil yet, but drilling for saltwater was common, and Drake wanted to study how it was done. Saltwater wells frequently became contaminated with petroleum, which was sometimes collected and sold. Historians disagree about who first thought of *drilling* for oil. Townsend claimed it was his idea; so did Drake. Whoever thought of it, Drake had it in mind on his first expedition out to Pennsylvania.[48]

Our ancestors used salt not merely to flavor their food. Until the introduction of household refrigeration in the 1920s, salt curing was one of the few means available for preserving food, meat in particular. Salt taxes were common across the centuries precisely because salt was a commodity vital to food preservation.

There were two ways to bore a salt well, just as there had been in probing for coal in earlier times: kicking down a well using a spring pole, the old hard way; or chiseling a borehole with a drill string supported by a derrick and lifted and dropped by a small steam engine.

When Drake arrived in Titusville with his family in May 1858, he discovered that Townsend had prepared the townspeople by addressing him as "Colonel" Drake in letters and documents sent ahead to await him there, anticipating that the postmaster would spread the word. He did, and Drake was greeted respectfully with his presumed title. He had served with the Michigan militia as a young man; his wife would claim later that he had earned the rank, and perhaps he had.[49] Townsend, a banker to his core, demonstrated no such charity where Drake's salary was concerned. Across the next year, the Seneca Oil Company's superintendent of oil production frequently had to dun the company for his pay, borrow money locally on his good name, and once even return to New Haven to appeal to the board of directors for money.

It took Drake more than a year to find someone to drill his well. One salt-well driller after another refused him or failed to appear when promised.

Steam-powered rig for salt (and oil) drilling; essentially a powered version of a spring pole.

His standards were high. "Although I arrived at Titusville on the 15th day of May," he recalled, "it was the 20th day of July before I found a Man that would contract to bore a five-inch hole 1,000 feet deep and agree to forfeit or lose the pay if by his own negligence or Carelessness he did not succeed." That man, like others to come, never showed up. One recruit died on the way to Titusville.[50] "Salt borers," as Drake called them, tended to be a drunken lot in any case, "thirsty souls [who] preferred Whiskey to any other liquid for a steady drink." But most or all of them also thought the tall, bearded stranger was insane.[51] "We knew there was oil there," one local explained. "But that didn't count for much with us because the oil didn't seem to be good for much."[52]

Drake filled the days improving the skimming trenches, building an engine house, ordering and installing the steam engine, and riding back and forth to various towns looking for a well driller or buying supplies. The winter of 1858–59 was miserable. "It rained at least four days in each week all winter," Drake remembered.[53]

Finally, in April 1859, a friend wrote him from Tarentum, Pennsylvania, about eighty miles south of Oil Creek, "that he had found me a Man, one that would suit me precisely. He could recommend him as he had been in his service several years."[54] Drake rode down to Tarentum to meet William Andrew Smith, "Uncle Billy," a compact, bearded blacksmith. The area's salt borers relied on him to solve any problems they encountered drilling wells. Uncle Billy had been planning to retire from blacksmithing and take up farming. He agreed to drill Drake's well for $2.50 a day ($72 today, or $9 an hour), a good price, especially since three of Smith's sons and one of his daughters (to cook and clean) would eventually join him in the work at no extra charge. Learning that Drake hoped to buy drilling chisels and other tools for the project in Tarentum, he proposed making them himself before he closed down his forge.

The Smiths arrived in Titusville on 20 May 1859, without Uncle Billy's wife. His daughter, Margaret, said her mother thought Titusville "seemed too much like 'back woods' and she was afraid something might happen to us."[55] The family soon moved from the local hotel into the engine house, next to the well site. Work on the well began immediately with construction of a new engine house and a wooden derrick twelve feet square at the bottom and about thirty feet high, nailed together on the ground and then lifted into place like a barn raising by a crowd of local volunteers in late May or early June. Many of them had never seen a derrick before. One wag told a farmer who asked about it that a New York man was erecting it as a monument to mark the location of the biblical Tower of Babel. Locals called the project "Drake's Folly." In June as well, Mrs. Smith summoned her courage and moved to Titusville with the rest of her six children. She was prescient; during their time in Titusville, the Smiths lost their youngest daughter, Lida, to diphtheria, shortly before her seventeenth birthday.[56]

Lacking any other way to locate oil underground—dowsing and consulting spiritualists would come later—Drake chose to drill in the middle of the narrow island formed by Oil Creek on one side and, on the other, the water-powered sawmill's millrace (a channel to divert water to a mill wheel).[57] In August 1858, while he was still trying to find a well driller, he had hired a crew to dig for oil in that location. Their spades soon struck oil, but they struck water at the same time, flooding the hole. Drake had decided then that it would be cheaper to bore.[58] What remained of that trial was an eight-foot hole cribbed with logs. That was where Smith began work, intending to crib on down to bedrock and then to bore a hole.

Spring rains had saturated the island soil. Water welled up as Smith and his son began digging. Working submerged to their waists in icy water and taking breaks five or six times an hour to warm up, they managed to dig through sand and clay down to about sixteen feet. But like the old one, the new hole filled with water, collapsing the walls. "There was enough water in it to drown us all," Uncle Billy recalled.[59]

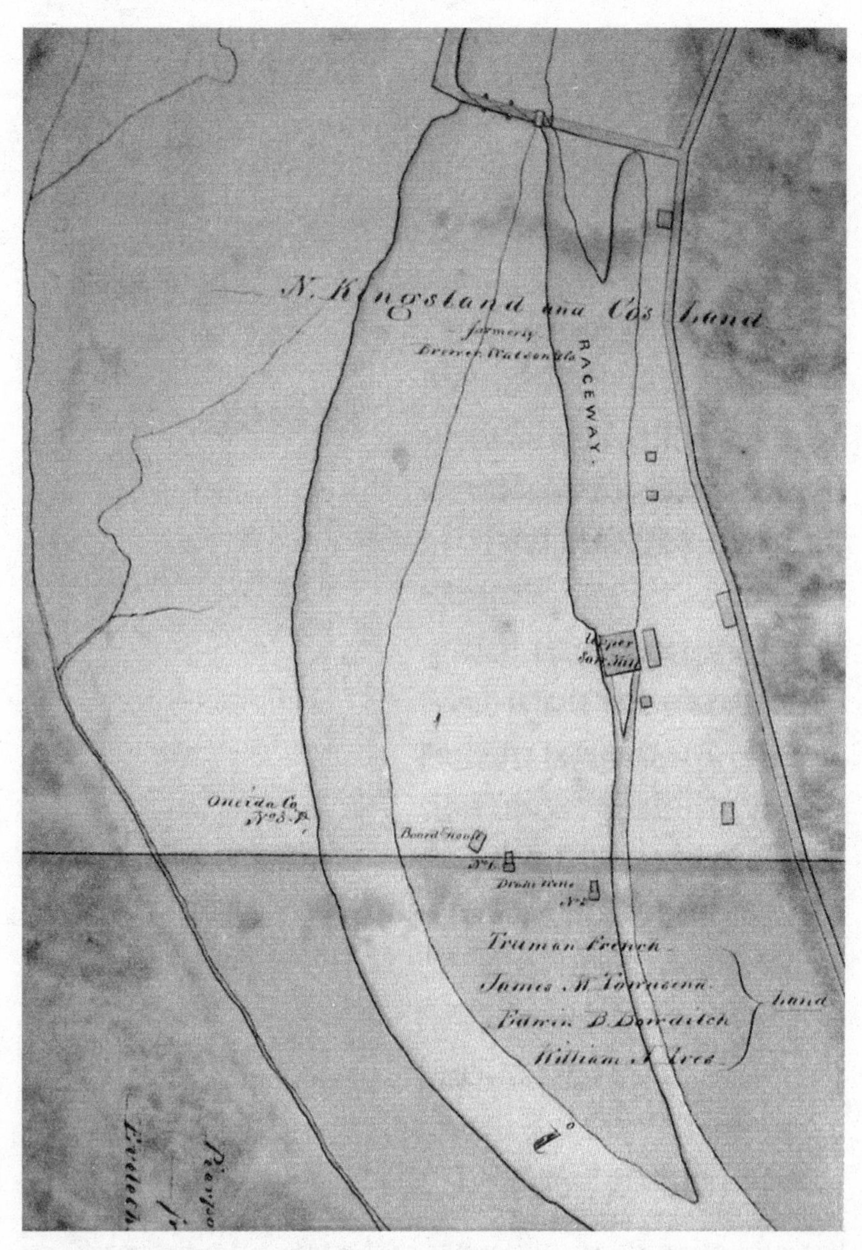

Contemporary map of island formed by Oil Creek and millrace. Drake's first well, captioned "no. 1," is located below the center of the map on the map fold line that divides the island horizontally. (Well no. 2, on the right, closer to the millrace, was drilled later.)

Drake, or Uncle Billy, or both of them, decided they should drive a pipe—a well casing, it came to be called—down through the sand and clay to bedrock. Drake located two lengths of cast-iron pipe each about nine feet long. Using a windlass to crank a battering ram, they rammed down the first pipe, but the second crushed the upper end of the first, after which Smith drew up a pattern for a thicker-walled pipe and Drake had it cast in ten-foot lengths. ("I could not have suited myself better," Drake would praise Uncle Billy, "if I could have had a man made to order."[60])

The new pipe arrived around 2 August 1859. By 9 August, the ram had driven it down to bedrock at forty-nine feet, eight inches.[61] Uncle Billy went to work with a chisel on the drill string, using a rope pulleyed over the derrick and hauled by the steam engine to lift and drop the drill string. When the hole filled with stone chips, he had to pull the drill string, replace the chisel with a tubular scoop, and scoop out the hole. The dulled chisel had to be sharpened regularly as well, which he usually did on site. Across the month of August, he worked away, chipping down about three feet a day, to fifty and then to sixty feet.

Back in New Haven, the investors had given up on Drake's Folly—all but James Townsend, the banker. "You all feel different from what I do," Drake appealed to the investors. "You all have your legitimate business which has not been interrupted by the operation, while I staked everything I had upon the project and now find myself out of business and out of money."[62] By then, two of Drake's friends had cosigned a $500 note for him at a Pennsylvania bank to see him through. That August 1859, even Townsend decided that the venture was a lost cause and moved to shut it down. He wrote Drake to that effect and enclosed a money order for $500 to pay any final bills. Fortunately, the mail was slow.

On Saturday, 27 August, near the end of the day, Uncle Billy's chisel broke through a rock layer at sixty-nine feet and dropped a further six inches. Both he and Drake reserved Sunday for church and rest, keeping the Sabbath, so he pulled the drill string and called it a day. He must have wondered about the breakthrough during church on Sunday morning, because on Sunday afternoon, he walked over from his house to the well and looked in. Oil had filled the hole up to four feet of the opening. It shimmered in the afternoon light.[63]

TEN

WILD ANIMALS

Oil, Edwin Drake would say, was hard to sell (a common if surprising theme with new sources of energy). "It takes time and work to introduce it," he cautioned New Haven banker James Townsend a few frustrating months after his historic August 1859 strike.[1] No one seems to have anticipated that challenge. The first response of the people of Titusville to the Seneca Oil Company's success in drilling an oil well was a mad rush to acquire mineral rights along Oil Creek and drill more wells. Drake's well was a pumper, not a gusher. Uncle Billy Smith had attached an ordinary hand-operated well pump to the drill casing on that first Monday after the strike while Drake scouted barrels.

"They had to begin with so simple and elementary a matter as devising something to hold the oil," writes Ida Tarbell, the Titusville native and contemporary reporter. "There were not barrels enough to be bought in America, although turpentine barrels, molasses barrels, whiskey barrels—every sort of barrel and cask" were put to use.[2] Townsend, a shrewd businessman, soon bought a Titusville cooperage—a barrel factory—to profit meeting the demand. He opened a bank in Titusville as well. A year after the Drake strike, oil flowed from a crowded field of wells at a thousand barrels a day, which made

Early Pennsylvania oil wells.

about five hundred barrels of lamp oil. Only a small part of that volume was ever shipped, and much was lost to leakage along the way.[3]

Petroleum was hard to sell because it was costly to transport by wagon and smelled terrible in its raw state. Refining it required moving it from Oil Creek to one of the many refineries, most of them small, that sprang up between Titusville and Pittsburgh. But swampy or frozen dirt trails slick with spilled oil played hell with caravans of up to one hundred wagons loaded with thousands of gallons of barreled oil.

"Fresheting" became a desperate alternative: raising the creek level with a wave of water—a freshet—to float the oil in flatboats perilously down Oil Creek to the Allegheny River. Flatboats holding up to eight hundred barrels of oil lined the creek banks waiting. Word went upstream to the first of more than a dozen private mill dams ponding the water on the creek and its tributaries. Crews stationed at the dams opened them in succession. The swell of water from the first dam, farthest up the creek, ran down to the second dam and so on, one after another as the freshet advanced. When the Kingsland Dam opened, three miles below Titusville, enough water had been released to cause the creek to rise two feet or more.

A fresh breeze would announce the approaching flood. It was a nice trick to know when to push off into the swell. "Inexperienced boatmen generally

One freshet run drove flatboats into a bridge near Oil City.

cut loose their boats upon the first rush of water," writes historian Paul Giddens, "only to be grounded and battered into kindling by those coming later. An experienced boatman waited until the water commenced to recede, then cut loose his lines, throwing himself upon the mercy of the swift current." The full concourse was no small event, Giddens continues:

> A pond freshet afforded a most unusual sight, for here were 150 to 200 flatboats, little and big, loaded with 10,000, 20,000, or 30,000 barrels of oil, either barreled or in bulk, floating along endways and sideways on a rushing flood and wildly fighting their way down Oil Creek, which was only twelve rods [198 feet] wide and very crooked as it wormed its way through steep hills. It required all the skill and strength of some 500 boatmen to avoid collisions with other boats, the rocks, and other obstructions.[4]

Even a successful freshet run didn't guarantee delivery. Often as much as a third of the oil was lost to barrel leakage before the run began, and another third leaked away during the float down the Alleghany to Pittsburgh.[5] With such limited and inconsistent delivery, petroleum was not an immediate threat to the coal-oil business: by 1860, coal-oil refiners had raised their production to 262,500 barrels a year.[6]

Conditions began to change in 1861. In April, the month that the Civil War began with Confederate Brigadier General P. G. T. Beauregard firing on Union forces defending Fort Sumter in South Carolina, the first gusher came in on Oil Creek. Shallow pioneer wells such as Drake's had tapped into oil too close to the surface to confine natural gas. But on 17 April, around sunset, on a farm at the lower end of Oil Creek, a column of oil propelled by natural gas roared out of a deeper drill hole and gushed sixty feet into the air.

Within a half hour of its release, time for a small crowd to gather, it found an errant flame, caught fire, and exploded. Black smoke billowed up and oil fell to earth as burning rain. A sheet of fire swept across the surrounding ground and ignited two more wells, several open vats of oil, a barn, and a legion of oil barrels stored nearby. Nineteen people were killed; dozens more badly injured. The well burned for three days before drillers succeeded in smothering it with dirt and manure. Once under control, it flowed at an astonishing three thousand barrels a day.[7] Other gushers followed, beginning with one in May that flowed three hundred barrels a day for fifteen months before abruptly running dry, earning, for the lumberman who had bought the farm site for $1,500, about $2.5 million, or $67.8 million today.

The first rail connection reached Titusville in 1862, eventually superseding the artificial-freshet mêlée. In its first fourteen months, the new Oil Creek Rail Road carried away more than 430,000 barrels of oil and delivered more than 459,000 empty barrels to the oil well sites; sixty thousand passengers traveled in and out of the region by rail during the same period.[8] The ultimate improvement would be pipelines to move the oil from the wellhead to the railroad. Those came in various gauges from two to six inches, the oil flowing by gravity or pumped by steam, beginning in 1863. For a time, the teamsters who had monopolized the oil-hauling business holed the pipes with pickaxes and ignited the spilled oil. Eventually armed guards discouraged them, and they moved on—1,500 in one week in the summer of 1866.[9]

Before then, all the coal-oil distillers making kerosene had switched their feedstock from expensive cannel coal to petroleum, far cheaper at less than $1 per gallon. No adequate records survive of the increase in crude oil production during the Civil War, but Giddens found a suggestive parallel in the increased taxes collected by the federal government on refined petroleum: for the last third of 1862, about $237,000; for 1863, $1.2 million; for 1864, $2.3 million; for 1865, $3.05 million.[10]

At the same time, the war choked off three other important sources of lamp fuel. The Union blockade of Southern ports interrupted shipments of turpentine to northern distilleries, accelerating the adoption of petroleum-derived kerosene in the North.[11] An ill-considered new federal tax made grain alcohol unprofitably expensive. And Confederate predators hunted whaling ships, not whales, to near extinction.

Earlier in American history, to finance the government of the newly established United States, Congress had taxed grain alcohol, a tax repealed in 1817. By 1860, Americans were consuming more than thirteen million gallons of grain alcohol annually, 80 percent of it as a component of burning fluids.[12] That large market was an obvious source of revenue when the Union moved to raise funds to fight a war with the breakaway Confederacy. Between the beginning of the Civil War, in April 1861, and 1 July 1862, Congress authorized a system for levying and collecting taxes, An Act to provide Internal Revenue to Support the Government and to pay Interest on the Public Debt.[13] The act established a new government agency, the Internal Revenue Service, within the Treasury Department, and assessed several new taxes and licensing fees. The most destructive of these was a "duty of twenty cents on each and every gallon" of grain alcohol, collectible at the distillery. The duty increased across the Civil War to a final level in 1864 of some $2 per gallon, or $30 today.[14]

The alcohol tax was intended to assess *beverage* alcohol, but it failed to exclude industrial and illuminating alcohol. The high tax, raising the price of those alcohols to about $2.50 per gallon, drove the fuel out of the market just as petroleum-derived kerosene was entering it.

Camphene, the other component of burning fluid, similarly increased in price, because of its blockaded scarcity, from 35 cents per gallon prior to the Civil War to $3.80 in 1864.[15] Not for the last time, a petroleum product—kerosene—rode into the marketplace on a government subsidy—exclusion from a punitive tax—and crowded out other fuels. By 1870, camphene and alcohol had all but disappeared from the market, while petroleum-derived kerosene sales had reached 200 million gallons annually.[16]

The other lighting product competitive with kerosene was sperm oil. The whale fishery, economically strong before the war, fell into decline across the war and after.[17] A major cause of the decline was predation. Ironically, as the whalers hunted and killed whales, so also did rebel cruisers of the Confederate navy hunt down and destroy Yankee whaling ships.

The captain of the most notorious Confederate cruiser was an unprepossessing former US naval officer named Raphael Semmes, fifty-two years old in 1861, a Maryland native, lawyer, and naval journalist known for his prickly disposition and indifferent performance, married improbably to a daughter of Cincinnati Methodists with antislavery convictions. Somehow his contrarieties synchronized with the challenge of war. He became a relentless, deadly hunter for the Confederate cause, whom a Philadelphia newspaper christened "a wolf of the deep."[18]

Semmes's first raider, a slow steam and sail hybrid named the *Sumter,* captured no fewer than ten Union merchant ships in its first month of operation, June 1861.[19] Semmes burned only one of them, a merchant out of Maine sailing in ballast. The other nine carried foreign-owned cargo, limiting him to unloading them in foreign ports and confiscating them for the Confederate cause.

Three months followed with only two prizes. The *Sumter* sailed well enough, but its screw propeller was fixed to its hull, causing drag that slowed the ship when operating under sail alone. Nor could it carry enough coal for more than eight days of steaming. Its first flurry of captures had already given Semmes a fearsome reputation in the Yankee press, leading merchant ships to hide in port or avoid the Caribbean, where his ship's limitations had confined him. "It is of no use to chase sails anymore in these waters," he decided in December; "the Yankees have nearly all disappeared."[20]

He took two final prizes and then sailed northeast into the Atlantic. There, in deep water, he caught and burned his first whaler, the *Eben Dodge* out of New Bedford, before a hurricane turned his makeshift warship leaky and crank.*[21] Semmes was proud of the *Sumter* despite its inadequacies. "She cruised six months [and] captured seventeen ships," he wrote in his war memoir of his 1861 patrols.[22] Reluctantly, he abandoned the *Sumter* at Gibraltar to be sold.

Seven months later, stopping in the Bahamas on his way home after leave

* "Crank": a nautical term for top-heavy.

in England, Semmes received orders to proceed to Liverpool to take up a new command. That ship, the thousand-ton, all-black *Alabama,* still under construction by the Scottish firm of John Laird, Sons & Company, had everything the *Sumter* had lacked. It was rakishly long (220 feet) and narrow (32 feet wide), three-masted, with a 300-horsepower horizontal two-cylinder steam engine and a twin-bladed brass screw that could be retracted and stowed, eliminating one of *Sumter*'s worst features. The smokestack could be lowered as well, making the cruiser appear an innocent clipper ship. It was fast compared with the ships it would hunt, capable of ten knots* under sail alone and more than thirteen knots with steam and sail combined.

It departed Liverpool without armament to meet British neutrality restrictions. Once armed off Terceira Island in the Azores, in the mid-Atlantic west of Portugal, it racked six muzzle-loading thirty-two-pound† cannon fixed in position pointing port and starboard and two powerful pivot-guns positioned amidships to sweep either direction—one a long-range hundred-pound menace with a rifled seven-inch bore, the other an eight-inch smoothbore. Most important of all, the *Alabama* could carry 350 tons of coal, enough to operate at full steam for eighteen days, and with a condenser attached to its steam engine to make potable water from seawater, it could stay out hunting for months at a time.

Semmes took command of the *Alabama* in the Azores on 24 August 1862 and began prowling for Yankee whaling ships as soon as its weapons were fitted and its decks recaulked. Sperm whales chased schools of anchovies and sardines in the Azores in spring and summer, up to the beginning of October. It was rich pickings, with pairs of whales herding the fish into swirling balls and then breaching up through them with their mouths open to feed.

The first six ships the *Alabama*'s lookouts sighted were either out of range or foreign, but on 5 September, Semmes writes, they encountered a ship "lying to . . . with a huge whale, which she had recently struck, made fast alongside, and partially hoisted out of the water by her yard tackles." Encumbered by the whale the crew was flensing, and trusting the United States flag the *Alabama*

*Knots—nautical miles—are sea measures of time and distance combined; 1 knot equals 1.15 miles per hour.

†A cannon's pound measure refers to the weight of the shell it can fire.

displayed, the *Ocmulgee* of Edgartown, Massachusetts, was easy pickings for the Confederate raider.[23]

Several days later, having prepared his rowdy British crew for duty by reading them the punitive articles of war, Semmes chased a whaling brig that turned out to be Portuguese, to his disappointment. "This was the only foreign whaling ship that I ever overhauled," he comments; "the business of whaling having become almost exclusively an American monopoly." He considered the population of the North to be mongrelized and inferior, he says elsewhere, but here he makes an exception for whaling, "the monopoly . . . resulting from the superior skill, energy, industry, courage, and perseverance of the Yankee whaler, who is, perhaps, the best specimen of a sailor, the world over."[24]

That same afternoon, Semmes took the *Ocean Rover*, a whaling ship heading home to New Bedford, fat with 1,100 barrels of oil, and the next morning the *Alert* of New London, Connecticut, only sixteen days out of port and loaded with clothing, fresh provisions, and the recent newspapers Semmes used for shipping intelligence. He burned those ships and many more besides in the weeks that followed, sailing north to the Grand Banks of Newfoundland when his prey thinned out in the Azores, no fewer than twenty prizes across two months of hunting. Collectively, from 1861 until its defeat and sinking by the USS *Kearsarge* in the English Channel off Cherbourg, France, in 1864, the *Alabama* captured sixty-five Union ships worth more than $5 million ($95 million today). Most of them were whalers.

In the last year of the war, another and equally deadly Confederate raider, the *Shenandoah*, sailed and steamed up into the Bering Strait between Siberia and the Russian colony of Alaska looking for Yankee whalers to burn. It arrived almost two months after the war officially ended on 9 May 1865, but neither it nor the fleet of whalers hunting in the Bering Sea knew of the Confederate surrender. *Shenandoah* had already captured and burned four ships and avoided one stricken with smallpox—a death sentence for the Union crew so distant from any aid. Now the raider, faster than any whaler, entered the narrow strait, bided its time in darkness, and, when the light came up in the morning, counted ten sails within range.[25]

Realizing the whalers were trapped, *Shenandoah* first chased and overhauled a ship attempting to escape southward: the New Bedford whaler *Waverly*. When

Edouard Manet, *The Battle of the* Kearsarge *and the* Alabama.

they burned it, the conflagration alerted all the other ships in the strait, but there was nothing the unarmed whalers could do. In all, *Shenandoah* captured fifty whaling ships that week, forty-six of which it burned. The loss was estimated at more than $1 million for the ships (more than $14.6 million today) and $400,000 ($5.9 million today) for their cargoes of oil and bone.[26]

The American whale fishery never recovered. From a total of 722 ships in 1846, the fleet declined to 124 by 1886.[27] In September 1871, pursuing increasingly wary whales—bowheads—through the Bering Strait into the Chukchi Sea off Alaska's North Slope, thirty-three of some forty whale ships, twenty-two from New Bedford, were trapped in pack ice and had to be abandoned. By then, kerosene derived from petroleum had replaced sperm oil as the fuel of choice for lighting, in America and across the world. Whaling survived, greatly diminished, by shifting its emphasis from oil to whalebone—for corset stays, umbrella ribs, and other uses that today are met with flexible metal or plastic. Sperm oil continued to serve as a refined lubricant—it was the favored lubricant for machine guns during both world wars—until the 1960s, when first jojoba oil and then synthetic ester lubricants replaced it.

The unexpected bounty of oil in the early years at Oil Creek accounted in part for the appalling waste in spillage, fire, and freshet riding. A more indirect—but in the long run, more destructive—cause of waste and loss was the legal status of underground resources.[28] By the law of that era, petroleum underground was analogous to underground water, was ferae naturae ("of a wild nature"), a wild animal of sorts and subject to the same rules of property. No one understood yet that water saturated porous rock underground; it was believed to flow in underground streams much as surface streams and rivers do. The only way to fix the location of underground water was to dig a well. Then the water that filled the well, having been located in a particular place, became the property of the person who owned the land on which the well was dug or the person to whom the owner had leased the rights.[29]

An early American case of this legal theory involved two men in dispute about who owned a fox: Mr. Post, the man pursuing it, or Mr. Pierson, the man who, knowing Mr. Post was on the chase, caught the fox and killed it anyway. (Neither man owned the land upon which he was hunting—a beach, in fact.) Mr. Post claimed possession of the fox by right of active pursuit and sued Mr. Pierson for trespass. Mr. Pierson asserted in his defense that the fox, which Mr. Post had not yet brought under his control by wounding, catching, or killing, was not Mr. Post's property; that it became Mr. Pierson's property when he caught and killed it.

Mr. Post won at trial, but the judgment against him was reversed on appeal. "In order to obtain title to a ferae naturae, a person must take it," the Massachusetts appeals court ruled in 1805. "The 'first to kill and capture' is the superior rule of law. Had Post mortally wounded the animal, it would have been sufficient to show possession, since this would have deprived the animal of its natural liberty. However, the plaintiff was only able to show pursuit and therefore acquired no property interest in the animal."[30]

British and American courts agreed that underground fluids also have their natural liberty, just as foxes do. The clearest statement of the theory in nineteenth-century American law occurs in a Pennsylvania Supreme Court judgment dating from 1889. "Water and oil," the court held, "and still more strongly gas, may be classed by themselves, if the analogy be not too fanciful, as minerals ferae naturae. In common with animals, and unlike other minerals, they have the power and the tendency to escape without the volition of the owner. . . . They belong to the owner of the land, and are part of it, so long as they are on or in it, and subject to his control; but when they escape, and go into other land, or come under another's control, the title of the former owner is gone."[31]

Some minerals—iron ore, coal—remain fixed where they are found and can be counted as property. Others—water, oil, natural gas—move underground in unknown channels, sometimes to the detriment of other potential users. In an important 1843 British case, *Acton v. Blundell*, responding to the frustration of a cotton mill owner whose mill water well had gone dry when a nearby coal mine was deepened, an English court concluded that underground sources were unpredictable and might change at any time. "No proprietor knows what portion of water is taken from beneath his own soil," the court concluded: "how much he gives originally, or how much he transmits only, or how much he receives: on the contrary, until the well is sunk, and the water collected by draining into it, there cannot properly be said, with reference to the well, to be any flow of water at all." Since the mill owner didn't own the underground water except as it filled his well, the court awarded him no relief when his well went dry.[32]

As with water, so with petroleum. The surface of the earth belonged to those who held title, but the oil under the surface belonged to no one until it was found and taken, and whoever took it first made it his property. This common-law principle, called the rule of capture, establishes a condition that the biologist Garrett Hardin, in a historic 1968 paper in the journal *Science*, called "The Tragedy of the Commons."[33] The tragedy of the commons—of any resource held in common by a community—is that each user is motivated to use as much of the resource as possible without regard for its depletion or despoiling. With petroleum, the tragedy of the commons meant that each well owner was motivated to pump as much oil as possible as quickly as possible, before other wells drained away the common supply. "Ruin is the destination toward which all men rush," Hardin warned, "each pursuing his own best interest in a society that believes in the freedom of the commons. Freedom in a commons brings ruin to all."[34]

The ruin in the case of Oil Creek oil production was less depletion than environmental destruction of the valley through which Oil Creek flowed and the waterways adjoining and downstream. When Ida Tarbell was three, in 1860, she moved with her family into what she calls a "shanty" on a tributary of Oil Creek so that her father, a carpenter, could make wooden oil tanks for the burgeoning trade. She grew up in Venango County, attended high school in Titusville, and saw the negligence that the law of capture encouraged. "If oil was found," she wrote in her autobiography, "if the well flowed, every tree, every shrub, every bit of grass in the vicinity was coated with black grease and left to die. Tar and oil stained everything."[35]

The tragedy of the commons that Tarbell witnessed was local, but the pollution attending the production of petroleum extended far beyond the oil fields themselves. By 1870, investment in the US oil industry had reached $200 million, the equivalent of almost $4 billion today. Annual production in Pennsylvania alone totaled more than 4.8 million barrels. Only cotton accounted for more US export dollars.[36] As output increased, refineries accumulated byproducts for which they had no use. They dumped them onto the commons. They ran off the volatile lighter distillates—gasoline in particular—into pits or onto open ground to evaporate. Or they flushed them into creeks and rivers to join the industrial and slaughterhouse waste and raw sewage fouling American waterways. Rivers ran opalescent with them; creeks burst into flame.

It remained for the next century to confront the tragedy of the commons on a national and then an international scale. But first technology would confront the challenge of transmitting power at a distance, and a competition would emerge between competing forms of electricity.

ELEVEN

GREAT FORCES OF NATURE

Benjamin Franklin was accustomed to Philadelphia's muggy summers. He'd run away there in 1723 from Boston, his birthplace, when he was seventeen, and made it his refuge for the rest of his life. By his mid-thirties, he was financially secure from publishing *Poor Richard's Almanac,* his annual review for general readers, and began devoting his time to invention and experiment.

Electricity caught his attention beginning in 1743, and more intensely in 1747, when a London friend, Peter Collinson, sent him a glass tube for generating static electricity, the only kind of electricity then known. "I never was before engaged in any study that so totally engrossed my attention and my time as this has lately done," Franklin wrote Collinson appreciatively.[1] He was still engrossed with electricity at the end of April 1749, when he told Collinson that "with the hot weather coming on," he and his friends planned to end the winter experimental season with an electric picnic on the banks of the Schuylkill River.[2]

Franklin said he imagined the picnic party igniting flammable "spirits"—alcohol distilled from wine—on the far side of the river by sending a "spark . . . from side to side through the river, without any other conductor than the water," an experiment they had performed previously. They would use an electrical shock to kill a turkey for dinner, roasting it on an "electrical jack"—a spit turned by an electrostatic motor of Franklin's devising—before a fire kindled with a discharge from a Leyden jar. They would drink the health of the famous electricians of England and Europe from "electrified bumpers"—full glasses tingling with static electricity to shock the drinker's lips—while firing guns ignited by an "electrical battery."[3]

Franklin's electrical battery was a connected row of Leyden jars. He had coined the term by analogy with a battery of cannon, a number of cannon lined up so that they could be fired together to *batter* a target. The chemical battery had not yet been conceived; when it was, it would borrow Franklin's collective noun.

The Leyden jar was the first electrical storage device, invented independently in Pomerania and in Leyden, Holland, in 1745. It stored electrical charge generated by contact, the kind we call static electricity today. In his famous kite

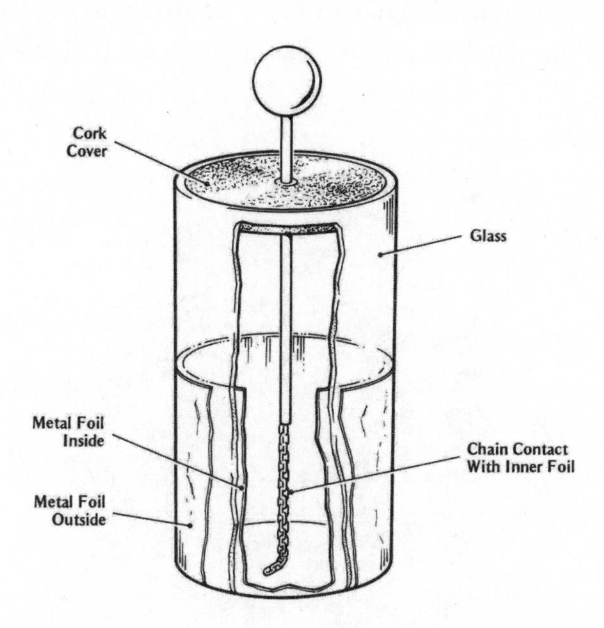

Cutaway view of a Leyden jar, the first battery. Inner and outer foils, separated by nonconducting glass of jar, store static electricity. Ball and chain allow circuit to be closed between foils, discharging them.

experiment of 1752, Franklin flew a kite in a thunderstorm to collect electrical charge, which he transferred from his wet kite string into a Leyden jar. The experiment demonstrated that the modest sparks and shocks of static electricity were identical with the great bursts of lightning that split the sky in storms. For such "discoveries in electricity," the Royal Society of London elected Franklin to membership in 1753 and awarded him the Copley Medal, its highest honor.

Announcing the award, the Earl of Macclesfield described electricity as a subject "which, not many years since, was thought to be of little importance . . . nor was anything much worth notice expected to arise from it."[4] Now, as the contemporary chemist Joseph Priestley would affirm, "The greatest discovery which Dr. Franklin made concerning electricity, and which has been of the greatest practical use to mankind, was that of the perfect similarity between electricity and lightning."[5] Franklin's kite experiment advanced electricity from the status of a parlor-magic curiosity to that of serious science and practical application.

The practical application was Franklin's lightning rod, an invention that seems almost trivial today, forgotten behind electricity's vast range of uses. But before electricity could compete with existing sources of power—animal

Benjamin Franklin conducting his kite experiment to demonstrate that lightning is electrical. Note Leyden jar at boy's feet, to be charged from kite key.

power, wind and water mills, sails, the steam engine—a large problem had to be solved: finding a way to generate it continuously, cheaply, and in sufficient quantity to use it to drive machines capable of accomplishing useful work. In Franklin's day, such applications had not been imagined yet. Electricity was something to be protected against and something to be explored. It was particularly valuable as an investigative tool for chemists. Priestley isolated and identified oxygen with it as well as nine other gases, including nitric oxide, nitrous oxide, sulfur dioxide, ammonia, nitrogen, and carbon monoxide, work for which he also received a Copley Metal, in 1773.

Electricity was a harder problem than steam had been. Though it was conceived first as a fluid, it wasn't something that could be boiled up by heating a mass of liquid and released in controlled volumes to push a heavy piston to turn wheels. It was not a prime mover—a machine such as a windmill or a steam engine, which converts a natural source of energy into mechanical energy—it was a transfer agent. A Leyden jar discharged intermittently, in multiple bursts, each successive burst weaker than the last. A conductor—a wire, a river—could carry the charge, but when it reached the end of the wire or the other side of the river, it discharged all at once. Franklin's electrostatic motor was powerful, yet it lacked a source of energy beyond a workman rubbing a sulfur ball to charge a Leyden jar. Assigning a workman to turn the spit by hand continued to be both simpler and less expensive. Again, unlike steam, electricity lacked obvious applications, however mysterious and fascinating it might be. With enough equipment—cat skin and amber, Leyden jar mounted with a spark gap—you could use it to light a candle. A few did, another parlor trick, but most continued to borrow a flame from the fireplace or strike a light with flint and steel.

The first major breakthrough toward harnessing electricity for power came in 1800, a decade after Franklin's death. It followed from the confusing discoveries of an Italian surgeon and physiologist working at the Academy of Sciences in Bologna, Luigi Galvani, who was studying the effects of electricity on the muscles of animals. No one knew what caused muscles to contract. Some speculated that the motive force was electrical, but it was difficult to understand how electricity could operate in such a saline, conductive environment as a muscle. Galvani wondered if the nerves might carry the electric charge, with their greasy tissue somehow insulating it from its wet surroundings.

Galvani's wife, Lucia Galeazzi, worked as his assistant. Lucia was the daughter of Galvani's old physics and anatomy professor, whose death in 1775 had opened up the position he held at the Academy of Sciences. About two months along in his electrical studies, on 26 January 1781, in the laboratory he maintained in his home, Galvani had dissected a frog and splayed its legs with the ends of their central crural nerves exposed on his dissection table. For generating electric charge, Galvani used a Dollond, a machine of the latest design, manufactured in England. Turning the crank on the Dollond rotated a large glass disk against wool pads on the ends of joined iron arms. The arms carried the charge produced by the contact to a terminal from which current could be drawn.

Galvani had tried many electrical experiments with frog legs in the preceding months. This time, as before, he was preparing to connect the central crural nerves in the legs to the Dollond, as experimenters had been doing for years. An assistant cranked the machine, which began discharging sparks from its nearby terminal. Previously, the frog legs had reacted when the experimenter touched a wire from the Dollond to the frog's exposed nerves. This time, however, *before* Galvani completed the connection, he writes, "one of those who were assisting me gently touched the point of a scalpel to the medial crural nerves of this frog. Immediately all the muscles of the [frog's] limbs contracted . . . in violent tonic convulsions."[6]

Galvani's Dollond electrical machine. Hand at upper right demonstrates how to elicit a spark.

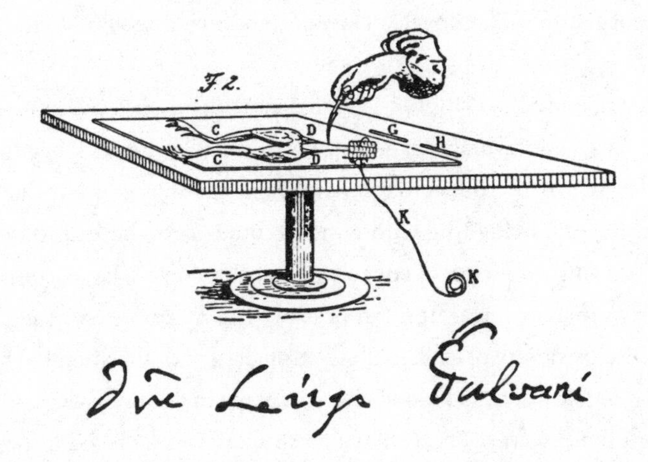

An assistant "gently touched the point of a scalpel to the medial crural nerves of this frog." The signature below the schematic is Galvani's.

The frog's legs had contracted *without any direct electrical connection* to the Dollond. Of such unexpected and seemingly marginal events are scientific discoveries made. By his own account, Galvani failed to notice this unexpected response. He was distracted, he recalled, "wrapped in thought . . . and pondering something entirely different." The one who noticed the unusual event was his wife. As soon as Galvani understood what Lucia had seen, he grasped its radical provocation.

Across the next months and years, Luigi Galvani explored this discovery in detail, using both Dollond and Leyden jar discharges—"artificial electricity," as it was called at that time—as well as Franklin's "natural electricity" of stormy skies. He published a full report of his work in Latin in 1791: *Commentary on the Effects of Artificial Electricity on Muscular Motion* (known in brief as the *Commentarius*).

The Italian natural philosopher Alessandro Volta read the *Commentarius* in March 1792. A native of Como, on the southwestern end of the lake of the same name, Volta was professor of physics at the University of Pavia. He had already investigated static electricity and invented an improved way to generate it: a simple device called an electrophorus. During a two-year period when he was studying gases, he had discovered methane, first collecting it from a swamp on the edge of Lake Maggiore. His discoveries and inventions made him known throughout Europe. The Royal Society of London elected him to membership in 1791.

Volta responded to Galvani's 1792 report with enthusiasm and set to work repeating his countryman's experiments.[7] By April, he had begun to have doubts. Trying to reproduce one of Galvani's experiments—connecting the back and leg of a living frog with a metal conductor—he had found that he needed two *different* metals in contact with each other ("a key, a coin . . . but of an entirely different metal from tin or lead"[8]) to make the frog's leg contract. He pondered a variety of explanations, struggling with the accepted belief that animals generate a special kind of electric charge in their muscles and transmit it through their nerves. This "animal electricity" was believed to be different from static electricity or lightning and unique to each animal. Galvani believed his experiments had elicited animal electricity rather than ordinary, non-vital electricity.

Volta doubted Galvani's claim: If so, why did a living animal's leg respond only when he linked spine to muscle with a bimetallic conductor? Why not a conductor made of only one metal as well? Volta suspected that the electric charge demonstrated by the frog's response came not from the frog but from the contact between the two dissimilar metals—a conclusion that offered an exception to the animal electricity theory and thus challenged its validity.

To confirm his suspicion, Volta tried another experiment. He prepared a frog leg with a long piece of nerve extending out of the thigh. To the bare nerve, but not touching the thigh, he then clipped two leads from a weakly charged Leyden jar. The leg muscles contracted. The current from the Leyden jar had stimulated the leg muscles even though it had circuited only through the nerve. To Volta, that meant the current didn't originate in the muscle, as the animal-electricity theory proposed.

Volta then administered his coup de grâce: he prepared another frog leg with a similarly exposed crural nerve, but instead of connecting the nerve to a Leyden jar, he clipped on an arc of two dissimilar metal strips, one of tin leaf and the other of brass. "Instantly the entire limb will be excited into convulsions and kicks," he wrote, "yet the limb has not been touched, and it is inconceivable that it could be reached by the electrical fluid, which has traveled only between . . . two adjacent parts of the nerve."[9]

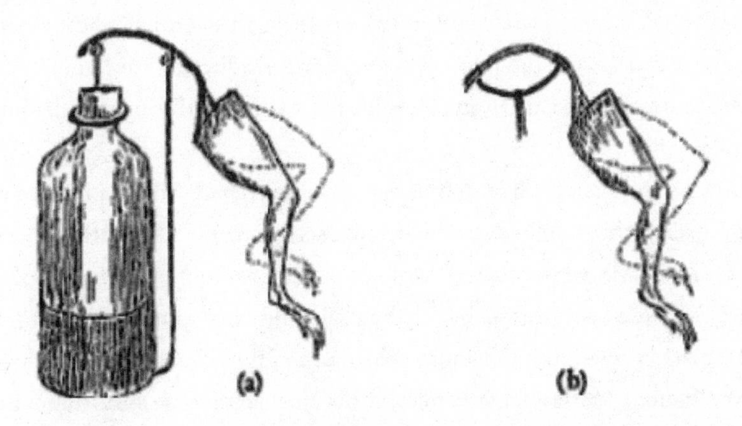

Volta's "nerve only" experiment: (a) with Leyden jar, (b) with bimetallic conductor.

If he could make the frog leg jump by attaching a bimetallic conductor, Volta concluded, "then there is surely no reason to assume that a natural, organic electricity is at work here."[10]

Volta was generous enough, or sensitive enough to the politics of science, not to try to crush Galvani entirely. He praised him instead with a qualified superlative: "Galvani's great discovery of a full-fledged animal electricity," he announced, "remains solid and stable; nonetheless, it must be limited to fewer phenomena, and nearly all his suppositions and explanations collapse."[11] Volta went on to prove that no animal tissue need be part of the apparatus to make electricity.

Years intervened between Galvani's 1792 report and Volta's ultimate response, years of debate across scientific Europe and of further experiments and reports. Napoleon invaded Italy in 1796 and demanded a loyalty oath of public officials, including Galvani. The physicist refused because the oath would have committed him to atheism, and he was a devout Christian. He lost his appointments and his income and died in poverty in 1798. Volta swore Napoleon's oath and continued his research. In 1799 he invented the ultimate instrument of his rebuttal. He introduced it to the world in a letter to the Royal Society of London on 20 March 1800.

Volta called his new instrument a *pila*: a pile. It stacked together disks of dissimilar metals—copper or silver, tin or zinc—separated by saltwater-saturated pieces of cardboard. With no animals involved, Volta's pile generated electric charge. It did so continuously, not in isolated bursts.

Volta had invented what we now call a battery, an electrochemical generator of electric charge. For the first time, a machine produced a continuous flow of electric current. The elusive and invisible force had been brought under human control.

Across the next decade, the battery was improved. Volta's pile tended to lose power as the saturated cardboard spacers dried out. So batteries were constructed of rosin-waterproofed wooden boxes containing a solution of electrolyte—dilute acid worked better than saltwater and soon replaced it—with metal plates slotted into the liquid. The more of these "cells," the more powerful the battery. With a large battery, it became possible to heat fine wires to incandescence and even melt diamonds. Humphry Davy used a battery of two thousand cells to demonstrate the intense blue-white brilliance of an electric arc at one of his public lectures in Britain's Royal Institution in 1809.

An eyewitness to Davy's demonstration described how "the spark, the light of which was so intense as to resemble that of the sun, struck through some lines of air, and produced a discharge through heated air of nearly three inches in length, and of a dazzling splendor."[12]

Carbon-arc lighting, too bright and hot for household use, would become important in industrial, lighthouse, street, and retail lighting in the decades ahead. Curiously, neither Volta's battery nor Davy's arc light would stimulate further scientific discovery for another twenty years—as if comprehending the novelty of current electricity took time.

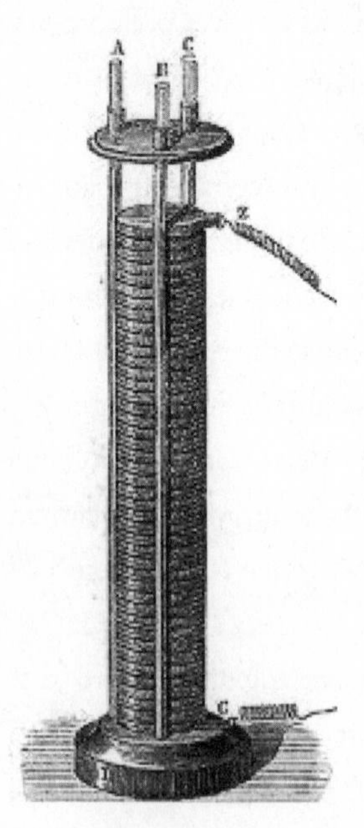

Volta's pile.

That further discovery involved a connection between electricity and magnetism so fundamental that the two phenomena would ever after be understood as a unity: electromagnetism. The shortcomings of batteries today—low power density and limited storage—are the same shortcomings batteries suffered in 1810, nor was any alternative technology available throughout most of the nineteenth century for generating electricity. Knowledge of electromagnetism eventually opened the way to

The two-thousand-cell Davy battery assembled in the basement of the Royal Institution.

the modern world of electric power, where electricity is available in almost unlimited supply, and power can be accessed cleanly and efficiently within homes and factories and transmitted across cities and continents.

The Danish physicist Hans Christian Oersted's historic discovery of electromagnetism has long been attributed to a mere accident, as if Oersted had tripped over it as he might have tripped over a loose brick. All discoveries are partly accidental, by definition, but those who make them are seldom unprepared. Like almost every other natural philosopher in Europe, Oersted had responded to reports of Volta's pile by building one himself. He used it to experiment with the varying effects of acids and alkalis. He had learned the elements of chemistry in his father's pharmacy in a small Danish town on the island of Langeland, where he was born in 1777, about one hundred miles southwest of Copenhagen. By 1800,

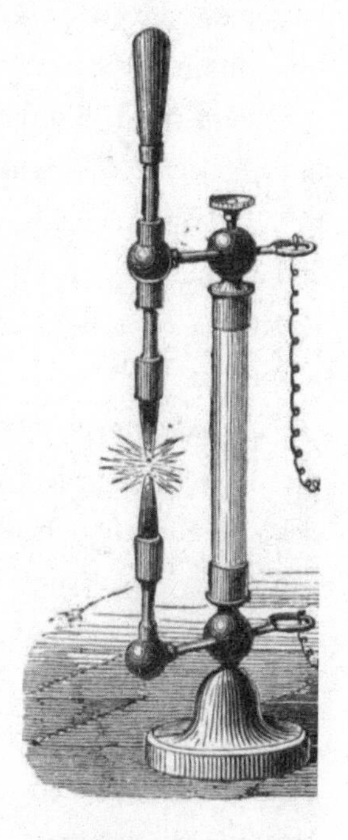

A laboratory instrument demonstrating Davy's simple carbon arc.

after attending the University of Copenhagen with his brother, Anders, he had earned his doctorate in physics and was lecturing at the university. The two brothers were talented. Anders, who studied law, would serve for two years in the 1850s as prime minister of Denmark.

Oersted's experiments with acids and alkalis led him to discover the advantages of acid as a battery electrolyte simultaneously with Humphry Davy in 1803. By 1806, he had earned an appointment as a professor of physics at Copenhagen. Long drawn to the Romantic idea of the unity of nature, he believed that physics expressed itself as an underlying unity among the great physical forces of electricity, magnetism, heat, and light. He explored this question in a theoretical paper he wrote while he was traveling abroad and published in Paris in 1813, "Studies in the Identity of the Forces of Chemistry and Electricity."[13] Here he first raised the question of a possible connection between electricity and magnetism.

"Men have always been inclined to compare magnetic forces with electrical forces. The great resemblance between the electrical and magnetic attractions and repulsions and the similarity of their laws must necessarily lead to this comparison." But there was a problem with the comparison, Oersted continued: electricity seemed to act upon magnetic materials without responding to their magnetic fields, which had led other theorists to argue that electricity and magnetism were separate, independent forces. Experiment would have to resolve the issue, he concluded, but such intertwined forces would be difficult to sort out.

The experiment, when Oersted prepared it, proved to be surprisingly simple. But because he had anticipated difficulty, he allowed teaching and work to delay his effort for a half decade.[14] He found time to invent an electric fuse for detonating explosives, a thin wire heated to incandescence by electricity. He conducted an expedition at the order of the king of Denmark to the Danish island of Bornholm. He invented the mercury-vapor lamp with one of his expedition colleagues, the attorney Lauritz Esmarch, and with the same colleague, an old friend, substituted copper for wood for the troughs of his liquid batteries, greatly increasing their output of electricity.

Finally, in spring 1820, when Oersted was delivering a series of lectures to an audience of science students, he once again thought through his old conviction of the identity of electrical and magnetic forces. This time he "resolved to test my opinion by experiment."[15]

Oersted prepared the experiment on a day when he had an evening lecture. He meant to try it prior to the lecture. Something interrupted him and he decided to postpone it. During his lecture, his mind still on the experiment, he realized it was likely to succeed. He interrupted his lecture then and ran the experiment. It was, a science historian observes, "the one case known in the history of science when a major scientific discovery has been made before an audience of students during a classroom lecture."[16]

The experiment might easily have failed. It consisted of a magnetic compass, a battery array, and a length of wire connected to the battery to carry electric charge—simple enough. But Oersted believed the wire needed to be heated to incandescence for maximum effect, so he used what he calls a "very fine platinum wire" for his conductor to add light and heat to electricity to see if those forces together would affect the compass position. He didn't yet know that a heavier wire would have generated a more powerful magnetic field. His thin wire limited the effect.

It revealed itself nevertheless. Oersted prepared his students by reminding them of the way a thunderstorm can make a compass needle swing. He proposed that a magnetic needle placed close to an electrified wire might behave the same way. Displaying the experimental apparatus, he closed the circuit. Current flowed through the platinum wire, and the compass needle deflected.[17]

Excited though he must have been, Oersted thought the experiment encouraging but not yet sufficiently convincing—at least, not to the larger scientific community that would judge it. The apparatus was "feeble," he writes, the effect "unmistakable" but "confused."[18] He decided he needed a more powerful battery. Across the next several months, he and his friend Esmarch built one of twenty copper troughs with zinc and copper plates suspended in a solution of dilute sulfuric and nitric acid. That July, with Esmarch and another friend as witnesses, Oersted tried the experiment again with his bigger battery and a heavier wire.

"A very strong effect was immediately obtained," he writes in his autobiography. With a blizzard of more than sixty experiments, Oersted discovered that the kind of conducting wire didn't matter: he tried platinum, gold, silver, brass,

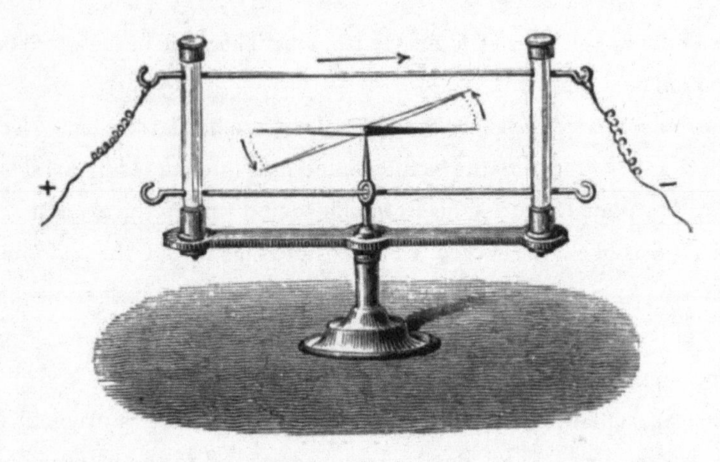

Oersted's historic experiment demonstrating an electric field deflecting a magnetic compass.

iron, ribbons of lead and tin, and even a trough of mercury, all "with equal success." The electrified wire affected the compass needle even when separated from it by glass, metal, wood, porphyry, or stoneware, even "when the needle was included in a brass box filled with water." Needles of brass, glass, or lacquer showed no response. But significantly, if he strung the electrified wire *under* the compass needle instead of above it, the effects remained the same, but the needle's direction of movement was reversed.

That discovery gave Oersted the law he was looking for, because the needle's positional reversal could mean only that the magnetic field which the electric current generated filled the space adjacent to the wire in circular form around it.

In July 1820 Oersted published a brief summary of his experiments in the form of a four-page report in Latin that he had printed and mailed out to friends and fellow experimenters throughout Europe. The report established his priority of discovery. He then prepared a longer account in French detailing his many experiments, which was received in Paris in September. Within three years, a Russian diplomat in Saint Petersburg who was an amateur experimenter, Baron Pavel L'vovitch Schilling, had begun designing a telegraph system based on Oersted's discoveries. Schilling demonstrated the system to Czar Alexander I sometime before the Czar's death in 1825. In the 1830s, the Russian government planned a telegraph link between Kronstadt

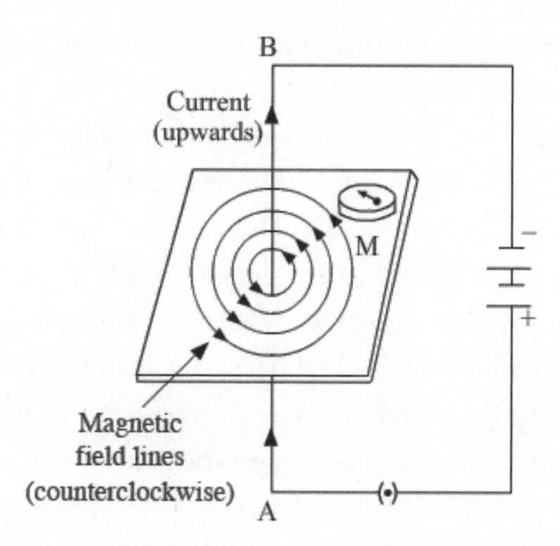

By positioning a compass above and below an electrified wire, Oersted discovered the circular form of the magnetic field surrounding the wire.

fortress on its island at the entrance to Neva Bay and the palace in Saint Petersburg, but Schilling died in 1837 before his system could be commercialized.

Besides electromagnetism itself and the presence of an extended magnetic field around a current-carrying wire, one more fundamental discovery was necessary before engineers and inventors could begin to see how to apply current electricity to the electrical uses of the modern world. That discovery was the work of an extraordinary British chemist named Michael Faraday, a protégé of Humphry Davy's at the Royal Institution in London.

Faraday was born into poverty in London in 1791, his father a blacksmith in poor health. With only a few years of schooling, he began working at the age of thirteen as an errand boy for a bookseller and bookbinder, who apprenticed him the following year. "There were plenty of books there," Faraday said later, "and I read them." He also began studying the new science of electricity in the pages of the *Encyclopædia Britannica* and annotating a four-volume work of chemistry. In 1810 he joined a lecture society organized by a group of young London men interested in self-education. A literate new friend in that society committed two hours a week to helping Faraday improve his grammar and spelling, a program they continued together for seven years. Another friend, a banking clerk, sharing Faraday's enthusiasm for chemistry, discussed new discoveries and solved chemical problems with him.

When Faraday's bookbinding apprenticeship ended in 1812, he faced the prospect of full-time bookbinding work that would engulf the hours he had been devoting to study. By then, he had begun attending public lectures at the Royal Institution, particularly those of Humphry Davy. In October 1813 Davy injured his eyes during an experiment when a mixture exploded in his face. A friend who had seen Faraday's exceptional penmanship in the bookstore recommended him to Davy. That brief employment as a secretary prepared the way for the young man to appeal to the great scientist for laboratory work. Davy arranged to have Faraday hired at a guinea a week plus candles, fuel, and a small flat in the Institution attic. Faraday stipulated an apron supply and laboratory privileges as well. The deal was struck.[19]

Across the next two decades, Faraday revealed himself to be an experimenter of increasing skill and originality. He became essentially the manager of the Royal Institution while exploring chemistry and then electromagnetism. Oersted's discovery that an electrified wire generated a magnetic field around it drew Faraday to search for the reverse phenomenon: electricity induced in a wire wrapped around a magnet. When that experiment failed—he was one of many who tried it—he began a long series of experiments trying every conceivable arrangement of wires and magnets.

In 1831 Faraday first heard of powerful electromagnets which two scientific competitors, one of them an American researcher, Joseph Henry, had constructed by winding hundreds of turns of insulated wire around soft iron cores. When they ran a battery current through the wire winding, the iron core acted like a conventional magnet many times its size. Extrapolating from this experiment, Faraday acquired an iron ring like a large doughnut and wrapped two coils of wire separately around opposite halves. The leads from one coil he attached to a battery, the leads from the other coil he attached to an instrument for measuring electric current (a galvanometer, named in honor of Luigi Galvani).

When Faraday connected the battery, the galvanometer needle deflected—but then swung back and forth and steadied. When he disconnected the battery, the needle deflected again in the opposite direction, oscillated, and steadied.

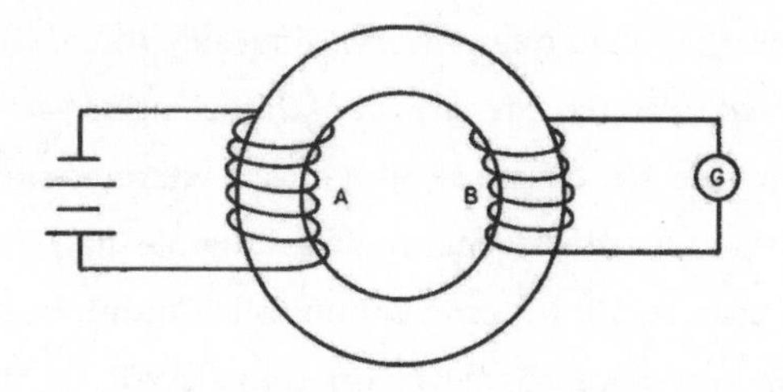

Schematic of Faraday's induction experiment. Stacked lines on left indicate a battery; circle on right with G inside indicates galvanometer.

Joseph Henry had noticed this effect before. Sparks had flashed when he disconnected his battery, which he had called "sparks from a magnet." But because Henry was looking for a steady current, he had failed to explore further.[20]

Once again the honor of discovery would go to the investigator who noticed and followed up a seemingly trivial side effect: Faraday noticed, paid attention, and ran a series of further experiments. "In one setup," writes the science historian Michael Schiffer, "he learned that the needle [of the galvanometer] would also deflect when the second coil was moved toward or away from the first. In addition, Faraday found that thrusting a permanent bar magnet into a hollow coil, or removing it quickly, briefly stirred an attached galvanometer. In further experiments with permanent magnets and coils, he obtained a consistent result: only when the magnet was moved briskly in relation to the coil did the galvanometer's needle move."[21]

Like his competitors, Faraday was looking for a steady electric current from magnetism. He held back reporting his results. Instead, he borrowed the design of an experimental apparatus from a French researcher and used it to explore further. The instrument Faraday built consisted of a twelve-inch copper disk mounted on a brass axle between the two poles of a horseshoe magnet powerful enough to lift thirty pounds. A wire attached to the brass axle, and a wire heading in a wire brush pressed against the copper disk, made a circuit with the galvanometer. When Faraday cranked the copper wheel through the magnetic field, the galvanometer registered a flow of induced electric current that continued so long as the disk was cranked.

So electricity produced magnetism, and magnetism produced electricity.

The two forces were indeed one powerful, invisible force: electromagnetism. That deep proof won Faraday the respect of the scientific world. It also cleared the way to generate electric charge steadily, in any volume, without the need for batteries. Faraday's copper-disk generator was a simple magneto, a first example of a mechanism that would become common in automobiles and other machinery, a way to convert mechanical work into electricity. An arm could perform such work, turning a crank. So could a steam engine. So could a waterfall.

TWELVE

A CADENCE OF WATER

Niagara, the falls and the river, take their name from a Mohawk Indian word, *ohnyá·kara î*, meaning "neck of land [between the lakes]."[1] The river and the falls move vast quantities of water across a neck of land from one of the Great Lakes, Lake Erie, to another, Lake Ontario, and thence to the Gulf of Saint Lawrence and out to sea. Lake Erie, at 572 feet above sea level, is elevated more than 300 feet above Lake Ontario. Niagara, falling 170 feet, accounts for about half of the descent between the two lakes. Lake Erie contains some 115 cubic miles of water. If Erie were not refreshed regularly with rain and snow, Niagara Falls would still need about one hundred years to empty it, which means about a cubic mile of water passes over the falls annually.

Falling water is the oldest source of industrial power other than muscle. Niagara Falls attracted commercial interest from its earliest days, but harnessing it was well beyond the capability of early engineers. The exploring Franciscan Louis Hennepin visited the falls in 1679 and described it in a report as "a vast and prodigious Cadence of Water which falls down in a surprising and astonishing manner, in so much that the Universe does not afford its Parallel."[2] There are higher falls elsewhere in the world, but none that flows so reliably from season to season. The Great Lakes form Niagara's reservoir, a collecting area of

Niagara Falls in 1679: an engraving from Louis Hennepin's *New Discovery of a Very Large Country.*

almost eighty-eight thousand square miles. The seasonal variation in lake level rarely exceeds two feet.[3]

Above the falls, where Lake Erie narrows into the Niagara River, early settlers built waterwheel systems along the riverbank or on small loop canals. These supported the common constructions of pioneer settlements: a sawmill and blacksmith's shop, a grist mill, a rope walk, a tannery, the first houses. A settler cleared an island in the river. Stocked with goats, it acquired the name Goat Island. By the 1830s, the river supported a line of modest industries built along its banks: a nail factory, a paper mill, a flour mill, another sawmill, and a woolen factory. Two hotels, the Eagle and the Cataract House, offered one hundred rooms between them for visitors and residents.[4]

In 1841 two American engineers calculated the energy available from the falls for turning waterwheels at 4.5 million horsepower. The US Army Corps of Engineers, surveying the Great Lakes in 1868, estimated Niagara's total available energy as about 6 million horsepower. But cutting a millrace through the eighty-foot thickness of Niagara's hard limestone shelf would be costly, and tunneling below the limestone through the soft shale would require lining the tunnel with multiple courses of expensive hard brick. Niagara was twenty miles from Buffalo, the nearest city, with no technology at hand to deliver power that far away. Nor did a population of workers live near the remote falls. Until the last decade of the nineteenth century, Niagara Falls was a powerful natural engine not yet in gear.

"All machines for the conversion of work into electricity," a lecturer told the London Society of Arts in 1881, "are founded on Faraday's great discovery of the induced current, derived from the relative motion of a magnet and a coil of wire."[5] Yet Michael Faraday's early experiments in electromagnetism had not led immediately to practical application. "Nearly all important discoveries pass through a stage of neglect or obscurity," Professor W. Grylls Adams went on to explain. "Either the public attention is already preoccupied, or the discoveries come at a time when the public are not prepared for them, and they are disregarded."[6]

Which was what happened with electricity. The almost magical applications of electricity that did emerge preoccupied the public: Samuel F. B. Morse's telegraph (1837), Alexander Graham Bell's telephone (1876), and Thomas Edison's phonograph (1877) and electric light (1879). The frictional generator with its Leyden jar and the chemical battery continued to be the primary sources of electricity until late in the nineteenth century. Both were feeble, limited, and expensive compared with the products of the development of steam, the broad-shouldered steam engines that powered factories, raised water, propelled ships, and hauled trainloads of passengers and freight. On a smaller but complementary scale, horses moved goods and passengers within the city and generated power directly or by turning sweeps on the farm. The fuels most in demand for heating and to power machinery were wood and coal. United States energy consumption reached 70 percent wood in 1870, shifting to 70 percent coal by 1900.[7] Kerosene was a cheap lighting fuel where coal-derived town gas wasn't available, and petroleum increased its share as its use for lighting and lubrication grew.

Revolutionary though steam was, the prime mover of the industrial revolution, it had significant disadvantages. Since its output was mechanical motion, it was inherently local. No one had found a way to deliver it at a distance except by cable, as with San Francisco's and Pittsburgh's cable cars, or pressurized air. It was inconvenient, requiring up to two hours to raise a working head of steam and a stoker's regular attendance to maintain—hardly an arrangement that craftsmen or small businesses could afford in time or money. Its smoke was foul, and because steam was local, its foul smoke polluted cities.

But if electricity had only begun to enter commercial service, Professor Adams told his London audience, such discoveries may still "pass through a stage of quiet development in the laboratory."[8] Laboratories in the mid-nineteenth century systematically investigated the properties of electromagnetism. That research paralleled the development of the electric generator and its reverse, the electric motor. Spin the device with mechanical power from a steam engine or a waterwheel, and it output electric current. Input electric current into the device, and it output mechanical power, turning machines such as looms or the wheels of electric trolleys.

The *kind* of electric current mattered. Batteries produce a continuous flow of charge: direct current (DC). Generators—coils rotating past the opposing poles of a magnet—produce a regularly reversing flow of charge: alternating current (AC).

Since batteries were the only source of current electricity until late in the nineteenth century, almost all the commercial developments of the era worked with the low-voltage direct current that batteries produce, much as mobile phones and laptops do today. Following that line, Thomas Edison pursued direct-current generation to supply electricity for the electric lighting system

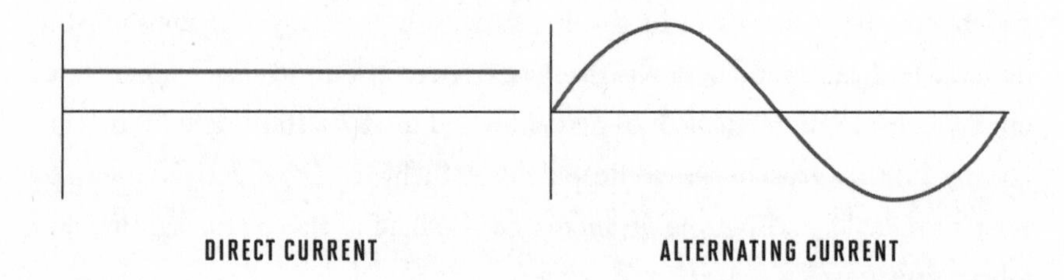

Two kinds of electric current. The midlines indicate zero charge, with positive above and negative below.

he was devising. Lacking experience with alternating current, engineers were skeptical of AC systems. "They cannot make it safe," one of them claimed in a public lecture as late as 1888, *after* AC had become available commercially. "They cannot make it reliable. . . . They cannot make its lamps even [i.e., glowing steadily, not flickering]. They cannot make its lamps last. They cannot make it sell by meter [i.e., meter it]. They cannot make it run by motors."[9] But they could, and they did.

Alternating current had a massive advantage over direct current: it could be transformed easily into a higher or lower voltage. Voltage, like water pressure, moves electric charge. Amperage, like water volume, delivers more charge. The two qualities interact inversely. Stepping up voltage reduces amperage. Stepping up voltage allows alternating current to flow on wires of smaller diameter without encountering as much energy-sapping resistance.

Not so direct current. "It was the common saying of the day," the American inventor William Stanley Jr. remarked of the period before AC was introduced, "that, if one should attempt to light Fifth Avenue from Fourteenth Street to Fifty-Ninth Street [with direct current], the conductors required would be as large [in thickness] as a man's leg."[10] (Conductors of small-diameter wire would heat up and even melt under such conditions, much as old-fashioned single-use house fuses used to do.) In consequence, direct current was physically and financially limited to local distribution. "The cost of copper wire for transmission," notes a contemporary observer, "would be prohibitive for distances of more than a few hundred yards."[11]

Thomas Edison had to invent much more than the electric light. As do all innovators of new technologies, he faced the larger problem of developing and deploying the infrastructure required to support his inventions. Behind the steam engine, a network of mines and distribution systems supplied coal for its operation. Local generating plants and networks of underground pipes sustained gas lighting. When Edison planned his direct-current system of electric lighting, not wanting to run wires as thick as a man's leg, he envisioned neighborhood-scale generating stations—steam engines turning direct-current generators—modeling his system on the gas-lighting system and even running his wiring, like gas, in pipes underground.[12]

Given direct current's limitations, the gas-lighting model fit it reasonably well, at least locally. But how could electricity be extended from city to city? How could powerful sources of energy such as Niagara Falls deliver the electricity they would generate to the distant centers of population where it was needed? Two engineers discussing electrical power transmission in 1879 cited "a certain electrician"—probably Edison—"who asserts that the thickness of the cable required to convey the current that could be produced by the power of Niagara would require more copper than exists in the enormous deposits in the region of Lake Superior."[13]

Edison knew he had a problem: direct current didn't travel well. It had to be used at low voltage. It wasn't easily transformed. The standard method of transforming direct current from one voltage to another involved (a) using DC to run a DC motor that (b) turned a generator to generate AC, which (c) a transformer then raised to a higher voltage, after which (d) a device called a rectifier converted the AC back to DC. At the other end of the line, converting higher-voltage direct current back to low voltage required another pass through a motor-generator system. Such a complicated process was inefficient, with losses of efficiency at each stage and correspondingly increased expense.

Edison tried at least to alleviate the bottleneck. To extend the range of his service, he developed a three-wire system, which made it possible to deliver direct current at reasonable cost to an area around a generating station about three times as large as his previous two-wire system had allowed.[14] But long-distance transmission still escaped him. The problem wasn't one of design but one of basic physics.

The inventor whose work led to the application of alternating current to long-distance transmission was an eighth-generation New Englander named William Stanley Jr., born in Brooklyn in 1858. Growing up in Great Barrington, Massachusetts, his family's ancestral home in the Berkshire Mountains, Stanley had demonstrated a distinct technical gift, repairing watches, installing a telegraph line between his house and the house of a friend, and maintaining a small steam engine a relative had given him.[15] Stanley's father, a prominent attorney, wanted his son to enter law. The twenty-year-old made a start at it, matriculating at Yale in 1879. Other possibilities distracted him, however, including a romantic dream of going west. "I did not study," Stanley said later, "and I left college at Christmas time."[16]

Electricity was not yet a professional field in 1879. "It is difficult for engineers of the present day to appreciate the conditions of that time," Stanley told a lecture audience in 1912. "Please remember that there were but few books on electrical engineering, no formulae, except those hidden away in scientific papers, no nomenclature, and there was hardly any information about alternating-current phenomena available."[17] These opportunities, rather than dreams of going west, are probably why the young inventor dropped out of Yale. "Have had enough of this," Stanley wrote his parents abruptly from New Haven that holiday season, "am going to New York."[18] His father was not pleased.

A month earlier, in November 1879, Thomas Edison had filed the fundamental patent for his "electric lamp."[19] To learn the new field, Stanley signed on as a novice electrician in a telegraph-key and fire-alarm factory. A year later, having reconciled with his father, he borrowed $2,000 ($50,000 today) from him and bought a nickel-plating shop, where he learned electrochemistry. He repaid his father's loan within a year and found work with Hiram Maxim, later the inventor of the Maxim machine gun but at that time chief engineer of the Electric Lighting Company of New York, an Edison competitor.[20]

"He was very young," Maxim recalled of Stanley, who was twenty-two in 1880. "He was also very tall and thin, but what he lacked in bulk, he made up in activity. He was boiling over with enthusiasm. Nothing went fast enough for him."[21] In 1881 Maxim asked Stanley to install the first extended incandescent lighting system in New York City, in the Caswell-Massey drugstore in the Fifth Avenue Hotel on Broadway, powered by a Maxim no. 20 generator with a capacity of a hundred Maxim lamps. Stanley ran an insulated wire from the Maxim machine shop at Sixth Avenue and Twenty-Sixth Street, where the steam engine and generator were located, down the avenue to the hotel roof and into the drugstore to its chandeliers, grounding the return line on the store's gas pipe.[22]

Maxim went off to France in 1881 to represent his company at the Paris Exhibition of that year. While there, he made the first drawings of his machine gun, a formidable weapon that could fire more than six hundred rounds a minute and would make him rich and famous when it was purchased in various calibers by most of the major armies and navies of the world.[23] From Paris, he moved on to London; he would not return to the United States for years. Probably because Maxim was gone, Stanley left the Electric Lighting Company in

1882 and relocated to Boston to work in product development for Swan Electric Light, which allowed him time for his own experiments as well.

Stanley's drive and impatience once again short-circuited his employment. Within months of his arrival in Boston, he resigned from Swan and moved in with his parents, who lived at that time in Englewood, New Jersey. There he set up his own laboratory. For a time, he worked unsuccessfully on a storage battery. He turned then to inventing a generator that would solve one of the fundamental electrical-generating problems of the day: the flickering of electric lights when the generator that powered them responded to changes in electric load. Most of us have had the experience of lights dimming when a refrigerator or an electric heater switches on. With the primitive generators of Stanley's day, the problem was common and frequent, one reason that many hesitated to replace their gas lights with electric lighting. Edison's lamps had to be adjusted by hand whenever other lamps or motors in the circuit were switched on or off.

In 1883 Stanley conceived the idea of an alternating-current generator with a secondary feedback circuit: a pair of small extra coils that automatically cut in or out of the circuit to stabilize it whenever the voltage dropped. He was riding a train one day, full of his new idea, when George Westinghouse's younger brother, Herman, happened to sit down next to him. They began talking; soon Stanley told Herman about his idea for a self-regulating alternating-current generator.[24] Herman knew a good idea when he heard one. He connected Stanley with George, the successful developer of the air brake and other railroad machinery that made long trains and long-distance transportation practical. George was just then considering entering the electric-lighting field, pursuing alternating-current technology rather than direct current. He had recruited a team of young engineers to build a knowledge base for him, but he wasn't yet fully committed. Stanley's work won him over. Early in 1884 he hired the twenty-five-year-old to develop a complete AC system, from generators to motors and lighting.[25]

Westinghouse had hired Stanley to work as his chief engineer. The young inventor's first project, in 1884, was setting up a full-scale factory in Pittsburgh to produce incandescent lights based on his own designs. There he moved beyond silk filaments to a filament made of leucine, a white crystalline form of amino acid derived from animal spleen, pancreas, or brain. It could be extruded through a die like pasta into a uniform fiber, cut and shaped as desired, and then carbonized.[26] Stanley continued to think about long-distance transmission of electricity, however. "During this busy year," he reflected in 1912, "I was carrying in my mind the old problem of distribution. It had become my major ambition by this time—my secret ambition, if I may confess it. Several times during the year, I thought to get at it, but I could not until fall." Having not yet had time to develop his self-regulating alternating-current generator—his "alternator"—he experimented with converting direct current into alternating current using a form of induction coil. "It worked," he recalled, "some, but oh how it sparked!"[27]

In the spring of 1885, Stanley fell ill. The record is silent on the specific nature of his illness, but he speaks of his waning "ability to withstand Pittsburgh and its work." His doctor advised him to move to the country, which suggests that the noxious smoke pollution from Pittsburgh's booming iron and steel industry was affecting his lungs. He was depressed as well by his lack of progress toward developing alternating-current technology: "I was rather discouraged, for the surroundings were uncongenial, the work hard, and the results meagre."[28]

Then Westinghouse intervened with an idea that helped Stanley break through. "One day, when the experimental work was troubling me, Mr. Westinghouse told me that he could get an option on the work of Gaulard & Gibbs"—a European electrical partnership—"and suggested that he send for their alternating (Siemens) machine and their induction coils."[29]

The Siemens "alternating machine" was a Siemens alternator, a version of what Stanley was trying to develop. Driven by a steam engine, it generated high-voltage alternating current. That current was then fed into a bank of induction coils to reduce its voltage sufficiently to power electric lights. But because the induction coils were wound to be connected in series, like old-fashioned Christmas tree lights, they shared a common current, which meant their voltage varied with demand, making lights flicker and risking dangerous overloads should any component fail.

Stanley's breakthrough came in studying the Siemens system. He realized, he said, that if he could make an induction coil—a "transformer," he called it now—wired in parallel rather than in series, each coil would operate independently. That arrangement would keep the current steady whatever the demand and even if a component failed. "The problem," he exulted, "would be beautifully solved."

The solution was analogous to the system used in some types of electric motors to regulate their speed. "I saw this analogy faintly at first," Stanley recalled, "but soon with strong and clear conviction. I was very much excited by it. It seemed too simple and too easy to be true. I was almost afraid to believe or speak of it, for I had experienced a good many disappointments and was in a nervous and overworked condition; but as my convictions grew and strengthened, I gained courage. Then I clearly saw that the solution was found."[30]

Westinghouse was not yet the expert on electricity he would become. He was still preoccupied with railroad safety systems and with natural gas. He asked an expert adviser, Franklin Pope, a patent attorney, the coinventor with Edison of the stock ticker, and the editor of *Electrical Engineer* magazine, to review Stanley's ideas on his behalf.

Stanley tried to explain to Pope that connecting his transformer in parallel, like connecting a motor in parallel, automatically adjusted the current. "I waxed eloquent over the automatic regulation of the system of parallel connection,"

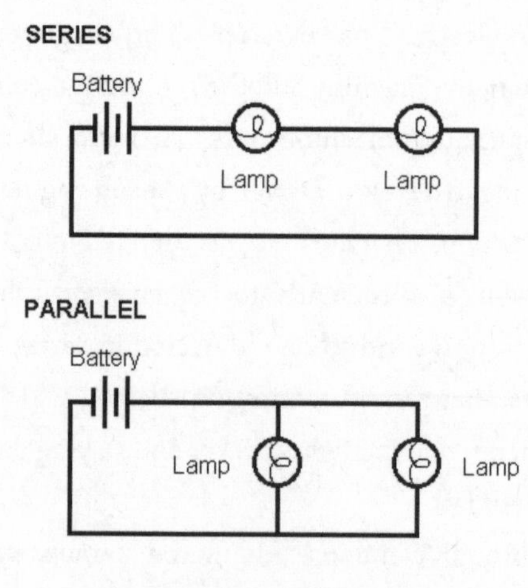

Series versus parallel circuits.

he recalled, "and tried my best to picture the phenomena clearly to him. But Pope's conviction came slowly—more slowly than my patience could stand."[31] Stanley's insight was that his parallel-connected transformer could be used to step up the voltage from an alternator while stepping down the amperage. The higher voltage and lower amperage would eliminate the need for thick, heavy wires and make it practical to transmit electricity efficiently across long distances. At the other end of the transmission line, the same process in reverse could step down the voltage to local requirements while boosting the amperage to useful power levels again.

Pope, it seems, could hear only the number for the high voltage Stanley proposed to use, evidently not understanding that it is amperage, not voltage, that kills. (An electric fence uses a 9-volt battery to charge a capacitor to about 8,000 volts, at the same time reducing the amperage to about 0.1 amp. An encounter with an electric fence feels like a brief, unpleasant, but harmless muscle cramp. A US household electrical circuit operates at 120 volts and 15 amps, which, because of the higher amperage, is much more dangerous, though usually survivable.) Pope didn't understand. He told Stanley that stepping up a current to 500 volts would risk starting a fire or fatally shocking users. "He held back," Stanley concludes, "and did not at first approve of the plan."[32]

Because of Pope's disapproval, George Westinghouse withheld direct support for Stanley's proposal to demonstrate that long-distance transmission of alternating current was practical. Ill as Stanley was that spring, he had money of his own to invest and believed in his breakthrough. He sold Westinghouse half his Westinghouse stock and agreed to use the proceeds to experiment.

Assuming the Stanley system worked, that was a good deal for Westinghouse—the arrangement, Stanley said in 1912 with the temperance of long hindsight, seemed "a trifle one-sided, as I did not benefit by it excepting as a stockholder in the company." And one-sided it was, but Stanley was so happy to leave the smoky city to return to Great Barrington, where the demonstration would take place, that he remembered it as "the best bargain of my life. . . . We shook the dirt of dreadful Pittsburgh from us, and hastened to the green fields of Berkshire, to build a laboratory and succeed or perish in our work."[33]

Before Stanley and his family left Pittsburgh in the early summer of 1885, he mustered enough energy to design and build several transformers for parallel connection. One of them, he recalled in 1912, wound with a 500-volt primary coil and a secondary that would reduce that voltage to 100 volts, became "the prototype of all transformers since made." That project and the move to Great Barrington prostrated him; he was too ill throughout the rest of the summer to work. "But when the cool September nights came on, my health came back, and I started in to equip my laboratory."[34]

For his laboratory and power station, Stanley leased a deserted rubber factory at the north end of town. He bought a 25-horsepower steam engine for a power source and had it installed in the old factory. To his frustration, the steam engine proved to be almost diabolically balky: "I have frequently met, in a long and stormy life, serious and obstinate difficulties," he recalled, "but I have never encountered any mechanism, of any kind whatsoever, that possessed so profound a genius for going wrong as this engine." It took a month to get the steam engine and its boiler operating in unison. Another "interminable delay" followed as he waited for delivery of a Siemens alternator that Westinghouse had ordered for him from London.[35]

Stanley used the delay to begin building what would eventually total twenty-six transformers, six of which he installed in the basements of the Great Barrington buildings he was supplying with power. (The others went to Pittsburgh.) At the factory power station, the Siemens alternator would produce 12 amps of electric current at 500 volts. One of his transformers would then step the current up to 3,000 volts, dropping the amperage accordingly. From the old factory, Stanley had workmen string a pair of insulated copper wires down the main street of the town: four thousand feet of line, about four-fifths of a mile, supported on insulators nailed to the old elm trees that lined the sidewalks. Distribution lines then led the current into the building basements, where the transformers installed there dropped it back down to 500 volts.[36]

With everything in place and the power station up and running, the transformers lit thirteen stores, two hotels, two doctors' offices, one barber shop, and the telephone and post offices. "At last the town was lighted," Stanley concludes, "and we had ocular evidence of our success. We made a gala night of it. The streets and stores were crowded with people, the big 150-candle-power lamps were running at about double their candle-power, and my townsmen, though very skeptical as to the dangers to be encountered when going near the lights, rejoiced with me."[37]*

Electric service delivered regularly from a central station began in Great Barrington, Massachusetts, on 6 March 1886. Stanley's townsmen might have rejoiced, but no one came from Westinghouse Pittsburgh to see. Deeply frustrated, Stanley took the train to New York and looked up an old friend and Westinghouse ally, Henry M. Byllesby, a young but talented manager whom George Westinghouse had lured away from the Edison Electric Light Company in late 1885 when he was forming the Westinghouse Electric Company. Byllesby noticed then—in November, just as Stanley was lighting up Great Barrington—that "there was substantially no one in the organization, except Mr. Westinghouse himself and dear old Frank Pope, who had any real expectation of anything commercial coming out of the alternating-current system."[38] Byllesby liked Stanley, however, and when Stanley looked him up in New York that March 1886, he listened to him:

> Stanley came down to see me in New York on a Friday, and impressed me with the fact that he actually did have a small alternating-current central station running at Great Barrington, and he quite pathetically implored me to go back to Great Barrington with him to look at it.
>
> This I did, and spent the following Saturday there. I found he had a complete system, barring, of course, the meter and the motor, that it was actually performing, and performing well, and, with relatively slight modification, could be put upon the market.[39]

From Great Barrington, Byllesby traveled to Pittsburgh to brief Westinghouse. Byllesby was "enthusiastic to the last degree," he says, but the Pittsburgh industrialist and his associates still hesitated.[40] A tour of the Berkshire installation on 6 April 1886 finally sold them on Stanley's system. "This visit," Stanley recalled, "determined Mr. Westinghouse to actively enter the

*One hundred-fifty candlepower is a little brighter than a 100-watt incandescent bulb or a 25-watt LED. Double that would then be somewhat brighter than 200 watts incandescent or 50 watts LED.

alternating-current field, as the novelty and scope of the system surprised him greatly."[41] Commercialization followed. By winter, a full-scale power station was delivering alternating current throughout Buffalo, and by summer 1888, no fewer than a hundred Westinghouse central stations powered electric lights in cities and towns across the eastern United States.

None of these early installations yet transmitted power between cities. Several DC intercity power systems had been in operation in Europe since the early 1880s, but the first operating AC intercity transmission system was installed at Cerchi, Italy, in 1886, to transmit 2,000-volt current seventeen miles to Rome. Two 150-horsepower steam-driven Ganz generators produced 112-volt AC, which was then stepped up to the higher voltage for transmission.[42]

Stanley decided in 1892 to find out if there were limitations on the level of high-voltage power that could be transmitted. He directed his assistant, Cummings C. Chesney, to organize a demonstration north of Great Barrington in Pittsfield, Massachusetts. "He instructed me to design and build . . . transformers and a line for 15,000-volt operation," Chesney writes.[43] To do so, Chesney set up transformers that converted the town circuit from 1,000 to 15,000 volts, erected a power line around a farm that led back to the same transformer station, retransformed the power down to 1,000 volts, and fed the power back into the town circuit. The little plant operated faultlessly throughout a New England winter. In the years that followed, Stanley would recommend even higher than 15,000-volt potentials. Long-distance electric power transmission today begins at 110,000 volts (110 kilovolts, kV) and can even reach as high as 765 kV.

The free energy of sunlight lifts water vapor into the air, from which it falls as rain or snow. Power from falling water harnesses that free energy: hydropower is an obvious first choice for generating electricity. The Willamette Falls Electric Company installed the first AC hydroelectric power station in the United States in 1889 to send power from Oregon City to Portland, Oregon, thirteen miles away.[44] The Telluride Power Company built a small hydroelectric station in Ames, Colorado, in 1890 to power the machinery of a gold mine three miles away. Similar installations followed in San Bernardino, Pomona, and Redlands, California, and in Hartford, Connecticut.[45] "With these plants," write two early historians, "began the era of hydro-electric power transmissions in the country, and statistics show that nearly three hundred plants were in actual operation about 1896."[46]

The largest and most significant of these was at Niagara Falls. Plan after plan had been advanced since the 1850s to harness the immense waterpower of the falls of the Niagara River, from a hydraulic canal scheme upriver to drive pulp and flour mills, to a Buffalo combined waterpower and sewage-disposal system. With extraordinary disregard for the scenic qualities of the area, the Buffalo plan would have discharged the sewage from a city of 256,000 people through a tunnel under the falls into the river below, where the combined sewage and water might power thousands of mills and factories.[47]

The hydraulic canal was built eventually, upriver and out of sight of the falls (the sewage discharge, fortunately, not). What the artist Frederic Church, in an 1869 letter to the landscape architect Frederick Law Olmsted, called "the rapidly approaching ruin of the characteristic scenery of Niagara" roused popular enthusiasm for preservation.[48] By then, every foot of riverbank on the American side above and below the falls was owned privately and would have to be purchased, something the New York State legislators in Albany were loath to appropriate the funds to do. From 1879, when popular agitation began, it took seven years to persuade them. The owners of the riverbank properties claimed ownership to the middle of the river—and thus of the river's power potential—and valued their holdings accordingly at some $30 million. The New York State Supreme Court disagreed about the reach of their ownership into the river, which reduced their claims to $4 million. Commissioners appointed by the state legislature then evaluated the land to be taken for public use at $1.4 million ($37.5 million today). The state took it, and the Niagara State Reservation came into existence in 1885.[49] Within two years, the number of visitors traveling to the falls on excursion trains in season had already reached 166,000.[50]

In 1886 a group of investors distributed a prospectus for *Power from Niagara Falls by Electricity*, which indicates how ambitious the prospect was of generating power on the Niagara scale. It was "entirely practicable now," the prospectus claimed, to illuminate Buffalo with power from Niagara Falls, "and the opinion is rife among scientific men that ways will be found in the near future for transmitting this power to much greater distances." Ways had not yet been found in 1886—the Stanley system of AC power transmission had not yet emerged—but, the prospectus continued, "an application has already been received from a manufacturer in Birmingham, England, for . . . conveying power by means of compressed air." Compressing air with electric pumps at Niagara Falls and piping it to Buffalo to replace steam in operating steam engines might have worked, however exotic or desperate it sounds today, but farther transmission by so inefficient a system was unlikely.[51]

From 1887 to 1896, Westinghouse and Thomas Edison fought what has come to be called the War of the Electric Currents, with Westinghouse championing alternating current and Edison championing direct current. That story has been told in great detail elsewhere, somewhat exaggerating the role of the inventive Serbian engineer Nikola Tesla, whose only important contribution to the "war" was the alternating-current electric motor. It culminates in the development of hydroelectric power generation on a large scale at Niagara Falls.

A crucial turning point came in 1890. A power system had been planned in the late 1880s that would avoid intruding into the area of the Niagara Reservation. That area included the shoreline of the Niagara River from the American Falls back past Goat Island about a mile to Port Day, as well as the river below the falls to the Suspension Bridge, a railroad bridge two and a half miles downriver. The plan involved digging a tunnel two and a half miles long and fourteen feet in diameter from the riverbank east of Port Day underground by the shortest line of descent past the falls to the lower river, where it would discharge below the Suspension Bridge. Connected to this raceway by more than three miles of cross tunnels, 238 wheel pits would hold that number of 500-horsepower waterwheels to power 238 mills. A new industrial community could then establish itself outside the Reservation to process and manufacture goods by waterpower.

The plan might have worked, given enough investment. But tunneling through more than five miles of hard limestone would be expensive, and developing an industrial city with hundreds of mills in what was still a semiwilderness would have been, in the words of the first historian of the Niagara development, Edward Dean Adams, "visionary, requiring a generation in time and fortunes in expenditures to create."[52] The plan pointed backward to older technologies at a time when new technologies were almost at hand for generating electricity and transmitting it to distant cities, Buffalo first of all.

Edward Dean Adams turned around the Niagara Falls project. Before he became its informal historian, he was an engineer and investment banker who specialized in reorganizing failing companies, railroads in particular. A "financial observer" told *Time* magazine of Adams's early career, "Mr. Adams [was] one of the shrewdest and most close-mouthed young financiers in New York."[53] Adams was a small man, the middle child of ten of a Boston grocer and his wife, a science graduate of Norwich University in Vermont and an engineering student at the Massachusetts Institute of Technology (MIT) in its first years of operation. Besides holding executive positions with a number of railroads in the last two decades of the century, he served on the board of the Edison Electric Lighting Company in the early 1880s when Edison was wiring up New York City.[54]

Thomas Edison had developed his distribution system for lighting and only incidentally for power. Niagara offered the opposite challenge, developing a distribution system for power and incidentally for lighting. As a partner in Winslow, Lanier & Co., a New York private banking company, Adams became involved in the Niagara hydropower project in the summer of 1889. Three investors who had acquired the rights to a previous Niagara Falls development company had formed a new organization, the Cataract Construction Company of New Jersey, but had been no more successful than their predecessors in raising the necessary capital. "Three such projects had already failed," *Time* recalled; "$800,000 had been thrown away." In August Cataract Construction had offered Winslow, Lanier a half interest in the project. Adams asked for a six-month option to give him time to investigate.[55] One of his first decisions was to resign from Edison Electric, to free himself from any conflict of interest as he set about reviewing the conditions at Niagara and the state of the electric industry.

The first question Adams posed to the experts he began consulting, in September 1889, was how practical and economical would it be to transmit large amounts of power across long distances. One of the consultants Adams retained was Dr. Henry Morton, the president of the Stevens Institute of Technology. Morton answered that the problem had not yet been solved: "Large amounts of power have been transmitted to distances of 1 or 2 miles, and small amounts of power have been transmitted for long distances, such as

30 miles, but the combination of large amounts of power and long distances has yet to be realized in practice."[56] Doing so, Morton thought, would require developing new electrical machinery.

Adams cabled Edison, then in Paris for the Paris Centennial Exhibition, for his opinion. The Sage of Menlo Park responded, "No difficulty transferring unlimited power"—meaning, of course, via direct current. "Will assist."[57] Edison had investigated Niagara in 1886, when the plan for local waterwheels powering local mills was still under discussion. Now he offered the alternative he had envisioned then. Instead of water-powered mills for nonexistent factories in remote areas, he proposed a system of direct-current electrical generation via a tunnel of water turning turbines connected to generators, with insulated and waterproofed electric cables laid in the riverbed to carry the power upriver to Buffalo.

Edison's reputation encouraged the investors at Cataract Construction to consider rejecting the waterwheel plan in favor of generating electricity at Niagara. Edison wanted to conduct a new survey before he committed further. In November 1889, having done so, he offered an estimate for producing direct-current electricity at Niagara, but not for distributing it, of $5.2 million ($134.6 million today). Pending a response from Adams and his partners, Cataract held off deciding whether or not to accept Edison's direct-current plan.[58]

Cataract's board understood that any plan it adopted would require working around the Niagara Reservation, which was off-limits to industrial development. Since the water rushing over the falls could not be harnessed directly for power, power would have to be drawn from water bypassing the falls through a large tunnel dug through the hard limestone and underlying shale from the river above the falls to the river below the falls. Such a tunnel would be necessary regardless of the kind or volume of electricity generated and whether or not it was used locally or transmitted beyond. Nor would the volume of water drawn off by even a large tunnel affect the appearance of the falls: such a tunnel would divert little more than 3 percent of the vast Niagara flow.[59]

Today it's not unusual to begin building some of the basic structures of a project before the final specifications have been determined. Such advance work was unusual, if not unique, in 1890. Cataract nevertheless committed to constructing the tunnel, contracting with a newly formed company, Niagara

Falls Power, and subscribing $2.6 million to meet construction costs. The board of the new company chose Adams as its president, later adding sixty-three-year-old Philadelphian Coleman Sellers II as chief engineer.[60]

Adams went to Europe in February 1890 to consult with leading scientists and hydraulic engineers. He traveled first to Switzerland, which because of its mountains had more experience with waterpower than any other country in the world. Oerlikon Machine Works outside Zurich received special attention. It had won the only Grand Prix awarded for dynamos at the 1889 Paris Exhibition, and it manufactured the largest motors and generators in the world. In France, Adams found electrical transmission replacing older methods of energy transfer such as shafting, rope, waterwheel, pressurized water, and compressed air. He collected and studied thousands of pamphlets, catalogs, photographs, journals, and reports from companies and engineering societies as well.[61]

By May, Adams had concluded that whether or not the power was used locally or transmitted away, only one powerhouse was necessary at Niagara, that the tunnel could be shortened, and that he should convene an International Niagara Commission of scientists and engineers to offer ideas and to study and endorse any plans. From Paris, he cabled New York accordingly, asking the Niagara Falls Power board to concur and to send Sellers to meet him in London. "Directors present approve your plan," the board responded. "Sellers sails Saturday with all papers."[62]

Sellers arrived in early June. "I met Mr. Adams in London," he recalled, ". . . and found that he was more enthusiastic than ever in the scheme."[63] Sellers and Adams visited Switzerland once more to confirm the decision to shift from running factories with waterpower to running them with electricity. Later in the month, the two men met in London with interested investors and organized the international commission Adams had proposed. William Thomson, Lord Kelvin, the eminent Scots-Irish physicist who formulated the second law of thermodynamics and directed the laying of the first transatlantic telegraph cable, became chairman.

For the next three years, engineers working under Adams and Sellers designed a system for harnessing the power of Niagara. A major challenge came at a meeting of the international commission when Lord Kelvin, a firm opponent of alternating current, offered a resolution to exclude it from further consideration. Sellers questioned the resolution, one of the commission members recalled, on the ground "that the commission's knowledge of the possibilities of alternating current at that time was not sufficient to justify action which would close the doors to that system."[64] The American engineer won the argument that day against one of the most distinguished scientists of the age. (Kelvin later apologized.) Then Westinghouse, having prevailed against Edison in its bid to light the 1893 Chicago World's Fair with alternating current, won the contract to build the power plant to generate alternating-current electricity from Niagara Falls and transmit it to Buffalo and beyond.

The power plant in its finished configuration was a marvel. McKim, Mead & White, the architectural firm that designed the Brooklyn Museum and the National Museum of American History, in Washington, DC, designed the transformer house, using limestone quarried in Queenston, Ontario, and laid up by Italian stonemasons. The generator and transformer houses were built at ground level upriver from the falls, above Niagara City. Water diverted into a channel from the river entered the powerhouse and dropped 140 feet down an array of ten vertical pipes called penstocks. The curved lower end of each penstock directed the rushing water horizontally through a large double-wheeled turbine. After spinning the turbines, the water flowed into a common tunnel that led to the main discharge tunnel, a brick-lined, shield-shaped conduit 17 feet wide by 21 feet high. That behemoth carried the discharge water 6,700 feet downstream to a partly submerged outlet into the Niagara River below the falls.

The turbines connected back up to the 5,000-horsepower generators aligned on the floor of the powerhouse above, turning the generators and generating electricity. The electricity was then transferred to a transformer house, stepped up to 11,000 volts, and sent on its way to Buffalo.

Power began flowing officially from Niagara Falls on 25 August 1895, first of all to a nearby aluminum-processing plant.* Within a year, it was being transmitted beyond Buffalo four hundred miles to New York City. By 1905, Niagara

*Aluminum, first purified by Hans Christian Oersted in 1825, was once so rare and expensive that Napoleon III reserved his set of custom aluminum dinnerware for occasions of state. The Washington Monument was similarly tipped with 6.28 pounds of the rare material; by then, 1884, the price had fallen to $1 an ounce. As electrical power became available to reduce aluminum from its ore, the light, lustrous metal became more common and less expensive.

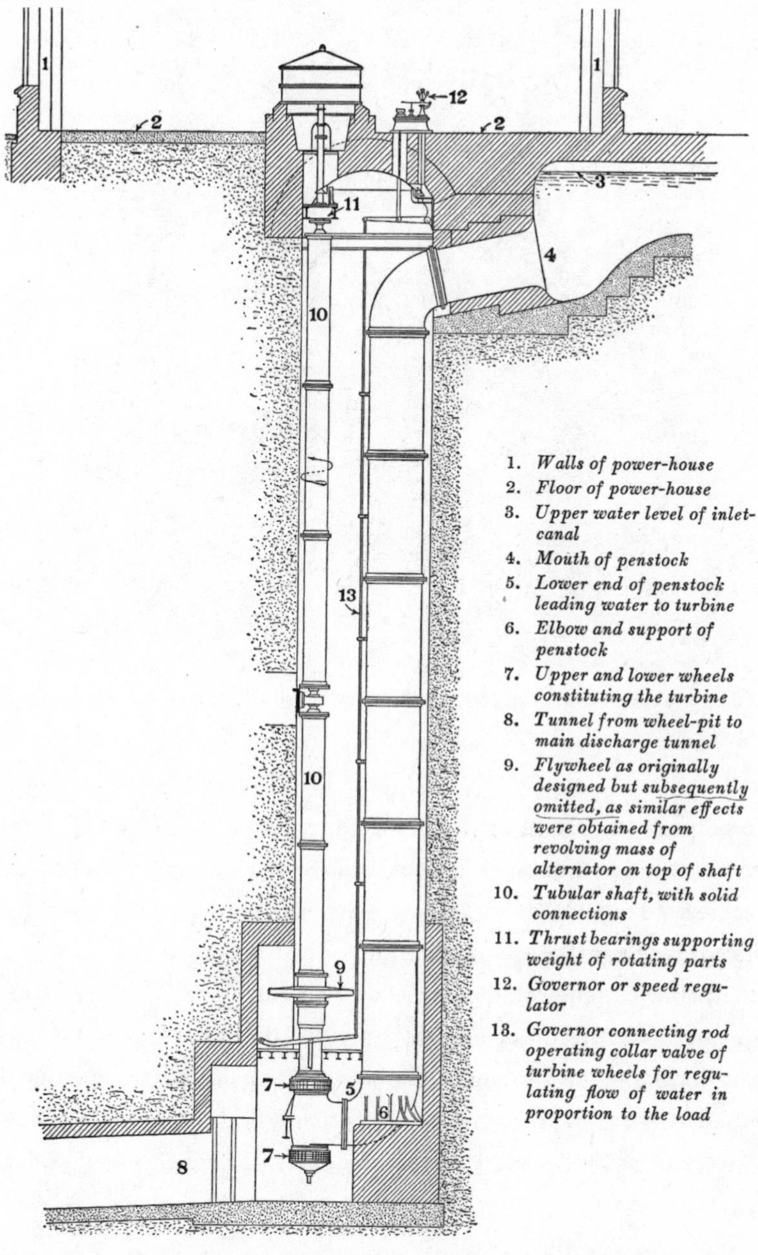

Cross section of Niagara generator unit. Water enters upper right (4), falls down penstock, spins turbine (7) at bottom of wheel pit, and exits into side tunnel (8). Turbine shaft (10) turns generator in turbine building at ground level above.

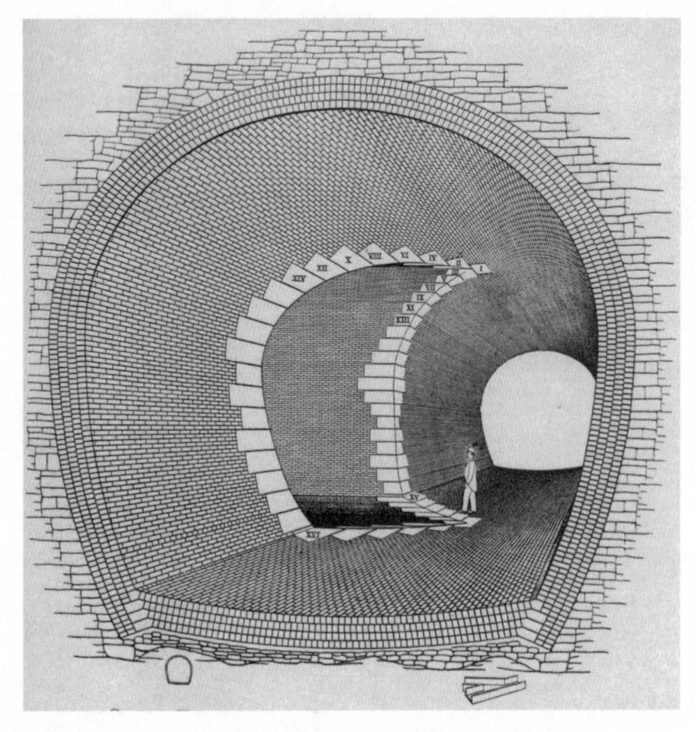

Main tailrace tunnel with wheel-pit tunnel side entry. For scale, compare size of man standing beyond side entry.

was producing 10 percent of all the electrical power in the United States.[65] When a second powerhouse began operation in 1904, Niagara power production aggregated to 100,000 horsepower—equaling the total power then being generated elsewhere throughout the entire United States.[66]

Alternating current won the War of the Currents. Even Edison grudgingly took up manufacturing the equipment to produce it. Motors large and small, all the way down to motors for individual sewing machines, began replacing the shafts and belts that transferred power inefficiently from steam engines. Country people still read and cooked with kerosene, but electric lights went on in the cities of the world.

THIRTEEN

AN ENORMOUS YELLOW CHEESE

Present on every street a century and a half ago, largely invisible today, the horse is a prime mover often overlooked in accounts of the history of energy. Yet Manhattan, as late as 1900, with 1.8 million people occupying an area of about twenty-three square miles, shared that narrow island with some 130,000 horses.[1] Except for the mounted police and for exercise in Central Park, few in the city rode on horseback. The horses of Manhattan, as in all the urban areas of America and Europe, pulled two-wheeled cabs and four-wheeled hacks, omnibuses and streetcars, carts and wagons. Besides passengers, they delivered milk, food, laundry, beer, ice, and coal, hauled fire engines and sprinkler trucks, and removed snow and their own voluminous manure from city streets.

In the eighteenth and early nineteenth centuries, short-haul stagecoaches served to deliver people from the countryside into larger American cities. (In smaller cities and towns, people went on foot. "The preindustrial 'walking city,' " writes historian Ann Norton Greene, "had remained geographically compact because the distance a person could commute on foot was approximately two miles."[2]) Stagecoaches evolved into larger omnibuses in the 1830s. Their enclosed carriages accommodated twelve to twenty-eight inside passengers protected from the weather, backs to the windows facing each other across

Gansevoort Market, New York City, 1907.

a central aisle. By 1852, in Lower Manhattan, some thirty companies operated more than seven hundred omnibuses. Rides weren't cheap: a 12-cent fare, at a time when workers earned a dollar and craftsmen only $2 a day, limited the omnibus to businessmen, young professionals, and their families.[3]

Setting omnibuses on rails increased the number of passengers that horses could haul and improved the ride. In 1856, when New York City's Common Council judged street-level steam locomotives to be dangerous and barred them below Forty-Second Street, horse-drawn street railways replaced them. Rails set flush with the pavement superseded the raised rails that had damaged private carriages and irked other drivers.[4] Line owners commissioned larger, lighter cars to increase ridership: accommodating more passengers per trip meant they could reduce fares. One such railway delivered more than three and a half million rides between Upper and Lower Manhattan in 1859.[5] Boston saw similar numbers despite its narrow, crowded streets. In every nineteenth-century American city, the absence of traffic controls made the confusion worse.[6] Traffic police and mechanical signals would be twentieth-century innovations.

Heavier loads required larger horses. Engineers today design efficient machines scaled to meet most human needs, from microchips to passenger jets. In earlier eras, animals were bred to such purposes: sheep for mutton and

Horse-drawn omnibus with upper deck. The ride on cobblestones was jarring.

sheep for wool; cattle for meat, cattle for milk, and oxen for hauling; dogs to a thousand purposes large, medium, and small, from herding sheep, to guarding property, to hunting deer or barn rats; and horses for transport, for racing, for hauling, or to generate stationary power by turning a sweep. The horses of the nineteenth-century city were living machines.[7]

Midwestern farmers bred the horses the city required—haphazardly at first, but with increasing expertise at selective improvement as the century advanced. By far the most popular workhorse in the United States was the Percheron, a breed that originated in the Perche region of France, about fifty miles southwest of Paris. Although it was long claimed that the Percheron breed was shaped in the Middle Ages when native Perche mares were bred with Arabian stallions brought back from the Crusades, no evidence other than oral tradition supports the claim. Some archeological evidence identifies the type as having Neolithic antecedents.[8] All modern Percheron bloodlines trace to a warhorse named Jean Le Blanc, foaled in Le Perche in 1823 when Perche breeders were breeding a heavier horse for the American trade.

The horse they evolved was gray or white, calm, powerful, and intelligent, weighing about two thousand pounds and standing six feet high at the shoulders. Edward Harris, a wealthy New Jersey farmer, horse breeder, and friend and patron of John James Audubon imported the first Percherons into the United States in 1839. A champion herd sire named Louis Napoleon, imported into Ohio in 1851, represented the breed standard. "The Percheron quickly became America's favorite horse," the breed association history claims. The US Census for 1930 found three times as many registered Percherons in America as the other four draft breeds combined.

Horses *increased* in number after the commercialization of the steam engine because horsepower filled the niche below steam power. A horse stood ready to pull a cart or plow a field on command, without the delay of building up a head of steam. Energy transitions are seldom so complete that they drive out every competitor. Much of the world still relies on animals for farm work and transportation: horses, oxen, camels, llamas, water buffalo, elephants, even fellow humans.

Feeding the urban fleet of horses hay and grain supported many thousands of farmers. An idle riding horse in New York City required about 9,000 calories of oats and hay per day. A draft horse in the same city working in construction required almost 30,000 calories of the same feeds. Annually, each draft horse consumed about 3 tons of hay and 62.5 bushels (1 ton) of oats. It took roughly four acres of good farmland to supply a working city horse that year's worth of feed.[9] At the beginning of the nineteenth century, when cities in America were limited largely to the East Coast, farmers seldom

Percheron cart horse, 1875.

transported bulky loose hay more than twenty to thirty miles to city markets.[10] The commercialization of the hay press in the 1850s, operated by hand or by horse-powered sweep, reduced the bulk and thus lowered the cost of shipping hay, while the opening of the Midwest's tallgrass prairies to settlement and farming in the intervening years met the increasing demand for horse feed. By 1879, national hay production totaled 35 million tons, a figure that had nearly tripled to 97 million tons by 1909. More than half the land in New England was devoted to hay by 1909 as well, and at least twenty-two states harvested more than a million acres a year of hay and forage.[11] The mechanization of American agriculture with horse-drawn or horse-powered machinery supported this vast expansion.

The volume of water and feed that city horses consumed was matched by their daily output of urine and manure. A working horse produced about a gallon of urine daily and thirty to fifty pounds of manure. That volume filled the New York streets daily with about four million pounds and a hundred thousand gallons of redolent excreta that had to be cleared away. When it wasn't, the streets mired up.

Urban manure, both human (night soil) and animal, was a valuable by-product of city living throughout the early and middle decades of the nineteenth century. Street-cleaning departments collected horse manure from stables and

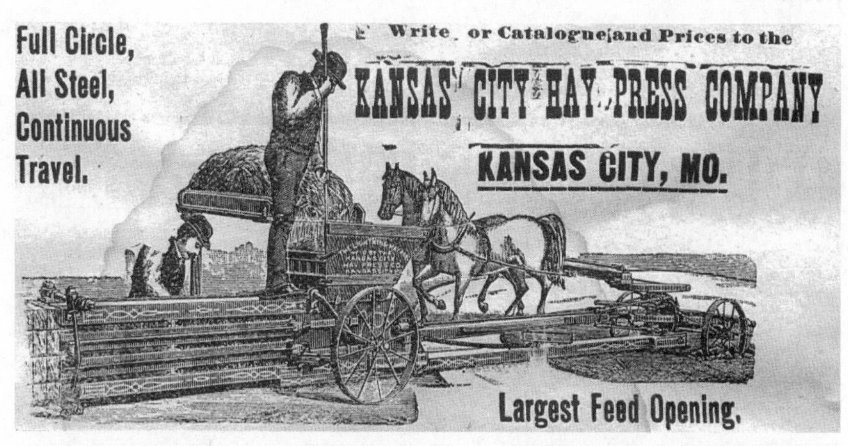

A hay press, an early version of the hay baler. A horizontal horse-powered ram compressed loose hay fed into the funnel into dense blocks.

Morton Street at Bedford Street in Manhattan's Greenwich Village in 1893, mired in uncollected horse manure.

streets and sold it to local farmers, who used it to fertilize the gardens, pastures, and fields where they grew food, hay, and grain for the city.

Competition for fertilizer arose first in Britain beginning in 1840, when another, less bulky, cheaper, and richer material than horse manure entered the market there. The Prussian explorer and naturalist Baron Alexander von Humboldt had alerted Europe to this material, guano—mineralized bird droppings—following his introduction to it in Peru in 1802 during an expedition to South America. Humboldt had encountered it on barges docked in the harbor at Callao. It smelled so strongly of ammonia that Humboldt erupted in a fit of sneezing whenever he walked near it. Mountains of guano, he learned, covered a number of islands off the Peruvian coast. Its name derived from the Quechuan Indian word *wanu.*[12] Humboldt couldn't believe that such a vast quantity of minerals was the work of birds. He speculated that the material might have been formed by a process analogous to the formation of coal.[13]

On his return to Europe in 1804, Humboldt sent a guano sample to two chemists in Paris for analysis. Independently, Humphry Davy received a sample from South America through the English Board of Agriculture in 1805 and experimented with it. He judged it "a very powerful manure," but thought that Britain's rainy weather "must tend very much to injure" it by leaching out its nitrogen.[14] The French chemists reported that Humboldt's sample contained a fourth of its weight in uric acid, saturated partly with ammonia, both sources of nitrogen; oxalic acid, combined partly with ammonia and potash; phosphoric acid; and small portions of potassium sulfate, potassium chloride, fatty matter, and rust-colored quartz sand.[15] Modern chemical fertilizers typically contain varying proportions of nitrogen, potassium, and phosphorus. Here was a concentrated natural fertilizer with a far higher content of elements vital to plant growth than horse manure. The function of fertilizer on plants was not yet well understood, however; nor was Peru, on the far side of South America, accessible to European markets. Humboldt passed on to other investigations, and the treasure of guano remained unmined except by Peruvian farmers, a usage that went back to the Incas.

Davy's lectures on agriculture, collected together in book form and published in 1813, became an international best seller, with editions published in London, Paris, Berlin, Milan, New York, and Philadelphia. Across the next decade, however, Napoleon I's wars with Spain and Portugal and the subsequent wars of independence in Spanish and Portuguese America delayed guano commerce until after their resolution in the late 1820s.

In 1824 the Maryland journalist John Stuart Skinner acquired a supply of guano. Skinner was owner and editor of a weekly journal, the *American Farmer,* and an American farming and fertilizer enthusiast. The USS *Franklin,* the seventy-four-gun flagship of the Pacific fleet, had visited the Peruvian islands before being laid up in ordinary in Philadelphia in August 1824. The captain of the *Franklin,* an amateur naturalist, sent Skinner a collection of seeds from Pacific island plants along with two barrels of Peruvian guano. Skinner divided and distributed the guano among his farming acquaintances for trial. One of them, a prominent politician, declared the substance "the most powerful manure he had ever seen applied to Indian corn."[16] But despite guano's power as a fertilizer, its time had not yet come in America. "The planters either neglected the samples sent them," Skinner's biographer reports, "or used the precious fertilizer so injudiciously as to realize no good effects."[17]

A Swiss physician and naturalist, Johann Jakob von Tschudi, exploring Peru in 1840 to study its antiquities, learned of the guano islands and visited them that year. He was astute enough to assess the guano phenomenon correctly:

> The immense flocks of these birds [guanay cormorants, *Phalacrocorax bougainvillii*] as they fly along the coast appear like clouds. When their vast numbers, their extraordinary voracity, and the facility with which they procure their food are considered, one cannot be surprised at the magnitude of the beds of guano which have resulted from uninterrupted accumulations during many thousands of years. . . . These birds are constantly plunging into the sea, in order to devour the fishes which they find in extraordinary masses around all the islands. When an island is inhabited by millions of sea-birds, though two-thirds of the guano should be lost while flying, still a very considerable stratum would be accumulated in the course of a year.[18]

Tschudi judged the guano stacks he saw to be thirty-five to forty feet thick, but he underestimated them. Careful measurements made in 1853, when the guano had already been exploited for a decade, found the maximum guano depth on the Chincha Islands to be 44.7 meters—147 feet.[19] "It was said," writes a historian, "that the . . . islands gave off a stench so intense they were difficult to approach."[20] Modern investigators have concluded that the stacks began to be laid down in the middle of the first millennium BCE.[21]

In 1840, in response to the years of investigation and publicity from Humphry Davy forward, a market for Peruvian guano began to develop in Britain. By 1844, Justus Liebig, a German chemist well known in England, could observe that guano, "although of recent introduction into England, has found . . . general and extensive application."[22] Tens of thousands of tons of guano annually began to be stripped off the three principal guano islands, North, Central, and South Chincha, each a flat granitic mass no more than a half mile long with high, aerodynamically shaped stacks of guano rising above them.

Working with pickaxes from the shore inward and the top down, the miners created Dover-like yellowish cliffs that looked from a distance like dirty snow. One German observer compared them to "a dark platter on which an enormous yellow cheese is being cut."[23] A survey conducted in 1854 concluded that the original mass of guano on the three islands had totaled no less than 13 million short tons (11.4 million metric tons), under hills covered with what one investigator calls "a dazzling coat of guano glass," a crust of calcium phosphate.[24]

North Chincha Island: Guano miners worked from the top down and the shore inward to harvest the guano stack, forming cliffs of Dover-like solid guano.

Another 1854 observer found the guano surface "like light dry earth and full of holes" that the birds scraped out for nesting and shelter. The surface was "difficult to walk upon, there being no certainty that every other footstep will not sink in nearly to the knee." A few feet below the surface, the guano "became compact," however, "and from thence through its whole thickness, it is of nearly the consistence of Castile soap."[25]

A terrible side effect of shipping guano to Europe was the transfer from Peru to Ireland of the late blight (*Phytophthora infestans*) that rotted Ireland's potatoes, the primary source of food for that impoverished country's eight million people. Between 1845, when the potato famine first struck, and 1860, more than a million Irish died of starvation. Another 1.5 million emigrated, most to the United States. Guano ships had carried the blight from Peru, probably in the form of infected potatoes taken aboard to feed their crews.[26]

By 1883, the Chincha Islands had been exhausted of their treasure, the great stacks of guano stripped away and shipped to Britain, Europe, and the United States. The powerful fertilizer, which began the agricultural transition from local animal manures to industrial chemicals, displaced the horse manure of British and American cities. In doing so, it turned a profitable by-product of urban transportation into an expensive nuisance that cities had to pay to have carted away. The burden increased as the cities' populations increased: London, from 3 million in 1860 to 7 million in 1900; New York City, from 516,000 in 1850 to 3.4 million in 1900; Boston, from 137,000 to 561,000; Philadelphia, from 121,000 to 1.3 million; and Chicago, from 30,000 to 1.7 million. Increasing volumes of urban trash and garbage in addition to horse manure added to the challenge.

Urban sanitary science in the last years of the nineteenth century, newly informed of the invisible vectors of disease, focused on filth as a signifier, flies in particular. Before then, flies had been considered charming and benevolent at best, useful scavengers at worst. "Our common houseflies seem jocund with mirth," an observer wrote playfully in 1859, "while they chase one another and dance their giddy rounds in the sunbeams."[27] By the end of the century, reports a historian of public health, the fly had been "transformed from a friendly domestic insect into a threat to health and hearth. . . . Its dangers were exaggerated so that at times it became as certain a killer as the mosquito that spread yellow fever and malaria. Health officials sought to develop the idea of germs into a practical and comprehensible weapon against disease. To do this, they portrayed flies as germs with legs."[28] And flies, of course, luxuriated in horse manure; their larvae fed on it. The head of the US Department of Agriculture's Bureau of Entomology estimated in 1895 that the ubiquitous manure of horses bred 95 percent of all America's flies.[29]

The real change away from horse-drawn transportation came with the advent in the late 1880s of the electric streetcar. Frank Julian Sprague, a West Point–trained electrical engineer, installed the first commercial electric streetcar system in Richmond, Virginia, in 1887. He summed up its virtues in a lecture to his fellow engineers the same year, pointedly mentioning the sanitary advantages:

> The riding of an electrical car is far easier than that of any cable or horse car, starting and stopping more easily, and being in a large measure free from lurching and oscillation. The cars are much cleaner. They can be brilliantly lighted, and they can be heated by electricity. There is no dust such as rises from the heels of horses. The sanitary conditions are entirely altered, and the health and comfort of the whole population is conserved. Stables with all their unsavory characteristics and the consequent depreciation of the value of adjacent real estate disappear.[30]

The fly cartooned as a disease-carrying "germ with legs." Its beak reads "torture"; its bloody knife, "typhoid"; its footprint, "filth."

The electric streetcar fostered a lucrative appreciation of real estate values: at twice the speed of horses, it extended the range of commuter transportation four times as far out into the countryside, offering profitable investment opportunities in suburban real estate. A year after Sprague's system began operating in Richmond, Massachusetts land speculator Henry Whitney and Charles Francis Adams II, the first president of the Union Pacific Railroad, began building an electric streetcar line connecting their Brookline real estate to central Boston. As a result, Whitney would report, the value of his Brookline holdings four years later had increased by $20 million ($524 million today).[31]

By the turn of the century, the electric streetcar had largely replaced the use of horses in public transportation. The animals continued to serve for general hauling, merchandise delivery, and small-scale energy generation. In fact, their urban numbers actually increased.[32] Only the development of the internal combustion engine and its application to power the truck and the automobile across the years 1900 to 1915 replaced the city horse with mechanical transportation. During the transition, historian Ann Norton Greene has discovered, animals that had once been represented as patient, humble servants of humanity, in the mode of the jocund housefly dancing his merry rounds, began to be depicted as "ungrateful and unruly. Horses repaid human care by kicking, bucking, and causing accidents." A popular new magazine founded in 1895, the *Horseless Age*, a champion of the automobile, claimed that "scarcely a day passes that someone is not killed or maimed by a wild outbreak of this untamable beast. . . . These frightful accidents can be prevented. The motor vehicle will do it."[33]

Eventually the motor vehicle did. Yet it was not the automobile but the electric streetcar and the interurban trolley that opened up the bedroom communities that began to separate middle-class suburbanites from the urban working class. The new century would be electrical, though the electricity would increasingly be generated with dirty coal rather than with clean hydropower. The twentieth century dawned not nearly so shining as optimistic newspaper editors predicted, a recognizably modern world but a world where smoke and smog still darkened the urban skies.

FOURTEEN

PILLARS OF BLACK CLOUD

Where there was fire, there was smoke. "There have been added to the domestic fires of our crowded towns pillars of black cloud which the manufacturers pour into the air as if the air were their own," John W. Graham, a British mathematician and Quaker activist, wrote in 1907.[1] In all the years since coal had begun replacing wood in home fires, in factories, power stations, and railroad engines, the air had been a commons, commonly used. Wood use peaked in the United States at 70 percent in 1870. (It had peaked about a century earlier in Britain.[2]) Thirty years later, in 1900, coal commanded that 70 percent of US demand, and wood use was declining.[3] "About four-fifths of our people live in towns under a smoke cloud," the British activist laments. The fraction was smaller in the still largely agricultural United States, but its major cities were smoky enough.[4]

It can be difficult today to visualize the extent of city smoke pollution at the end of the nineteenth century. Photochemical smog, the combination of nitrous oxides and volatile organics typical of the automobile age, was not yet the primary malefactor. The "smog" of the word's original 1905 coinage was a combination of smoke and fog. Smoke from coal burning, thick and brown or black, was the characteristic pollutant of day. In cities prone to weather inversions, it could turn day into night-like gloom.

Yet smoke was not at first generally perceived to be a toxic danger. For nineteenth-century industrialists and for many middle-class citizens as well, coal smoke was the price of progress, "a necessary and harmless corollary," one historian calls it. "Total freedom from smoke pollution was still regarded by many as a utopian goal, and those who pressed for abatement were often dismissed as irksome, interfering do-gooders: 'amiable and unpractical faddists.' "[5] Even Gifford Pinchot, the ardent conservationist who was the first chief of the US Forest Service, believed coal to be "the vital essence of our civilization."[6]

Chicago's colorful chief smoke inspector from 1894 to 1897, Frederick Upham Adams, a New York–born socialist reformer, inventor, journalist, and novelist, divided the Chicago population into two classes: "those who create a smoke nuisance" and "those who are compelled to tolerate a smoke nuisance." The creators, he said, maintained "that smoke is an irrepressible

A November fog in daytime London circa 1872. Street boys sold their services as torchbearers.

necessity—a concomitant of the commercial and manufacturing supremacy of Chicago," and that "smoke not only is not unhealthy, but that it is an actual disinfectant." Their opponents declared to the contrary that smoke "has resulted in an alarming increase in throat, lung, and eye diseases" and "point[ed] to ruined carpets, paintings, fabrics, the soot-besmeared facades of buildings, and to a smoke-beclouded sky." There were fifteen thousand steam boilers within the city limits at that time, Adams noted, at least twelve thousand of which burned soft coal. They were scattered over 186 square miles, and no two plants were alike.[7]

People died in greater numbers during smoke inversions, but such increases in mortality were more often ascribed to the cold winter weather than to the pulmonary effects of smoke. Cold seemed a far more likely angel of death than smoke, which people lived with intimately from day to day.[8] "Coal pervaded every sector of America's industrial cities," writes historian David Stradling. "Every class of resident saw it, handled it, purchased it, and smelled its dust. Residents knew good coal from bad and which coals burned best in their furnaces. They knew the names of mines and mining regions that labeled the black diamonds they produced. Middle-class homeowners bought coal by the ton and stored it in basement bins. The working poor bought it by the bucket and used it sparingly in their tenement stoves for heating and cooking. The desperately poor dug through ash dumps in search of bits of unburned coal or combed the railroad tracks for lumps that had fallen from rail cars."[9]

Coal and its smoke, like horses and their wastes, were daily encounters that have disappeared from modern life in mature industrial societies. For a view of the turn-of-the-century pall of coal smoke that blighted US and British cities, present-day Beijing, China, is a fair approximation. It's a reminder as well that societies develop first and then, as the more immediate needs of their citizens are met, clean up their pollution. During Chicago's 1893 World Exposition, the authorities temporarily switched to fuel oil (delivered by pipeline from Lima, Ohio) for power generation to reduce smoke pollution over the fairgrounds. Similarly, Beijing shut down factories to reduce smoke pollution at the time of the 2008 Summer Olympic Games.[10]

Americans mined two kinds of coal: soft, sooty bituminous, about 60

Daytime view of smoke in Toronto, Canada, in 1904, looking from the harbor into the city.

percent to 70 percent carbon; and clean, hard anthracite, 92 percent to 98 percent carbon. Because of its impurities, bituminous coal smoked; anthracite burned clean. The one anthracite region known in the United States before smaller fields opened in Colorado and New Mexico lay in eastern Pennsylvania. Once the residents of eastern cities learned how to burn anthracite in their fireplaces—it needed a raised grate to provide a draft to keep it burning—they were willing to pay a premium for the hard coal. The coal trade that developed in the eastern United States between 1820 and 1860 was predominately anthracite. Boston, New York, and Philadelphia were thus relatively cleaner than British cities, although pollutants from smelting, slaughtering, and other industries continued to foul their air and water.

Of American cities, Pittsburgh was the smokiest, partly because of its location, partly because of its heavy industry. The Victorian novelist Anthony Trollope, visiting Canada and the United States in 1861, judged Pittsburgh to be "the blackest place which I ever saw." The site was "picturesque," Trollope thought, "for the spurs of the mountains come down close round the town, and the rivers are broad and swift." Pittsburgh's setting at the foot of the Alleghenies, in a deep valley at the junction of two rivers, was one reason it collected smoke: "Even the filth and wondrous blackness of the place are picturesque when

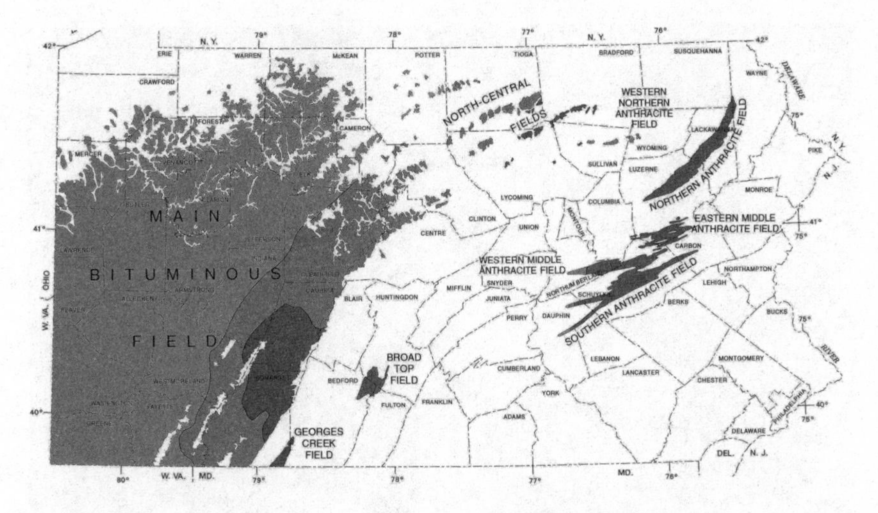

Coal regions of Pennsylvania. Gray field in west is bituminous; dark strikes eastward are anthracite, rarer but relatively smokeless.

looked down upon from above. The tops of the churches are visible, and some of the larger buildings may be partially traced through the thick, brown, settled smoke. But the city itself is buried in a dense cloud."[11]

Outcroppings of bituminous coal in those mountain spurs had led Pittsburghers to switch away from wood at the beginning of the nineteenth century, a half century earlier than the residents of other American cities. "Population was scarce," explains a Pennsylvania historian, "labor dear, and time precious in the busy western towns. Coal mined from Coal Hill, overlooking Pittsburgh, could be delivered by wagon to the door for as little as five cents a bushel. This was cheap enough, and far more convenient than cutting your own wood."[12] The result, for Trollope, was a grimy view that evidently amused him: "I was never more in love with smoke and dirt," he continues his description of the Pittsburgh scene in 1861, "than when I stood here and watched the darkness of night close in upon the floating soot which hovered over the house-tops of the city. I cannot say that I saw the sun set, for there was no sun. I should say that the sun never shone at Pittsburgh, as foreigners who visit London in November declare that the sun never shines of the place more succinctly."[13] He called it "hell with the lid taken off."[14]

Pittsburgh seen from Coal Hill in 1910, almost obscured by smoke.

Charles Dickens, Arthur Conan Doyle, and Robert Louis Stevenson, among other writers, evoked smoke pollution to color the atmosphere of their late-Victorian London fictions. The attorney who narrates Stevenson's horror story *Strange Case of Dr. Jekyll and Mr. Hyde,* in London on his way to Jekyll's house to discover a murder, passes through "a great chocolate-colored pall lowered over heaven."[15] The convention of relating smoke to moral disorder bloomed in the United States when middle-class activists sought a reason to justify abatement at a time when more urgent ills—air- and water-borne epidemics of typhoid, dysentery, and other diseases—took priority with health officials.

Mrs. John B. Sherwood, president in 1909 of the Women's Club of Chicago, was certain of the connection. "Chicago's black pall of smoke," she told club members in a speech that year, "which obscures the sun and makes the city dark and cheerless, is responsible for most of the low, sordid murders and other crimes within its limits. A dirty city is an immoral city, because dirt breeds immorality. Smoke and soot are therefore immoral." For Mrs. Sherwood, as for many Americans today, homicide loomed larger than its numbers justify: the three leading causes of death in the United States in 1900 were pneumonia

(202 deaths per 100,000 population), tuberculosis (194 per 100,000), and diarrhea/enteritis (143 per 100,000).[16] In contrast, the US homicide rate in 1900 was only 1.2 per 100,000 population, a historic low.[17] (The rate in 2016 was 4.9 per 100,000.)

Pittsburgh won a glimpse of a different future in the last decades of the nineteenth century when natural gas emerged to compete for a short period with coal. Gas had been a frequent by-product of saltwater and petroleum drilling going back to the 1859 Drake well and before. Other than for local applications such as salt boiling, no one knew what to do with it once it had served to push petroleum out of the ground. Producers typically burned it off. "Natural gas wells were often abandoned when oil was not discovered," writes historian David A. Waples, "left to blow freely. If ignited purposely or accidentally by friction, lightning, or careless open flames, they would burn for months or years."[18]

The problem with natural gas was always how to deliver it from the wellhead to the customer, at the right pressure. Manufactured gas—town gas—had the advantage of its controlled production and nearby location, from which it could be piped like water and its pressure controlled. Natural gas required pipeline delivery from wellhead to point of use, and the technology of constructing leakproof pipelines that could deliver gas of varying natural pressure across miles of intervening land developed only slowly. Early gas pipelines in Pennsylvania were cast iron, screwed together and caulked. Wrought iron replaced cast iron in the last decade of the century.[19] But natural gas was far cheaper than coal and its heat more uniform, qualities which made it desirable for process heating in Pittsburgh's many glass factories and steel mills. The Great Western Iron Company was the first in the Pittsburgh area to use natural gas, around 1870.[20] Many other manufacturers followed. By 1885, Pittsburgh had acquired about five hundred miles of natural-gas pipelines.[21]

Town gas manufacturers, whose gas cost more to make than natural gas did to prospect and deliver, resented the competition. One tale they told the Pittsburgh gas commission to convince it to ban "high-pressure" gas from the city, Waples writes, involved a stableman who "struck up a match to light a lantern and leaking gas nearby blew [him] thirty feet through the air, killed a horse, and set a building ablaze."[22]

George Westinghouse entered the Pittsburgh natural-gas business in 1884, after he drilled a well on the grounds of his thirty-two-room mansion in the city's prosperous Homewood district. He was not the first, but by 1889, the company he'd formed to exploit natural gas was the largest producer in the nation. To take the lead in distributing a new source of energy, Westinghouse had to invent the necessary technology. He patented twenty-eight inventions in his first two years of effort, Waples notes, including "improved ways of drilling wells, meter measurement, methods to prevent gas leaks, a regulator for controlling air and gas in a steam furnace, and an automatic control which shut off gas when the pressure fell, extinguishing flames."[23] To adjust gas pressure from high and irregular wellhead values to a value low and steady enough for home heating systems, Westinghouse invented the gas-pipeline equivalent of a transformer: he increased the diameter of his pipelines in stages as they extended from wellheads to city distribution centers, so that the gas arrived transformed to larger volume but lower pressure, a concept he later applied to the transmission of alternating-current electricity. Like Edison, Westinghouse borrowed systems from one technology and applied them in modified form to another.

Natural gas reduced Pittsburgh coal consumption from three million tons annually in 1884 to one million tons later in the decade. Nearby gas supplies were limited, however, and as gas fields depleted, smoke pollution once again choked the city. "We are going back to smoke," a speaker told the Pittsburgh Engineering Society in 1892. "We had four or five years of wonderful cleanliness in Pittsburgh, and we have all had a taste of knowing what it is to be clean."[24]

A taste of cleanliness is all American cities enjoyed through the first decades of the twentieth century. Such limited smoke abatement as cities accomplished had to be suspended in favor of increased industrial production during both world wars. Only the development of long-distance pipelines for transporting natural gas after World War II finally cleared the air.

"The reduction of the death rate," the American public health pioneer Hermann Biggs wrote in 1911, "is the principal statistical expression and index of human social progress."[25] In 1900 life expectancy at birth in the United States was 47.3 years, lower than the 2015 life expectancy in Swaziland (49.18 years), the lowest a United Nations study identified anywhere in the world.[26] The new methods and technologies of the twentieth century, despite its terrible wars, would extend life expectancy in industrial nations by more than thirty years.

PART THREE

NEW FIRES

FIFTEEN

A GIFT OF GOD

Henry Ford worked for Thomas Edison before he built his first car. In 1891, when he was twenty-eight years old, Ford moved with his wife, Clara, from the Michigan farm where he was born into Detroit to work as a night engineer for the Edison Illuminating Company. His primary duty was maintaining the steam engines that generated electricity in downtown Detroit. From farm experience, and from an apprenticeship with a company that built steamships for the Great Lakes trade, he knew steam engines. By 1893, Edison Illuminating had promoted him to chief engineer.

Ford began building his first automobile after he moved to Detroit, in a workshop he set up in a brick shed behind his Detroit duplex. The quadricycle, as he called it, was more a four-wheeled, motorized bicycle than an automobile. With a two-cylinder, four-cycle, four-horsepower gasoline-fueled internal combustion engine installed under the bench seat, a tiller for steering, and no brakes, it weighed just five hundred pounds.[1] It took him three years to design and build, by hand. ("Ford was working in a world that contained no automobile parts," quips one of his biographers.[2]) He rolled the quadricycle out of the workshop—after enlarging the narrow brick doorway with a sledgehammer—at two o'clock on a rainy June morning in 1896.

Henry Ford posing on his first automobile, the five-hundred-pound quadricycle, rolled out in June 1896. The small engine, which drove the rear wheels, is mounted behind the seat.

Ford was far from the only inventor working on a horseless carriage in the 1890s. "It has always been my conviction," writes Hiram Percy Maxim, the automobile-building son of the inventor of the machine gun, "that we all began to work on a gasoline-engine-propelled road vehicle at about the same time because it had become apparent that civilization was ready for the mechanical vehicle."[3]

Only some three hundred automobiles traveled on United States roads in 1896, among their variety the Philion Steam Car, the Lambert Gasoline Buggy, the Gottfried Schloemer Motor Wagon, the Ellis, the Duryea, the Pioneer, the Black, the Perry Louis Electric, the Columbia Electric Runabout, the Balzar Quadricycle, the DeLaVergine Six-Passenger, the DeLaVergine Dos-a-Dos, the Hartley Steam Fourseater, the Benton Harbor Autocycle, the Riker Electric Tricycle, and the Hart Runabout.[4] As their names imply, some were electric or steam driven; others, internal combustion, fueled variously with town gas, gasoline, alcohol, kerosene, burning fluid, or mixtures thereof. Which fuel and which engine type would dominate had not been determined. The Stanley Steamer was the best-selling car in America in 1898. Two years later, notes the historian Rudi Volti, "of the 4,192 cars produced in the United States in 1900, 1,681 were steamers, 1,575 were electrics, and only 936 used internal combustion engines."[5]

None of these many and various machines was yet configured anything like what came to be the modern automobile. Their air-cooled engines were typically hidden under the seat or underslung behind. "The horse carriage was the standard," Maxim remembers, "and anything different was frowned upon."[6] Even thus disguised, they frightened the horses they encountered on city streets and country roads. Connecticut required the motor vehicle driver to pull off the road and stop his engine if an approaching horse driver held up his hand, and to wait to resume until horse and driver had passed. "I am certain that in those early days, we spent as much time fussing with horses as we did running," Maxim complains.[7]

The internal combustion engine evolved from the steam engine. Both use hot gases expanding within an enclosed cylinder to supply power. The gas in a steam engine is steam, generated externally by heating water in a boiler and introduced through a valve into the cylinder, where it expands and pushes on a piston connected to a rod that transfers the motion outside the engine to turn a pair of wheels. Early steam traction engines were heavy, but those designed for automobiles—particularly the engines of the Stanley brothers, inventive twins from Maine—were remarkably light and efficient: the boiler on a Stanley Steamer weighed only ninety pounds and its engine only thirty-five.[8]

The first steam automobiles were slow-starting, requiring ten minutes or more to generate a working head of steam. That changed at the end of the century, with the addition of a pilot light that maintained steam during stops. Early steamers also vented their waste steam into the air rather than recondensing it, requiring stops every twenty or thirty miles to take on water. Steamers could use the public drinking troughs that cities maintained for their many horses. Watering troughs were plentiful in the countryside as well.

Excursion drivers who enjoyed challenges found steam automobiles pleasurably complicated to operate. "The legendary Stanley Steamer was festooned with gauges that required regular attention," Volti reports: "boiler water level, steam pressure, main tank fuel pressure, pilot tank fuel pressure, oil sight glass, and tank water level. Just to get one started required the manipulation of thirteen valves, levers, handles, and pumps."[9] The Steamer's control systems, ironically, operated a much simpler machine than its internal combustion counterpart: the perfected Steamer had only seventy-two moving parts, including its four wheels.[10]

The Stanley brothers, identical twins, in their 1898 Stanley Steamer, a hundred pounds lighter than Ford's Quadricycle.

The internal combustion engine, in comparison, could be started by turning a crank to work its pistons and generate a spark. (Cranking was hard work, particularly in cold weather, one reason many women preferred electric cars.) As the engine turned over, fuel—gasoline or alcohol, or a mixture of the two—in a timed sequence sprayed into its cylinders, where it was compressed and then spark-ignited, causing it to burn, heating and expanding so that it pushed on a piston connected to a rod that, again, transferred the motion outside the engine to turn a pair of wheels.

The electric car was simpler: a box of batteries, an electric motor, and a sliding lever or pedal to control the motor's speed. Its problem, besides frequent and slow recharging, just as today, was its relatively low power: since batteries were heavy, higher power had to be traded for battery life. The electric was ideal for city driving, however, clean and quiet, the mechanical equivalent of a horse and buggy. But with little charging infrastructure outside the city, it was unsuited for pleasure driving or longer-distance travel.

One early Philadelphia electric-car builder and enthusiast, Pedro Salom, nevertheless thought the electric so superior to the internal combustion automobile that he predicted it would prevail. "Compare [the electric's] construction," he told an audience at the Franklin Institute in 1895, "with the marvelously complicated driving gear of a gasoline vehicle, with its innumerable chains, belts, pulleys, pipes, valves and stopcocks, etcetera, and then put the question: Is it practicable? Is it not reasonable to suppose, with so many things to get out of order, that one or another of them will always be out of order?"[11]

Salom condemned the internal combustion engine's pollution, "a thin smoke with a highly noxious odor." He asked his audience to "imagine thousands of such vehicles on the streets, each offering up its column of smell as a sacrifice for having displaced the superannuated horse, and consider whether such a system has general utility or adaptability!"[12] Pedro Salom had considered the question. He thought not.

Steam engines and electric motors could be run up smoothly from idle to full power without gearing. But to operate without stalling, internal combustion engines had to idle at a rate of at least 900 revolutions per minute, and, for maximum efficiency, at 2,000 rpm or more, which meant they required gearing to reduce the number of rotations delivered to their wheels.[13] With their gears as well as with their more complicated engines, they were more demanding (and expensive) to build and operate than the other systems. Nor was it easy to learn to coordinate a clutch and gearshift on what came to be called a standard transmission. At first, Maxim writes, "when someone rasped the gears, it was considered evidence that my design was faulty." Later, when enough people had mastered steering, clutching, and shifting gears at the same time, "anyone who forgot to push down the clutch and rasped his gears was in disgrace and considered not competent to drive."[14]

Each type of vehicle power system had its champions and its detractors. Why internal combustion prevailed is a subject that historians of technology have long debated. Most arguments relate to infrastructure: the necessary systems that surround a technology and support it. Electric cars were limited largely to city driving because the countryside, not yet electrified, lacked recharging stations. Steam and internal combustion engines could usually find fuel in the local paint or general store, partly because gasoline was used as a cleaning agent and solvent, partly because farmers had taken up stationary gasoline engines to operate everything from washing machines to grain mills.

For at least one steam carmaker, the Stanley Motor Carriage Company of Newton, Massachusetts, that advantage was lost in 1914, when an epidemic of deadly hoof-and-mouth disease among New England farm animals led veterinary officials to shut down the many public watering troughs along eastern roads where steamers had rewatered.[15] Massachusetts had already challenged steamers for the clouds of steam they discharged, which obscured the vision of the drivers behind them, especially in cold weather. To solve that problem, Stanley developed a recondenser for its engines, but sales collapsed in the meantime. The damage was done.

By 1914, the internal combustion engine had swept the field. The Stanley and other steamer companies built a total of only about 1,000 of their cars that year, compared with a total of 569,000 by conventional US automobile manufacturers.[16] There were 1.7 million registered motor vehicles in the United States by 1914, up from 8,000 in 1900. Automobiles outnumbered horses in New York City for the first time in 1912, and the difference widened across the decade.[17] By the 1920s, horses were disappearing into pet food at the rate of a half million a year.[18] Except for local delivery service, the horse was departing the city, never to return.

The substitution of the automobile for the horse left farmers poorer. "By using the power produced by gasoline instead of by corn- and hay-burning horses," a rural economist wrote in 1938, "we have deprived the farmer of a market for the crops from many million acres."[19] As unintended consequences, farmers lost a major source of income, while the world lost a renewable resource, replaced by a fossil resource of unknown extent and unanticipated effect.

It might have been otherwise. Power alcohol reentered the marketplace in 1906 when the federal government lifted the old Civil War alcohol tax that had made it uncompetitive with kerosene. It competed with gasoline as a source of automotive fuel across the first three decades of the twentieth century. Henry

Ford designed his first production car, the Model T, with a flex-fuel system: it could run on either gasoline or alcohol, a feature that Ford continued to offer until 1931.[20] A brass knob to the right of the Model T steering wheel allowed the driver to adjust the carburetor to accommodate either fuel. A spark-advance lever on the left side of the steering wheel then adjusted the timing of the spark plugs, which needed to fire at a different point in the engine cycle, depending on the fuel.[21] Farmers, Ford thought, could make their own alcohol. City drivers could buy their gasoline at one of Iowa inventor John J. Tokheim's patented measuring pumps at curbside filling stations.

Even without its punitive tax, alcohol cost more to produce than gasoline, particularly since it lacked a supply chain from farm to refinery nearly as efficient as the pipelines and railroad cars of the Rockefeller-monopolized petroleum industry. (By 1882, the Standard Oil Trust controlled 85 percent of the

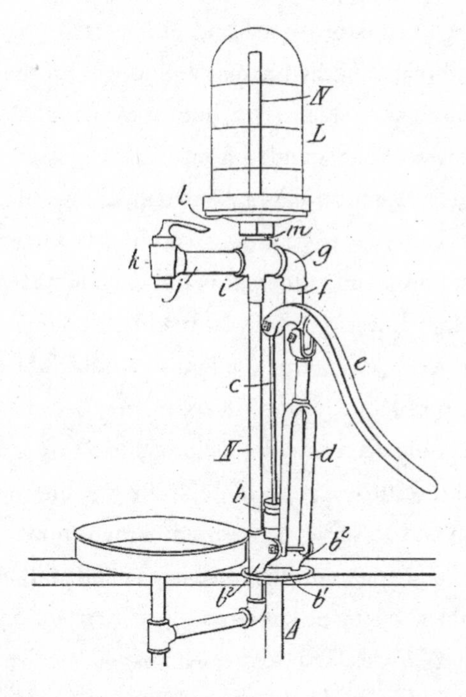

The Tokheim fuel measuring pump, patented January 2, 1900, pumped fuel stored in a tank safely underground into a clear glass dome where it could be measured and then spigoted into a gas can for transfer.

world oil market, one reason the US Supreme Court in 1911 ordered it broken up.[22]) As with wood compared with coal in Elizabethan London, the cost of delivery disadvantaged alcohol compared with gasoline. Alcohol would have required government support to develop as a full alternative. But it had immediate and valuable advantages as an additive.

A fuel's octane rating is a measure of how much it can be compressed before the heat of compression ignites it. Octane ratings at American gasoline pumps today, for example, range from 87 to 93. Pure alcohol has an octane rating of 105. Gasoline in the early years of the automobile era, called "white gas"—the plain distillate without additives—had an octane rating of only about 50. White gas was adequate for the first low-compression engines, but low-compression engines were inefficient and lacked power.

A higher-compression engine could power a larger car. It would give better mileage, saving fuel. Or it could be smaller, saving weight. But higher compression—squeezing the mixture of fuel and air injected into the cylinder into a smaller volume, so that it pushed harder and longer on the piston when it burned—caused white gas to break down, burn at the wrong time in the engine cycle, and waste power. What sounded from the driver's seat like a *ping* or a *knock* was, inside the cylinder, an aberrant and damaging explosion.

Engine knock became a serious problem around 1913, when the increasing demand for gasoline led oil refiners to maintain volume by distilling more crude into the product, lowering its octane further.[23] Engineers believed that engine knock was the result of premature ignition of the fuel—that compression alone was the problem. No one knew for sure, because it was difficult to know what was going on inside the cylinders of an operating engine firing at thousands of revolutions per minute. If the gasoline-powered internal combustion engine was to serve as the predominant power source for the automobile, its fuel and engine-design problems needed to be addressed.

In 1916 the problem of engine knock engaged the attention of a highly creative engineer named Charles F. Kettering, who would later become vice president for research at General Motors. Born in 1876 on a farm in northern Ohio to a family of Alsatian and Scotch-Irish descent, Kettering was tall, strong, and smart but beset with eye problems that almost kept him from attending college. He made it through with the help of classmates who read to him when his eyes failed. He was twenty-seven years old by the time he graduated from Ohio State University with a degree in electrical engineering.

Kettering was a good experimenter and an excellent inventor. For National Cash Register, he developed the first electric cash register, powered by a small, high-torque electric motor. For Cadillac, in 1911, piggybacking on his cash register experience, he invented the electric starter—another high-torque electric motor—which opened the automobile market for the first time to the many women who found it difficult to start a car with a hand crank.

In 1916 Kettering and his young associate Thomas Midgely Jr., a mechanical engineer, attacked the knock problem head-on. They built a window into the cylinder of the one-cylinder engine they used in the laboratory for research and recorded engine knock with a homemade high-speed strip camera. No one had ever looked inside a working cylinder before. They discovered that it behaved in miniature much as did the cracking* tower of an oil refinery. "I want you to keep this in mind," Kettering told an audience of automotive engineers in 1919, "that, if heating under pressure and high temperature produces disintegration of [petroleum] molecules in a cracking still, the same identical thing may happen when you maltreat the fuel in a gas engine."[24]

Crude oil is a varying mixture of a number of different oils and fats, as might be expected of a material transformed from the bodies of zooplankton and algae deposited in ancient shallow oceans and on continental shelves and covered over with sediment that became sedimentary rock. In the second half of the nineteenth century, when petroleum was first refined for kerosene and lubricants, it was distilled like whiskey: boiled in a still, the boiling vapors run off through a cold pipe to cool them back to liquids, the useless gasoline thrown away, and the kerosene collected. Simple distilling was inefficient: in 1910, when refineries now wanted the maximum output of gasoline for the burgeoning market in automobile fuel, the best they could get from their petroleum was only about 13 percent.[25] (The yield today is about 40 to 45 percent.)

The demand for higher gasoline output from limited supplies of American crude encouraged innovation. A petroleum engineer named William M. Burton, who trained at Johns Hopkins University and began work in

*"Cracking" simply means breaking down, as in cracking a nut.

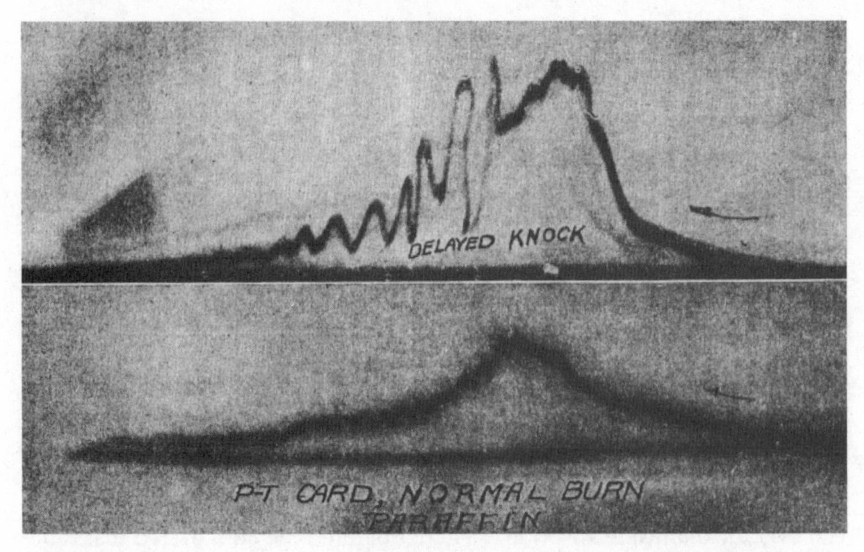

Recording graph of engine knock (top) versus normal burn (bottom). Graphs read from right to left.

petroleum refining in 1890, developed the first major advance for Standard Oil of Indiana. The Burton process, thermal cracking, moved beyond simply boiling the petroleum and piping off its distillates to boiling it at much higher temperatures and pressures. Standard Indiana opened the first commercial refinery using Burton's thermal-cracking process in January 1913. "Burton's still was operated at 95 psi and 750° F," write a team of oil historians, "and just about doubled gasoline yield."[26] Greater yield, and different products, would become possible with the use of catalysts to break down some of the distillates separated by thermal cracking and reassemble them in new molecular combinations.

Kettering and Midgely worked on the problem of engine knock for six years. Once they understood what caused it, they realized that the primary solution was not in redesigning the engine but in raising the octane of the fuel. The obvious way to improve automobile fuel was to mix alcohol, with its high octane rating, with gasoline. Alcohol and gasoline don't mix unless the alcohol is free of water, but they mix freely with the addition of a third ingredient, benzene, a hydrocarbon distilled from coal. A 30 percent or 40 percent mix of alcohol with gasoline and a little benzene made engines purr. Kettering and Midgely called that solution to their problem the "high-percentage additive."[27] They found it by 1918, patented it, but then set it aside.[28] It would function as a backup, a reserve, in case they found nothing better. Excellent additive though it was, alcohol had its own problems.

Fuel alcohol was available after 1906 for development as a gasoline additive. Then as today, however, it was distilled primarily from corn (not the kind we eat at table but the coarser kind grown to dry maturity and ground to cornmeal for animal feed). And corn prices generally increased from 1900 to 1917, crashed, and then increased again from 1920 to 1925, the years Kettering and Midgely were researching alcohol as an additive. Petroleum and corn prices tended to move up and down together, which meant adding corn-derived fuel alcohol to gasoline would increase an already increasing price at the pump.[29]

Nor were farmers much interested in growing corn for conversion into fuel alcohol when they were prospering growing corn for animal feed—animals to be fed in turn to the increasingly nonfarm human population as America urbanized. So prosperous was American farming in the first decades of the twentieth century that the years from 1900 to 1914 came to be called "the Golden Age of American Agriculture," a period when the value of farm products was higher than it had been in more than a hundred years.[30] Farmers saw alcohol production as an alternative in times of income decline, not in times of prosperity. That made grain unreliable as a source of raw material for automotive fuel.

Though he was interested in fuel alcohol, Kettering doubted if American farms could produce enough to meet increasing automotive demand. "Industrial alcohol can be obtained from vegetable products," he said in 1921, "[but] the present total production of industrial alcohol amounts to less than four percent of the fuel demands, and were it to take the place of gasoline, over half of the total farm area of the United States would be needed to grow the vegetable matter from which to produce this alcohol."[31] The alcohol-gasoline high-percentage-additive patent filed in 1918 had covered, but had not specified, fuel alcohol. It had specified benzene instead: 50 percent benzene and 50 percent gasoline.[32] But benzene was hardly more promising a source at that time than fuel alcohol. The entire 1920 US production of coal, if converted to benzene, would have been the equivalent of only one-fifth of that year's US gasoline supply.

More troubling to Kettering was the continuing decline in US oil reserves. Several members of Kettering's research team worked with a British organic chemist, Harold Hibbert, at Yale University in the summer of 1920.[33] From Hibbert, the inventor of ethylene-glycol antifreeze, they learned of the probability that domestic oil reserves might soon be exhausted.[34] If United States crude oil production continued at the 1920 rate of about 443 million barrels, Hibbert cautioned, the domestic supply would be exhausted by 1933. "Does the average citizen understand what this means?" Hibbert asked rhetorically. It meant, he went on to explain, that in ten to twenty years, the United States would be dependent entirely upon outside sources for a supply of liquid fuel, "for farm tractors, motor transportation, automobiles, the generation of heat and light for the thousands of country farms, the manufacture of gas, lubricants, paraffin, and the hundreds of other uses in which this indispensable raw material finds an application in our daily life."[35] There was no immediate solution in sight, Hibbert warned, "and it looks as if in the rather near future, this country will be under the necessity of paying out vast sums yearly in order to obtain supplies of crude oil from Mexico, Russia, and Persia."[36]

The Yale chemist had a long-term solution to offer, however: the substitution for liquid fuel of alcohol made "from corn and a variety of grains on the one hand, and from wood, on the other." Beyond waste wood, there were other potential sources of cellulose: corn stalks, flax, seaweed. And beyond alcohol, there was oil shale. Hibbert glanced hostilely at the Eighteenth Amendment, Prohibition, which had come into effect on 16 January 1920. He called it "pernicious and unenlightened legislation" that made research difficult by restricting access to alcohol (much as the illegality of marijuana later in the twentieth century would make research on the medicinal properties of cannabis difficult). He feared Prohibition might both "cripple existing chemical industries" and "endanger our national standing and security."[37]

Farsighted as Hibbert's proposal was—the world is reengaging with the same challenges today, from oil depletion, to oil shale, to renewable alcohol—he concluded by calling it a "scientific dream." And as with most dream reports, few listened.

Kettering and Midgley, having patented their high-percentage additive, with its complicating problems of supply, had moved on to researching a low-percentage additive. In 1921, after a discouraging several years when nothing seemed to work, they opened an investigation into compounds of several elements in group sixteen of the periodic table of the elements. Group sixteen, the oxygen group, includes oxygen as element 8, sulfur as element 16, selenium as element 34, and tellurium as element 52.

On a train to New York in late February 1921, Kettering read an Associated Press story about a Wisconsin chemistry professor's discovery that the compound selenium oxychloride was almost what the AP called "a universal solvent."[38] The story said it even dissolved Bakelite, the world's first synthetic plastic, which Kettering's Belgian American friend and colleague Leo Baekeland had patented in 1909. Baekeland had claimed in his announcement of the new plastic, made from carbolic acid (phenol) and formaldehyde, that it was "totally insoluble in all solvents."[39] The two men had long enjoyed a running joke between them about what could be dissolved—the old joke was that no container in the world would hold a truly universal solvent—which was why the AP story had caught Kettering's attention.[40] When he returned to Detroit, he suggested that Midgley test selenium oxychloride as an antiknock agent.

Oxygen and chlorine both increased knock, but such was Midgley's despair at that point of finding an antiknock additive that he agreed to try it. Exploring nitrogen compounds, he and his team had tried aniline, a coal-tar dye, which suppressed knock but produced exhaust that smelled like rotten fish. Selenium, it turned out, worked five times better than aniline but smelled like rotten horseradish. Moving on up this periodic table of the stinks, they tried tellurium. While it worked twenty times better than aniline, the element smelled, someone said, like "satanic garlic."[41] The foul odor, one of them recalls, "got into the men's systems and on their clothes. They couldn't wash it off, for water only made the odor worse. The smell was so bad that anyone working with tellurium was virtually a social outcast."[42] Fortunately for the noses of the world, selenium and tellurium were too rare to commercialize as additives in motor fuel.

Kettering, in one of his engineering talks, had spoken impatiently of his colleagues' tendency to conceptualize chemical compounds abstractly. "There has always been one trouble with all of our theoretical work," he cautioned them. "We take up [a] molecule as though it were simply something to think about and to talk about." To the contrary, he said, "these are real physical things, and the combination and decomposition of these things are real physical facts." Molecules were anything but abstract. For example, he said, "the more carbon atoms we get into the compounds, the heavier the fuel becomes."[43]

That practical perspective now informed the Midgley group's pursuit of an antiknock compound. One of their consultants, an MIT chemist named Robert E. Wilson, showed Midgley a periodic table he had constructed on a different principle of organization from Dmitri Mendeleev's original principle of chemical similarities. Wilson's table highlighted regularities important to organic chemists.*

"Tom was greatly interested in this," Wilson recalled, "especially since he believed that the antiknock properties of various agents were primarily properties of the elements, and he had some indications that the antiknock effect of an element varied predictably with its location in the periodic system."[44] Midgley's trail of increasingly heavy stinkers suppressing knock with increasing effectiveness was the crucial clue.

As they worked their way through more and heavier compounds in autumn 1921, they charted the antiknock effects on a large pegboard. "What had seemed at times a hopeless quest," Midgley recalls, "covering many years and costing a considerable amount of money, rapidly turned into a 'fox hunt.' Predictions began fulfilling themselves instead of fizzling."[45] In each of Wilson's groups—the fluorine, oxygen, nitrogen, and carbon groups—knock suppression improved with increasing atomic number. They tried tetraethyl tin at the end of October with good results, but further work with tin in November disqualified it—it caused another kind of knock, pre-ignition.

And then they came to the last element in the carbon group: lead. Lead isn't soluble in gasoline. Compounding it as tetraethyl lead improved its solubility. It was hard to make, but by the morning of 9 December 1921, a Friday, they had produced enough to test it. T. A. Boyd, the engineer who ran the test, remembered the morning vividly:

*To be specific, it was organized according to the elements' electronegativity or electropositivity: the degree to which they tend to gain or lose electrons in chemical reactions.

> With the men who were working on the endeavor gathered around the little [one-cylinder testing] engine, it was run on fuel containing a very small amount of the tetraethyl lead added purely by guess. And the engine purred along completely free of knock. An equal amount of untreated fuel was then poured in, cutting the concentration of the new compound in half. Still there was no sign of knock. The same process of halving the concentration of tetraethyl lead was repeated again and again and again, with the excitement of the observers mounted higher and higher with each dilution. In the end, it was found that as an antiknock agent, tetraethyl lead was *fifty times* as effective as the aniline on which so much work had been done earlier.[46]

Only when they had diluted their leaded gasoline by more than 1,000 to 1 were they able to produce knocking. Midgley rushed off to tell Kettering, who said later that day was the most dramatic of his entire research career.[47]

The new compound needed a name. For reasons never revealed, Kettering chose "ethyl," which confused it with ethyl alcohol and left out the significant fact that it was a soluble compound of lead, a substance long known to be poisonous.[48] To distinguish ethyl gasoline from unleaded white gas, they arranged to have it dyed red—partly, Midgley implied in a defensive 1925 report about the toxic effects of lead, to "specifically warn of the dangers of using it otherwise [than as a fuel], such as for washing motor parts, washing human body parts, for cooking purposes, etc."[49]

Much more development work followed, Midgley recalled. They knew very little about their breakthrough discovery. "We started spending more money, doing more research, and looking for other ingredients to go with tetraethyl lead, to make up a commercially practical compound."[50] In the meantime, the chemists guided it into early production. Ethyl gasoline was sold publicly for the first time at a filling station in Dayton, Ohio, on 2 February 1923.[51] Sales took off when the cars at that year's Indianapolis 500, which had been limited to smaller engines than had been allowed in past races, all burned ethyl gasoline, the winner averaging almost 91 miles per hour across the five-hundred-mile race.[52]

Standard Oil of New Jersey and General Motors both owned patents for methods of producing tetraethyl lead, a compound known since the 1850s. In 1924 they pooled their patents to form a new company, the Ethyl Gasoline

Corporation, to produce and sell the new product. Kettering became president; Midgley, second vice president and general manager.[53] By then, Standard Oil and DuPont were manufacturing tetraethyl lead at two semiworks: Standard in Bayway, New Jersey; DuPont in Deepwater, New Jersey. And trial runs were already producing casualties.

Lead has been a known poison since Classical times. It sickened Midgley himself and three coworkers in the winter of 1922–23, causing Midgley to spend a month in Florida recovering. The experience evidently failed to discourage him, nor did a lab explosion that blew lead into his eyes—lead he removed by flushing his eyes with mercury, which amalgamated with it and carried it clear.[54] In the article he published in 1925 defending his discovery, he noted that the symptoms of poisoning from tetraethyl lead specifically were "in order of their appearance, drop of blood pressure, drop of body temperature, reduced pulse, sleeplessness, loss of weight, sometimes nausea, sometimes tremor, and, in the most serious cases, delirium tremens."*[55]

The most extreme sign of tetraethyl lead poisoning is an agonizingly painful death. The pioneering American industrial-medicine physician Alice Hamilton and two colleagues described that harrowing extremity in a 1925 report for the *Journal of the American Medical Association* (*JAMA*), drawing on the recent rash of GM and DuPont cases:

> Symptoms of profound cerebral involvement appear, persistent insomnia, extraordinary restlessness and talkativeness, and delusions. The gait is like that of a drunken man, but there are no paralyses or convulsions. Finally, after a period of exaggerated movements of all the muscles of the body, with sweating, the patient becomes violently maniacal, shouting, leaping from the bed, smashing furniture, and acting as if in delirium tremens; morphine only accentuates the symptoms. The patient may finally die in exhaustion. In two fatal cases, the body temperature rose to 110 F. just before death occurred. One of these was a young man of fine physique who had been at work only five weeks. He is said to have suffered terrible agony. 'He died yelling.' "[56]

*That is, the DTs, more familiar as symptoms of alcohol withdrawal.

One DuPont Deepwater worker was stricken and died in September 1923, three more in summer and fall 1924. Kettering's and Midgley's laboratory in Dayton registered two deaths in June 1924. Five Standard Oil Bayway workers died from lead exposure in September 1924, and forty-four were hospitalized—a major catastrophe. Four more DuPont workers died in winter 1925.

"The *New York World* took up the crusade against this dangerous poison," Alice Hamilton recalled in her autobiography; "there was widespread panic lest the use of the blended gasoline involve risk to the public; several states hastily prohibited the sale of 'ethyl gasoline,' and foreign countries threatened to forbid its import."[57] The *New York Times* and other newspapers followed the story in detail through the middle years of the decade. Kettering reported the board of Standard Oil "in a blue funk over the whole thing. The directors were very much afraid about it. They didn't know what was going to happen to them."[58] Midgley complained to a colleague that "the exhaust does not contain enough lead to worry about," but that "no one knows what legislation might come into existence" as a result of the deaths. Any antiadditive legislation would, he was certain, be "fostered by [the] competition and fanatical health cranks."[59]

The US Public Health Service sought a scientific study to establish the toxicity of leaded gasoline, turning to the US Bureau of Mines because the study involved a petroleum product. Midgley and Kettering had approached the Bureau of Mines independently to perform a study as well, requesting that it call the product under study "ethyl" to avoid unnecessary publicity and that General Motors be allowed to comment, criticize, and approve the results.[60]

The bureau conducted an experimental study between December 1923 and July 1924, exposing a variety of animals—rabbits, guinea pigs, pigeons, dogs, and monkeys—to a low level of leaded-gasoline engine exhaust, the equivalent of exhaust level limits set to be allowed in the Holland Tunnel then under construction under the Hudson River between New York and New Jersey. The animals were exposed in a large, well-vented air chamber, free of lead dust, for periods of time ranging from three weeks to more than five months. Unsurprisingly, like Holland Tunnel workers, none died of lead poisoning. As a result, New York and New Jersey rescinded their bans on tetraethyl lead.[61]

In their 1925 *JAMA* report, Hamilton and her colleagues criticized the

Bureau of Mines study. "The period of exposure of the animals was too short," they concluded, "and the method used was not such as to produce convincing results." The study claimed, they wrote, "that 'the only danger of lead poisoning from products of combustion from ethyl gasoline seems to be confined possibly to the mechanic who is continually cleaning carbon from motors.'" They disagreed. They thought that the evidence showed a real danger of chronic lead poisoning from garage work and, more significantly, "a possible danger to the public from lead dust in the streets of large cities."[62]

Kettering sailed for Europe that winter to explore alternatives to his threatened additive. (A subsidiary of the German chemical giant IG Farben, BASF, showed him one, iron carbonyl, without identifying it. He recognized it and filed an American patent on it after he returned to the United States, which BASF understandably called "a rather sharp practice."[63]) Kettering met with Surgeon General Hugh S. Cumming in December 1924. Cumming now had the Bureau of Mines study in hand. The two men agreed that a conference of interested parties would be a good idea. Cumming called for a weeklong meeting in May 1925. More than one hundred industrialists, chemists, labor leaders, and physicians were invited.[64]

An interested party and president of the corporation whose product was under review, Kettering stood out among the principal speakers, most of whom were government officials.[65] He reviewed the history of engine knock and discussed alcohol and other additives but emphasized their limited supply. Frank Howard, Ethyl's vice president for research, later spoke more passionately. "Our continued development of motor fuels is essential to our civilization," he testified. "Now, after ten years' research . . . we have this apparent gift of God which enables us to [conserve oil]. . . . We cannot justify [it] in our consciences if we abandon the thing." Grace Burnham McDonald, the founder and director of the Workers' Health Bureau, countered that "it was no gift of God for the [workers] who were killed by it and the [many more workers] who were injured." Alice Hamilton entered a plea to the chemists "to find something else. . . . I am utterly unwilling to believe that the only substance which can be used to take the knock out of gasoline is tetraethyl lead."[66]

The organizers decided to cut the meeting short, insisting that its mandate extended only to tetraethyl lead, not alcohol and other alternatives.[67] The conference had been planned for a week but adjourned after only one day. A committee the surgeon general appointed to follow up met through the summer of 1925. It found "no good grounds for prohibiting the use of ethyl gasoline."[68]

As fuel historian William Kovarik notes in his study of the tetraethyl lead controversy, most of the boost in fuel octane ratings across subsequent decades came not from additives but from improved refining technologies and from blending. Midgley and Kettering were aware of the potential profit from developing a patented additive rather than adding unpatentable alcohol to motor fuel. "The way I feel about the Ethyl Gas situation," Midgley had written Kettering as early as March 1923, "is about as follows: it looks as though we could count on a minimum of 20 percent of the gas sold in the country if we advertise and go after the business . . . this at three cent gross to us from each gallon sold." By Kovarik's calculation, that minimum at that time represented some two billion gallons of gasoline, or $60 million per year ($837 million today).[69]

Both men were also conscious of the seemingly limited supply and declining quality of US petroleum. They had Harold Hibbert's formulation in mind. Kettering in particular had pushed for smaller, higher-compression engines operating on gas fortified with alcohol. Midgley, in his 1925 defense of tetraethyl lead, listed "conservation of petroleum" first among his reasons for having developed it.*

But their ideals, if such they were, conflicted with a larger movement in American life toward what Thorstein Veblen in 1899 had called "conspicuous consumption." With knocking no longer a problem, the internal combustion engine essentially perfected, General Motors could build more-efficient, higher-mileage cars or it could build larger, more-powerful cars. In its struggle for dominance over Ford, the proponent of efficiency, it chose power. No steam or electric automobiles were exhibited at the 1924 National Auto Show, and except for a brief flurry of interest in farm alcohol during the Great Depression,

*Such were Midgley's gifts for discovery and his long-term ill luck that he went on to develop the low-toxicity, low-flammability chlorofluorocarbons that replaced toxic ammonia or sulfur dioxide in refrigeration but were ultimately shown to damage the ozone layer of the atmosphere that protects humans and other living organisms from the destructive effects of solar radiation. Those, too, were eventually phased out.

fuel producers abandoned alcohol as a substitute or an additive for the next fifty years.

By 1936, 90 percent of all US gasoline was leaded. Domestic consumption of tetraethyl lead reached a high of 5.1 million pounds in 1956. In 1959 the US Public Health Service supported an Ethyl Corporation request to *increase* the lead content of gasoline from 3 cc to 4 cc per gallon—because refiners had reached a limit in improving fuel through refining and were now losing yield to keep up octane.[70] The Public Health Service did so despite a complaint by the committee of physicians it appointed to study the matter that "since the 1925 investigation, there have been no follow-up studies of large population groups on how adding TEL to gasoline affects the total body burden of lead."[71]

By 1963, more than 98 percent of US gasoline was leaded. When, a decade later, lead was finally ordered removed from the US gasoline supply, it was removed because it fouled the new catalytic converters mandated to fight smog, a different air pollution problem, not because it had been labeled a dangerous pollutant itself. Long before then, however, despite the extension of American oil supplies with a toxic additive, the United States had fulfilled British chemist Harold Hibbert's prediction that it would be "under the necessity of paying out vast sums yearly in order to obtain supplies of crude oil from Mexico, Russia, and Persia."[72] Persia—Iran—certainly, and Mexico as well, but most of all Saudi Arabia, a formerly sleepy kingdom of camels and sand.

SIXTEEN

ONE-ARMED MEN DOING WELDING

In late 1933 a small team of American petroleum engineers arrived in Saudi Arabia by arrangement with Abdul Aziz ibn Saud, the new king. Ibn Saud was a giant of a man, six foot three, broad shouldered and large handed, a camelryman and a warrior, immensely confident, committed to developing the impoverished kingdom he had won by intrigue and in battle across the past thirty years. Saudi Arabia became a country officially in 1932, an absolute monarchy: 850,000 square miles, more than three times the size of Texas, with a population of only about 2.5 million people, much of the land uncharted desert.

But across a narrow strait from its east coast, on the postage-stamp island of Bahrain, only thirty miles long and ten miles wide, oil had been flowing since June 1932 from Oil Well No. 1, at a site called Jabal al-Dukhan. Iran and Iraq had been the two important oil nations in the Middle East up to that time, their oil interests under British control. Bahrain was a minor exception, as Saudi Arabia would be a major. Through a series of sales and trades, motivated by what British engineers believed to be Bahrain's unpromising oil geology, the island's oil rights had devolved into the hands of one of the smaller international oil companies, Standard Oil of California (Socal). Besides freshwater springs, there were seeps of liquid bitumen underwater in the Persian Gulf (the Arabian

Saudi Arabia between Africa and Asia.

Gulf, the Saudis called it) north and east of Bahrain, and Bahrain itself had minor outcroppings of bitumen, a form of natural asphalt that is petroleum's denser first cousin.[1]

A Gulf Oil geologist, Ralph Rhoades, a Missouri-born ex-marine inevitably called Dusty, had pinpointed a promising structure: a rocky, dome-like formation called a *jabal,* on Bahrain in 1928, before Socal acquired it.[2] Then Fred Davies, a Socal geologist following up in 1930, had not only identified a well site on the Bahrain *jabal* but also had looked west across the strait to Saudi Arabia and spotted a cluster of *jabals* there as well. His identification would prove accurate: the Bahrain and Arabian *jabals* were related, both originally islands in the Gulf—the inland *jabal* now part of the land because sand had filled in the gap between it and the previous shoreline.[3]

Daniel Yergin, in his magisterial oil business history *The Prize,* describes in rich detail the long, complicated negotiations that followed between Socal and ibn Saud for oil rights in Saudi Arabia. Most of the country's revenues came from fees on and services to Mecca pilgrims. Pilgrimage numbers had declined with the financial disaster of the Great Depression—from about one hundred thousand per year up to 1930, down to only twenty thousand by 1933.[4] As a result, Yergin writes, "Ibn Saud was fast running out of money. . . . The kingdom's finances fell into desperate straits; bills went unpaid; salaries of civil servants were six or eight months in arrears. Ibn Saud's ability to dispense tribal subsidies constituted one of the most important glues bonding a disparate kingdom and unrest developed throughout his realm."[5]

An unlikely figure guided him: a rebel Englishman named Harry St. John Bridger Philby, known as Jack, a convert to Islam and ibn Saud's close adviser.* "The king," Philby writes, "in view of his recent troubles with the more fanatical leaders of the [fundamentalist] Wahhabi movement, was reluctant to open his country to the infidel." Philby says he told ibn Saud "that the king and his people were like folk sleeping over a vast buried treasure, but without the will or energy to search under their beds," adding a Koranic version of the universal saying "God helps those who help themselves": "Allah changeth not that which is in people unless they change what is in themselves."[6] The king still hesitated. Philby decided to act on the king's behalf, playing off British and American oil interests. His British counterparts, in any case, doubted if Saudi Arabia harbored any serious store of petroleum.

Ibn Saud, first king of Saudi Arabia.

From Socal's point of view, the time might not have been propitious. Global oil production, at 2.9 million barrels per day in 1925, had risen to more than 4 million bpd by 1929. The number of automobiles on the road in the United States alone had multiplied from 9.2 million in 1920 to 26 million in 1931.[7] Then the Depression killed commerce. US automobile registrations dropped by almost 2.6 million between 1930 and 1933. The world was awash in oil. In Oklahoma, you could buy a barrel of oil for 46 cents.[8]

*And the father of the notorious British atomic spy Kim Philby.

Some at Socal, however—people in the producing department, according to the novelist and historian Wallace Stegner, who wrote a Socal company history as a young man—"felt that the time to look for oil is when there is already plenty of it. That is when concessions and leases are easier and least expensive to get."[9] The Saudi Arabian concessions and leases were anything but easy to get, and they weren't inexpensive. Socal benefited, however, from a lack of interest by other oil companies, which often had contractual conflicts as well.[10] Even so, the initial negotiations between the American oil company and the Saudi king concluded only at the end of May 1933.[11]

The king had demanded payment of the contract's fees and long-term loans in gold: an annual rental fee of £5,000 in gold until a well came in, a loan of £50,000 (to be paid back only with oil revenues), and a royalty payment after the discovery of oil of four shillings gold per net ton of crude production. In return, a US government history reports, "The company received exclusive rights to explore for, produce, and export oil, free of all Saudi taxes and duties, from most of the eastern part of Saudi Arabia for sixty years. The terms granted by the Saudi government were liberal, reflecting the king's need for funds, his low estimate of future oil production, and his weak bargaining position."[12]

Just then, Socal's plans to pay ibn Saud in American gold stalled. One of Franklin Roosevelt's first acts upon taking the oath of office as president of the United States in March 1933 was to move the country off the gold standard and call in all gold coins and gold certificates. An appeal to the US Treasury Department to allow the exportation of $170,327.50 in gold (the dollar equivalent of £35,000, Socal's first installment) earned only a letter of rejection from a young Treasury undersecretary named Dean Acheson.[13] Socal met its obligation then from London by drawing a hoard of thirty-five thousand English gold sovereigns through its bank from the Royal Mint. Given the misogyny of the Wahhabi sect, well known then as now, the Mint agreed to lay Queen Victoria sovereigns aside and supply only sovereigns stamped with the images of English kings.[14]

The down payment arrived late—Saudi Arabia and Socal formally signed their deal on 14 July 1933—but on 4 August, seven wooden boxes of English gold sovereigns shipped from London to Jeddah, the fly-blown port town where the negotiations had been conducted, on the Red Sea coast about forty miles west of Mecca.[15] A Socal representative cabled the home office in San Francisco on 25 August, Stegner writes, "that he had counted them out on the tables of the Netherlands Bank at Jeddah under the eyes of Sheikh Abdullah Suleiman [ibn Saud's shrewd chief negotiator], and had Sheikh Abdullah's receipt."[16]

Now Socal had only to find the oil. It's obvious today that Saudi Arabia overlies one of the world's largest oil fields. It was not obvious in the 1930s, when Socal started prospecting, despite the Bahrain well. British interests, operating through their Iraq Petroleum Company (IPC), played catch-up in 1936 by negotiating a concession with ibn Saud for the Hejaz, the western part of Saudi Arabia. "The terms were much higher than those negotiated three years earlier by Socal," Yergin says. "The only drawback was the IPC never found oil in its concession."[17]

Socal assigned its Saudi Arabian concession to a subsidiary, the California Arabian Standard Oil Company (Casoc).* Casoc went to work immediately: its concession covered an area of roughly 320,000 square miles.[18] And as Henry Ford had built his first car in a world without automobile parts, so also did Casoc prospect a country unstocked with oil-equipment spare parts and few simple tools. Communications were equally limited. "Ibn Saud had conquered his Wahhabi followers' distrust of the telephone," Stegner reports, "by having passages of the Koran read over it, but the telephone so far linked only the cities of the Hejaz."[19] Two-way radios had to serve instead. Nor were there accurate maps other than a large-scale British War Office chart of the entire Arabian Peninsula. Despite these and other handicaps, by the end of September 1933, the first Casoc team had examined the *jabal* of limestone hills that Fred Davies had seen from Bahrain in 1930, decided it looked promising, renamed it the Dammam Dome, and established the first oil camp in Saudi Arabia nearby, four miles from the coast.[20]

The Casoc campaign to find commercial volumes of oil in Saudi Arabia comes in heroic colors or in realistic colors. The heroic version—Wallace Stegner's work for hire, commissioned in 1955 for public relations purposes and published only abroad (in Beirut) in his lifetime—depicts the Arabs and the Americans laboring side by side as equals and the king as a friendly cheerleader. The realistic version, a more recent scholarly work based on company documents and the unvarnished recollections of some of the participants, reveals

*Casoc became the Arabian American Oil Company (Aramco), in 1944.

pervasive racism on the American side: segregated housing, large pay differentials, Middle Eastern workers derided as "coolies" and "boys"—and a charming but avaricious king pushing relentlessly for money and other rewards.[21] Both portraits merge in their reconstruction of what it was like to seek and discover natural resources on the scale of Aladdin's lantern, oil gushing from the earth as if the great globe itself had been wounded.

The Casoc engineers completed detailing the Dammam Dome at the beginning of June 1934, but it was November before they sold the project to the Socal board back in San Francisco and began moving in a drilling crew. "In the news that the company planned to drill the Dammam structure," Stegner writes, "the [Saudi] government had the most welcome information it could have received. . . . The world was still sunk in depression, the hajj [Mecca pilgrimage] would be light again, the Saudi need of money was acute in spite of [a] second loan, which the company had made in advance of its due date. Needing income so badly, the government from the King down waved aside the warnings of the Americans that there might be no oil down there."[22]

By February 1935, lacking dynamite, they had built a derrick cellar for Dammam No. 1 the old way, by heating the surface rock with wood fires and dousing it with water to break it up. By mid-April, the derrick was in place. By the end of April, they had spudded in the well—bored down to bedrock, where the real drilling began—and started a 22½-inch hole.[23]

At the end of the first week of May 1935, they were down 260 feet through hard gray limestone. Then: 14 May, water at 312 feet; a slight showing of tar at 383 feet; still gray limestone, 496 feet. By 15 July, 1,433 feet, gray limestone. On 25 August, in a cable to San Francisco: "Slight showings of oil and gas at 1,774'. Not important but encouraging." By 18 September, at 1,977 feet, they had light oil surging at 6,537 barrels per day, which looked like success, but California advised caution: "These figures may need checking before jumping out of window." It was good advice. On 23 September, the drilling team reported, flow had steadied to only about 100 barrels a day. "It would have been an oil well in Pennsylvania," Stegner notes dryly, "but not out here." On 27 November, at 2,271 feet, they had a strong flow of gas but only a showing of oil. At which point they killed the well with mud. A month later, on 4 January 1936, they plugged Dammam No. 1 with concrete and started on Dammam No. 2.[24]

Oil had appeared at a deeper horizon on Bahrain, 2,832 feet. Looking for a parallel field, they decided to drill Dammam No. 2 deeper. By 11 May 1936, Stegner writes, down to 2,175 feet, it "was giving most encouraging indications." In a five-day test on 20 June it flowed 335 barrels a day. They decided to acidize it: to pump down hydrochloric acid at low pressure to dissolve open the pores of the limestone. That was safer than it sounded, since the limestone that the acid dissolved neutralized it. Acidizing No. 2 worked. Production went up to 3,840 barrels a day. Then they faced the same dilemma that Edwin Drake had faced seventy-seven years earlier when his first well came in: they ran out of storage. So they shut down the well. It had already confirmed what they and the Saudis had hoped: there was oil in Arabia.[25]

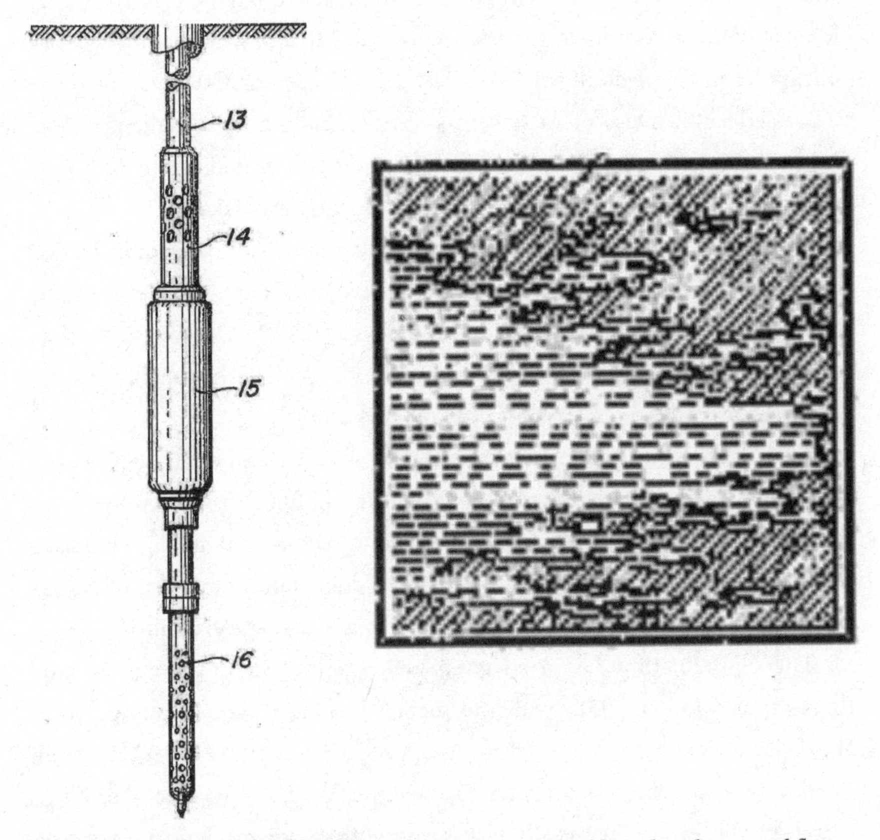

Injecting acid into an oil well opened channels into impermeable rock and increased flow—essentially an early form of fracking. *Left*: 1939 patent drawing of acidizer indicating injection holes; *right*: schematic of acidized rock.

Across the next two years, they drilled one disappointing hole after another. All that drilling swallowed water like a caravan of camels, 630 thousand gallons a day, the pumps never stopping on the submarine spring in the Gulf they drew from.[26] Dammam No. 3 never produced more than 100 wet barrels a day; they used the waterlogged product for road oil. Dammam No. 4 was a dry hole; they suspended drilling at 2,318 feet. So was Dammam No. 5, suspended at 2,067 feet. Discouraged, they got no further with Dammam No. 6 than digging its cellar and raising its derrick.[27]

Dammam No. 7, says Stegner, would be the first deep test hole. "What they would find was, by then, anyone's guess, but they all knew it had better show something. Time was running out."[28] They finished spudding No. 7 and began drilling on 7 December 1936. On 10 April they lost a drill bit. On 16 April they cleaned out the well with a larger drill bit down to 726 feet, when a bridge of rock collapsed, and a large boulder fell in. "Plugged with cement 200 sacks," Fred Davies cabled San Francisco, and then, reassuringly, "located top of cement at 704." Once the cement had hardened and entombed the cave-in and the boulder, they could redrill that section of the well, drilling through the blockage.

"By May 1937," Stegner writes, "everybody around Dammam admitted that the well was in bad shape and was going to be slow. There was a spurt in July that took them down to 2,400 feet, then delays again. On October 6 they had reached 3,330 feet. Tests then, as well as on the 11th and 13th at slightly greater depths, produced the same report: 'No oil, no water.' "[29]

Their first showing of oil emerged finally on 16 October at 3,600 feet—"about two gallons, in a flow of thin gas-cut mud." On that slim two gallons, the engineers faced a skeptical Socal board of directors back in San Francisco prepared to kill the well and stop the financial drain, tens of millions of dollars already invested in Saudi outreach or sunk into a hole in the ground.

Then, just in time, as in all good melodramas, Dammam No. 7 came through: on 4 March 1938, while the Socal board was still deliberating, No. 7, at a depth of 4,725 feet, started flowing at 1,585 barrels a day. Three days later, the flow was up to more than twice that volume, to 3,690 barrels, and to 3,810 barrels by the end of the month. It was sour with hydrogen sulfide, not sweet, but desulfurization treatment would remedy that. (Most of the sulfur produced in the world today is by-product sulfur from oil and gas operations.) With no storage at hand, they coupled No. 7 to No. 1 and flowed the oil back into the ground. By 27 April, Dammam No. 7 had produced more than 100,000 barrels.[30] Across the decades, until it was shut down in 1982, No. 7 alone produced more than 32 million barrels of oil.[31]

The deep play from which No. 7's oil flowed was named the Arab Zone. When Casoc extended Nos. 2 and 4 down into that zone, those wells also began producing in commercial quantities. The company built a pipeline then from Dammam to the port of Ras Tanura on the Gulf. The Saudis negotiated a much larger package of payments and royalties, in exchange for which Socal received a concession of 425 thousand square miles—an area equal to the areas of the United Kingdom, France, and Germany combined.[32] On 1 May 1939 ibn Saud visited the port with his retinue—nearly two thousand people, officials and courtiers, all arriving in some four hundred automobiles in great clouds of dust. A Socal tanker, the SS *Scofield,* was waiting to load, and they went aboard to celebrate. Then, with his kingdom now floating on oil, the king turned a valve on the harbor pipeline to load the first tanker of Saudi crude.[33]

The history of liquid energy is a history of pipelines. The natural gas field that George Westinghouse tapped to supply Pittsburgh with clean fuel, which reduced that city's annual coal consumption by two million tons, gave out in less than ten years. It may seem odd today that the city should then return to burning coal, but the technology did not yet exist in the 1890s to construct long-distance pipelines that might have drawn on more distant sources of natural gas.

Pipelines served at best for local and limited regional delivery. Constructed in sections with overlapping, riveted ends and caulked, they tended to leak. The technology that made possible long-distance pipeline construction was electric arc welding.

Humphry Davy had first demonstrated an electric arc at the Royal Institution in 1802—the demonstration that had required an entire basement full of batteries to support. Davy was demonstrating a new source of light, not welding. Welding was still something blacksmiths did by hammering together pieces of metal heated in a forge. An electric arc was more than hot enough at 6,500°F (3,600°C) for welding, but industrial electric welding pulled more current than batteries could sustain and awaited the development of the generator.

In the late 1870s, a Russian inventor, Nicholas Benardos, investigated using

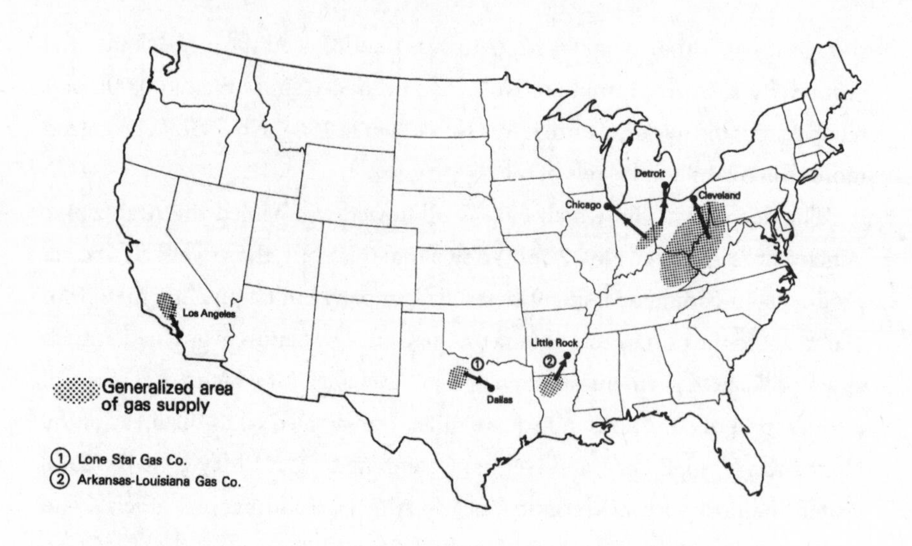

Regional gas transmission before 1925.

electric current to heat the edges of steel plates, much as blacksmiths did with their hot charcoal fires, after which the plates could be joined by vigorous hammering. Prolonged heating sometimes melted together the edges of the plates. Benardos noticed that such melting produced a monolithic connection stronger than hammering did. He demonstrated this technology at the great Paris International Exposition of Electricity in 1881, where Thomas Edison and Hiram Maxim, among others, exhibited their early electric lights. Benardos found a partner at the Exposition: a wealthy Polish engineer named Stanislas Olszewski, who supported Benardos's research and who shared his first, Russian patent on the process in 1885. A joint US patent followed in 1887.[34]

The Bernardos-Olszewski arc-welding system used a carbon rod clamped in a wired handpiece to deliver current to the pointed end of the rod. A clamp attached to the metal pieces to be welded carried the current back to the power supply. When the operator scratched the point of the carbon rod against the metal, he struck an arc, creating a hot, conductive plasma that completed the circuit. By moving the rod along the line where the two pieces touched, he melted and fused them together.

Using a carbon rod to form the arc limited this first electric arc-welding

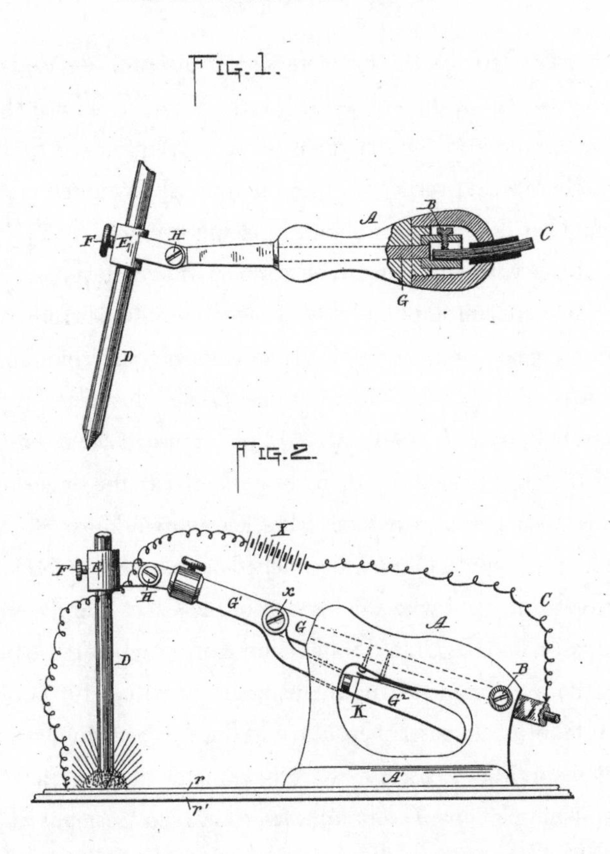

Two views of Benardos-Olszewski arc-welding system, patented in the United States in 1887. Battery symbol *I* in fig. 2 indicates power supply; the handpiece in fig. 2 sits in a stand.

system to melting metal parts to fuse them. If the operator wanted to add metal to the weld, he had to feed a separate metal rod into the hot arc, welding two-handed. Another Russian inventor simplified this process in 1888 by using a metal rather than a carbon welding rod. With that refinement, an operator fed the welding rod itself into the hot arc as he moved the rod along the line of the weld. The melt from the rod added metal to fill the weld. When the rod was nearly consumed, he unclipped and dropped the butt, clipped in a fresh rod and continued welding.

Other refinements followed at the beginning of the twentieth century. A hot metal welding rod interacted with gases in the air to produce compounds that embrittled and corroded welds, causing ship boilers, for example, to leak again after repair. To solve this problem, several inventors devised coatings that volatilized at the tip of the hot welding rod, producing a local shield of inert gas that excluded the air. The inert gas reduced dripping and sputtering, making it possible to weld overhead without undue risk of molten metal falling on the welder. That development allowed machinery such as ship boilers to be repaired in place without having to be taken apart and moved, saving time.

American merchant shipbuilding was nearly moribund before the Great War. One historian speaks of "the chronic inability of American shipbuilders to compete with foreign shipyards."[35] Most of the vessels other than proprietary passenger ships built between 1910 and 1914 were railroad-car and dump barges.[36] Steel ships were still riveted together, not welded, with the single known exception of one Great Lakes forty-foot icebreaker and workboat, the *Dorothea M. Geary,* built in Ashtabula, Ohio, and launched on Lake Erie in 1915.[37]

Conditions changed with the coming of the Great War. President Woodrow Wilson declared the United States neutral on 4 August 1914, but neutrality eroded as the war expanded and intensified. More than 100 Americans died among a total of 1,198 passengers killed in the German sinking of the British liner *Lusitania* in May 1915. A new German policy of unrestricted submarine warfare, and an intercepted communication between Germany and Mexico in January 1917 proposing a military coalition if the United States declared war, led America in April 1917 to join the alliance of nations fighting Germany.

With its 6 April declaration of war, the United States moved to intern all German and Austrian shipping then in American ports: twenty-seven German-owned vessels in New York Harbor alone.[38] "That afternoon," writes historian William Lowell Putnam, "a team of United States marshals marched through the cavernous warehouse piers of the North German Lloyd and Hamburg-America lines in Hoboken to formally seize everything within sight in the name of the newly belligerent United States of America." A few members of the German crews were allowed to remain aboard the liners as caretakers, an expensive mistake. The German caretakers responded with sabotage. A navy officer who inspected the ships found "breakages on all cast-iron parts."[39]

American shipyards turned to electric welding to repair the many ships the Germans had sabotaged. Despite the extensive damage, the repairs took only four months.[40] By March 1918, when the navy officer reported again, the ships had "made three or four voyages, without complaint."[41] Within eight months, the shipyards repaired more than one hundred ships, in time to carry munitions, supplies, and a half million American troops to Europe.[42]

That development in turn led to an investigation of welding for shipbuilding as the US Navy scaled up for war. "Ships, ships, and more ships is the call of the hour," Navy Secretary Josephus Daniels declared in February. "We must have more ships to win the war."[43] The British had faced so great a demand for oxygen and carbide for oxyacetylene welding since the beginning of the war that they had taken up electric welding as a substitute—and found it superior in most cases for welding mines, bombs, small vessels, and even a barge.[44] The US Shipping Board asked the British Admiralty to loan it an engineering officer knowledgeable about welding to investigate the state of US shipyard operations.

Captain James Caldwell of the Royal Navy arrived in the United States in mid-February 1918 and spent the next three months touring East Coast shipyards and factories in what he called "welding investigations."[45] Among other inspections, Caldwell toured the German liner SS *Prinzess Irene* at the Brooklyn Navy Yard on March 9. The ship had been transporting troops to Europe. Caldwell learned from its chief engineer "that no trouble had been experienced with the engine parts repaired by electric welding."[46]

Caldwell concluded that welding equipment and materials would cost no more than riveting, while the cost of labor would be reduced substantially. "One operator," he estimated, "would do the same work as now done by a squad of riveters (say 4 or 5 men) and a caulker."[47] And since plates could be welded butt to butt rather than overlapped, as riveting required, welded ships would be lighter as well. "Welding can be done by women as well as men," the *American Marine Engineer* acclaimed, seemingly straight-faced, after reviewing Caldwell's report. And, perhaps reflecting the terrible losses of British soldiers on the Western Front, "Even one-armed men can do welding."[48]

The US Navy ordered the construction of a fleet of 110 cargo ships and 12 troop transports to move war materiel and men to Europe. Electric welding, supplementing riveting, allowed for mass production, and though none of the ships launched until after the end of the war, the project advanced shipbuilding technology.

Shipbuilding stalled postwar—the Great Depression came early to the

Electric welding allowed mass production of American ships.

shipbuilding industry—but welding advanced, finding a major new application in pipeline construction. In 1925 the Magnolia Petroleum Company of Galveston, Texas, rebuilt a leaky two-hundred-mile bolted natural-gas pipeline with acetylene lap-welded pipe. After five more years of development—other companies followed Magnolia—electric welding replaced acetylene, eliminating overlapping, using less pipe, and cutting welding time in half. Alloy steels were also important to pipeline improvement, as were improved ditching machines and gas compressors. By 1931, pipeline workers were laying the first thousand-mile natural-gas pipeline from the Texas Panhandle to Chicago."[49]

Natural gas had advantages over town gas: it had double the energy content; it burned cleaner; and since it was drilled rather than manufactured, it was far cheaper—in 1930, about three times less expensive per million Btus,* meaning it promised greater profits to suppliers. Its disadvantages included the cost of pipelines necessary to move it from gas fields to customers (typically urban residential users) and uncertainty about the volume of reserves in gas fields—the problem that had forced Pittsburgh to abandon natural gas for a return to coal.[50] Adjusting tens of thousands of home appliances to handle natural gas's higher heat content required a large investment in service workers. Gas

*Btu: British thermal unit: the amount of heat necessary to raise the temperature of one pound of water by 1°F.

companies recouped some of their investment by training their service workers to sell homemakers additional appliances.

Pipeline technology advanced with increased demand, although collusion among gas companies in the wildcat days of the 1920s and 1930s complicated distribution. Reserve estimates yielded slowly to improved strategies, from drilling test wells in an expanding circle around a producing well to find the edges of a gas field, to estimating gas volume based on geological models. Major discoveries—the Panhandle Field in 1918 in North Texas, the Hugoton Field in 1922 around the conjunction of Kansas, Oklahoma, and Texas—eased early concerns about premature depletion. Panhandle and Hugoton together accounted for about 16 percent of total twentieth-century US natural-gas reserves, some 117 trillion cubic feet.[51] No fewer than twenty-nine larger American cities converted to natural gas between 1927 and 1936: from San Diego, Los Angeles, and San Francisco, to Phoenix, Denver, and Omaha, to Detroit, Memphis, and Atlanta, and eastward as far as Richmond, Pittsburgh, and Buffalo. Chicago, where financial interests were heavily invested in town gas, transitioned with mixed gas: town gas boosted to higher heat content with natural gas. By 1940, the national network of gas pipelines, though far from complete, spidered from Texas and Louisiana up through the Middle West and eastward into Pennsylvania.

Despite this expansion, huge volumes of gas went to waste. Wet gas—gas that flowed mixed with petroleum—was routinely vented into the atmosphere or flared off. Gas was often left to vent into the air, sometimes for years, when drillers abandoned dry holes. A 1935 US Federal Trade Commission report to Congress estimated that 20 percent more gas was *wasted* nationwide between 1919 and 1930 than was consumed: 4,375 billion cubic feet wasted compared with 3,520 bcf consumed.[52] A lack of pipelines to deliver the gas to markets accounted for part of the waste, but "Texas oil drillers," historian Christopher Castaneda concludes, "concerned only about 'black gold,' continued to vent trillions of cubic feet of 'waste gas' into the atmosphere."[53] Though the question of global warming had not yet emerged to public perception, natural gas—methane—is about thirty times more effective than carbon dioxide as a greenhouse gas. No one has calculated how much the vast waste of natural gas across the decades of the twentieth century—in the United States and

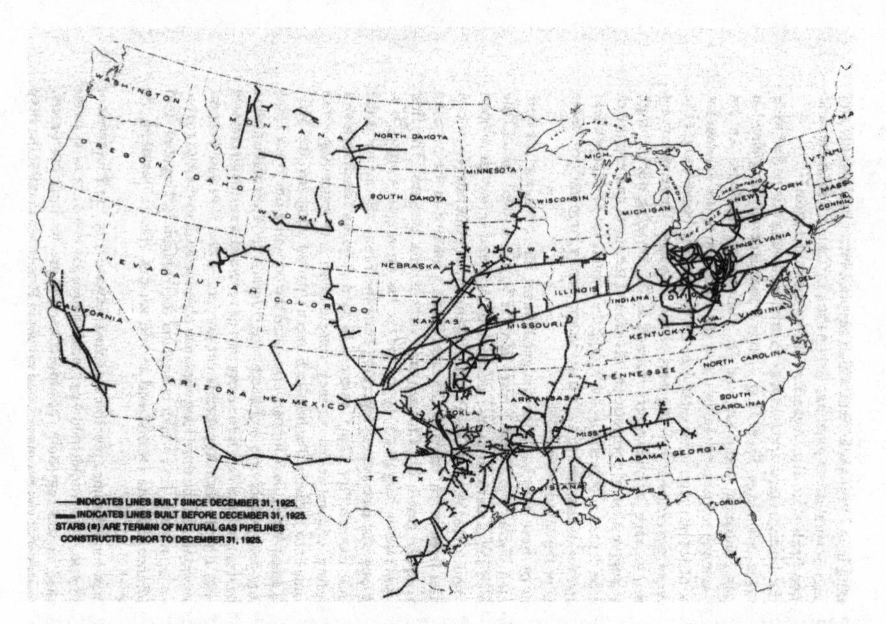

By 1940, pipelines tapping gas fields in Texas and Louisiana extended up through the Middle West and eastward into Ohio, Pennsylvania, and the Southeast.

throughout the world—contributed to global warming. The percentage was certainly more than zero.

Left out of 1930s natural-gas distribution was the Northeast, where town gas and coal continued to predominate. In the nineteenth century, New England, lacking other resources, famously supported itself selling granite and ice; its foundation of igneous and metamorphic rock wasn't conducive to oil and gas formation. As late as 2001, Pennsylvania and New York together had proven natural-gas reserves of only 2,093 bcf compared with the US total of 183,460 bcf, ranking them fifteenth and twenty-second respectively among states nationwide. The rest of New England and the Mid-Atlantic states—Connecticut, Maine, Massachusetts, New Hampshire, New Jersey, Rhode Island, and Vermont as well as Delaware and Maryland—produced little or no natural gas.

A different challenge emerged during World War II, one that would fortuitously remedy the problem of the Northeast's lack of natural gas. The war began in Europe with the German invasion of Poland on 1 September 1939. At that time, the United States accounted for more than 60 percent of total world oil production, with a surplus capacity of more than one million barrels a day.[54] That surplus would allow the United States to fuel its allies throughout the war. Even during the period of official American neutrality, from September 1939 until the Japanese attack on Pearl Harbor on 7 December 1941, US petroleum shipments supported the allied defense.

A fleet of oil tankers carried oil from the Gulf Coast around Florida and up the Eastern Seaboard to supply Eastern cities and for transshipment to England and Europe. With the German declaration of war in alliance with Japan on 11 December 1941, Germany sent a small submarine force under Admiral and U-boat Commander Karl Dönitz to attack the vulnerable tankers. Dönitz had asked for twelve submarines. Hitler, giving priority at that time to Mediterranean support of his campaign in North Africa, awarded the admiral only five. Dönitz chose the best crews, and in the six weeks between 11 January and 28 February 1942, his U-boats working the American East Coast attacked no fewer than seventy-four tankers, sinking forty-six of them and damaging sixteen more.[55] The submarines escaped unscathed. "Our U-boats are operating close inshore along the coast of the United States of America," Dönitz reported, "so that bathers and sometimes entire coastal cities are witness to the drama of war, whose visual climaxes are constituted by the red glorioles of blazing tankers."[56]

From a high of 1.4 million barrels delivered daily from the Gulf Coast to the Northeast in spring 1941, deliveries would drop to just 100,000 barrels a day in less than two years.[57] At a meeting in March 1942, the Petroleum Industry War Council Committee told the navy that if oil tanker attrition continued at such a high rate, it would be impossible to supply oil for the war effort past the end of the year.[58] Adolphus Andrews, the navy rear admiral who led the defense of the Atlantic coast, warned the secretary of the navy a month later that "the sinking of ships, tankers especially, on the coast is a serious matter resulting, if continued, in dire consequences to our war effort." Asking for more ships to take on the U-boats, Andrews emphasized the urgency of his request: "If such forces are not supplied in the near future, it is recommended that *consideration be given to the stoppage of tanker sailings* until escort vessels become available."[59]

The navy lacked sufficient armed escort ships to sustain a full-scale escort strategy. The admiral proposed instituting a temporary coastal convoy system. This Bucket Brigade, as he named it, escorted tankers around danger points such as North Carolina's Cape Hatteras and barricaded them in safe anchorages at night.[60] Dönitz increased the pace of his attacks, but the Bucket Brigade worked. Losses declined. Dönitz then moved his killer submarines into the Caribbean and continued destroying tankers. With the development of antisubmarine frigates and ship and airborne radar later in the war, the German submarine menace retreated. In the meantime, the United States went to work on a more effective protection for its northeastern oil deliveries: land pipelines, the largest and longest yet built anywhere in the world.

Pipelines to carry petroleum had been made larger and longer since the first two- and three-inch lines moved Pennsylvania Oil Creek crude from wellheads to the railroad in 1863. By the beginning of the twentieth century, pipe size had been standardized at eight inches because larger pipes tended to split at the seams. But an eight-inch line could deliver only about twenty thousand barrels a day, while by 1930, a major refinery could process up to 125,000 barrels a day. The initial solution to this bottleneck was "looping": laying down a second pipeline alongside the first.[61] In the 1930s, advances in steel technology allowed manufacturers to draw seamless steel pipes larger than twelve inches in diameter, which the oil industry called "big-inch" pipes. Lowered demand during the Depression didn't justify much construction, however. Only about ten thousand miles of pipelines were laid in the United States during that decade.[62]

To foil the German submarine menace, a consortium of eleven private oil companies called War Emergency Pipelines incorporated on 25 June 1942 to build a government pipeline to carry petroleum from East Texas to refineries in the Northeast. Construction began the next day.[63] The Big Inch pipeline would be twenty-four inches in diameter and 1,254 miles long, with pumping stations every 50 miles, capable of moving up to 335,000 barrels of crude daily, the largest and longest pipeline ever built up to that time.

Before welding, workers pulled a man lying on a cleaning pad through each pipe section while he wiped down the inner walls with hand rags, much as naked chimney sweeps in Georgian England wearing floppy felt hats had cleaned chimneys. After pumping a fifty-mile-long slug of water through the pipe to test it for leaks, the operators moved crude oil through the first section of the Big Inch on the last day of December 1942.

Big Inch twenty-four-inch pipe at a railroad crossing, waiting to be welded together, tarred, wrapped in protective paper, and laid in a 1,254-mile-long cross-country trench.

A second, generally parallel twenty-inch line, the Little Big Inch, had been proposed along with its big brother at the initial 1942 meeting of the Petroleum Industry War Council Committee. It would carry gasoline, kerosene, diesel fuel, and heating oil. Work began on the Little Big Inch, which would share the Big Inch's pumping stations, in February 1943. "Men dug a ditch four feet deep and three feet wide," a project history reports of the two lines, "and laid pipe over the Allegheny mountain range, through swamps and forests, under 30 rivers and 200 creeks and lakes, beneath streets, railroad right-of-ways, and through backyards, often during severe weather conditions. Total excavation was more than 3,140,000 cubic yards of earth,* and the whole job had to be done faster than anyone had ever laid a pipeline before." A trench blasted across the Mississippi River bottom allowed them to lay pipe there.[64] Tidal marshes in eastern New Jersey had to be filled with earth to create a raised bed, though

*Only about 10 percent less than the volume of the Great Pyramid of Giza.

pipeline construction never required a floating track like the one Robert Stephenson had laid across Chat Moss.

Fifteen days short of a year after construction began on the Big Inch line, on 19 July 1943, a ceremony near Phoenixville, Pennsylvania, celebrated the pioneering pipeline's final weld. "Oil had already begun flowing into the eastern extension five days earlier," the project history reports. "The line was filled at the rate of 100,000 barrels per day, moving eastward at 40 miles per day. It required 2.6 million barrels of oil to fill the eastern extension and a total of five million barrels to fill the entire line between Texas and the East Coast."[65] Gasoline began moving through the Little Big Inch on 26 January 1944, behind a slug of water injected for hydrostatic testing. (Solid rubber balls slightly smaller in diameter than the inside diameter of the pipe kept separate the batches of different liquids the Little Big Inch was designed to handle.[66]) The head of the gasoline stream arrived in New Jersey thirty-six days later, on 2 March 1944. A total of 185 million barrels of oil and product passed through the two pipelines in their first year of operation.[67]

When World War II ended with the defeat of the Japanese Empire in August 1945, billions of dollars' worth of military and industrial equipment became war surplus. The Big Inch and Little Big Inch pipelines were shut down and placed on standby. An intense debate followed among government, oil industry, and labor organizations about what to do with them. In 1944 an oil executive named Sidney A. Swensrud, an Iowan and a 1927 Harvard Business School graduate then working as an assistant to the president of the Standard Oil Company of Ohio, proposed an elegant answer: convert them to carry natural gas from the plentiful and underused Texas gas fields to the Northeast, barren of natural gas and still dependent on expensive coal-derived town gas for residential and industrial use.

"With a large unserved market at one end," Swensrud told a wartime meeting of petroleum engineers in New York City, "a large supply [of gas] at the other, and a potentially idle combination of lines in between, the possibility of using the . . . lines after the war obviously seems worth considering."[68] There were some fifteen million people who might be served, he said. They used town gas for cooking and hot water heating. But few heated their homes with gas. A half million existing houses in the region might convert to natural gas,

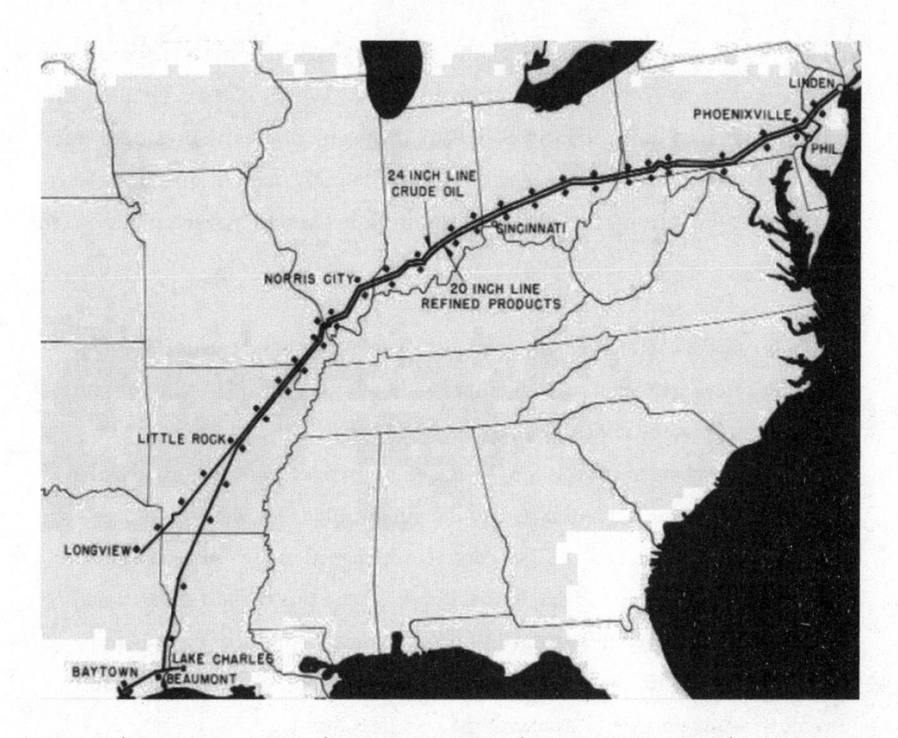

Big Inch (from Longview, Texas) and Little Big Inch (from Beaumont, Texas) oil pipelines, extending more than 1,200 miles across the United States. After the war, they would be converted to carry natural gas to the gas-starved Northeast. Black dots represent pumping stations.

Swensrud estimated. There would be a great expansion of housing when the war veterans came home, married, and settled down. A majority of new housing could be equipped for natural gas if it were made available at a reasonable rate. Commercial customers liked natural gas as well: "stores, offices, bakeries, hotels, restaurants, etc. . . . in the aggregate constitute a very substantial market."[69] And adjusting for natural gas's superior heat value, the gas delivered to the Northeast would cost less than the bare cost of merely manufacturing town gas.[70]

The most vociferous objection to converting the Inch lines to natural gas came from the United Mine Workers, America's coal miners, led by their pugnacious and effective leader, John L. Lewis. Born in Iowa in 1880, the son of a Welsh coal miner, with a great leonine head and a gift for fiery rhetoric, Lewis had led his six hundred thousand United Mine Workers out on strike as soon as wage and price controls lifted after the end of World War II. That venture had not endeared him to Harry S. Truman, who had become president upon Franklin Roosevelt's death in April 1945.[71] Truman made a note for the record, preserved among his papers, describing his version of what happened next:

> Lewis called a coal strike in the spring of 1946. For no good reason. He called it after agreeing to carry on negotiations without calling it. . . . He called one on the old gag that the miners do not work when they have no contract.
>
> After prolonged negotiation, I decided to exercise the powers under the second war power act and take over the mines. After they were taken over, a contract was negotiated. . . . The contract was signed in my office on the 5th of May, and Mr. Lewis stated for the movies* that it was his best contract and would not be broken during the time of Government control of the mines. . . .
>
> But Mr. Lewis wanted to be sure that the president would be in the most embarrassing position possible for the Congressional elections on Nov. 6. So he served a notice on the first day of November that he would consider his contract at an end on a certain date. Which was, in effect, calling a strike on that date. . . .
>
> The strike . . . lasted seventeen days, and then Mr. Lewis decided for the first time in his life that he had "overreached himself."
>
> [By then], it was a fight to the finish, by every legal means available, and in the end to open the mines by force if that became necessary. Mr. Lewis was hauled in to Federal Court, fined no mean sum for contempt. Action was started to enforce the contract, and I had prepared an address to the country. . . . [But] Mr. Lewis folded up. . . . He is, as all bullies are, as yellow as a dog-pound pup. . . . I had a fully loyal team, and that team whipped a damned traitor.[72]

(Lewis was further compromised when the *Washington* Post reported that he had converted his Springfield, Illinois, residence from coal to natural gas in 1945.[73])

*The newsreels, television's predecessors.

Lewis's November 1946 call for another strike energized those who had been promoting converting the Inch lines to natural gas, which replaced coal wherever it was introduced. The US House of Representatives held hearings into the lines' disposition. The Tennessee Gas Transmission Company temporarily leased the lines to deliver gas to Appalachia, a region where a winter energy crisis was brooding. And with a winning bid tendered on 8 February 1947, Texas Eastern Transmission Corporation bought the Big Inch and Little Big Inch pipelines for transmission of natural gas to the northeastern United States for $143.1 million, only $2.5 million less than the two lines' original construction costs. They continue to operate today.

By 1950, then, the three primary fossil fuels—coal, petroleum, and natural gas—all fed the large energy needs of the United States and, in various portion, the rest of the developed world. If coal share was declining worldwide, petroleum was approaching dominance, and natural gas had only begun to penetrate the world market. In those immediate postwar years as well, an entirely new source of energy, nuclear fission, the first potentially major energy source not derived directly or indirectly from sunlight, languished behind walls of secrecy, released as yet only to military use.

SEVENTEEN

FULL POWER IN FIFTY-SEVEN

On the cold winter afternoon of 2 December 1942, in a disused doubles squash court under the stands of the University of Chicago football stadium, the Nobel laureate physicist Enrico Fermi, a refugee from Fascist Italy, calmly initiated the world's first controlled nuclear-fission chain reaction. Other than hand-operated cadmium control rods, nothing visibly moved in the garage-sized graphite and natural uranium assembly Fermi and his crew had stacked up by hand over the preceding two months. (Fermi called the assembly a "pile" in amused reference to its stacked arrangement.) The reactor required no radiation shielding. The energy it produced by splitting—"fissioning"—uranium atoms, held to a mere 200 watts, was not even enough to warm the unheated court.[1] Yet the experiment was transformative, presaging both nuclear power and atomic bombs.

A nuclear reactor requires two basic materials: a fissionable element such as uranium and a moderator. The function of a moderator, as its name implies, is to slow down the neutrons that come bursting out of a uranium atom when it fissions, increasing their chance of encountering and penetrating another atom of uranium and causing another fission. For CP-1—Chicago Pile No. 1—the moderator was graphite. For most of today's power reactors, the moderator is

Assembling Chicago Pile No. 1 (CP-1) of graphite blocks and slugs of natural uranium. Note slug standing on end to right of the worker kneeling on top of pile.

CP-1 completed within wooden frame. The uranium-graphite pile first went critical in a doubles squash court at the University of Chicago on 2 December 1942.

water. Moderators slow down neutrons by giving them a target—for graphite, the nucleus of a carbon atom—to bounce off of repeatedly, losing energy with each bounce, much as billiard balls do.

Uranium atoms are unstable. They're as wobbly as water-filled balloons. The force that holds them together, called the strong force, is almost completely counterbalanced by the force that tries to push them apart: the positive electrical charge of the 92 protons that make up their nuclei. As the rule goes, opposite charges attract, but like charges repel. Uranium is the last natural element in the periodic table, element 92, because of this inherent instability. The elements beyond uranium—neptunium (93), plutonium (94) and so on—are all manmade by bombarding natural elements with neutrons or with other atomic particles.

Natural uranium is a mixture of two variant physical forms, called isotopes*: uranium 238, with 92 protons and 146 neutrons in its nucleus (92 + 146 = 238), and uranium 235, also with 92 protons but with only 143 neutrons in its nucleus (92 + 143 = 235). The isotope that both fissions and chain-reacts, releasing energy, is U235. But most of natural uranium is U238; U235 is only about 1 part in 140, or .07 percent—seven-tenths of 1 percent. Even more troublesome, the two isotopes are chemically identical, which means they can't be separated using chemical means. They're different physically: U238, with three more neutrons, is slightly heavier. So they can be laboriously separated by taking advantage of their slight difference in mass. Uranium isotope separation, also called enrichment, requires industrial-scale facilities such as factories full of centrifuges. There were no such factories in 1942. As Henry Ford had to build his first automobile in a world without automobile parts, so did Fermi have to work with natural uranium, somehow teasing a chain reaction from its scant content of U235.

The difference in nuclear structure between U238 and U235 makes a crucial difference in performance: U238 is insensitive to slow neutrons but absorbs fast neutrons and transmutes to a heavier element, neptunium, which in turn transmutes to plutonium.† If the goal is to fission uranium to release energy,

*Three, actually, but only a trace of the third isotope, U234, is present in natural uranium.

†Transmutation, the alchemists' dream—converting one chemical element to another—occurs in this case when a U238 nucleus captures a neutron.

U238 is a poison, soaking up neutrons with no energy release. The energy in a reactor comes from U235, which neutrons of any energy can fission. Fermi's challenge was to find a way to work around the propensity of U238 to absorb the neutrons he needed to produce a fission chain reaction in U235.

Fermi did that by taking advantage of U238's insensitivity to slow neutrons. He slowed down the fast neutrons the fissioning uranium produced by embedding the uranium in a matrix of graphite. Escaping the embedded slugs of uranium, the fission neutrons bounced against atoms of the surrounding graphite. They lost energy with each collision until they were slow enough to slip beneath the U238 absorption threshold, encounter a fresh U235 atom, and fission it in turn.

When an atom of U235 fissions, it releases energy—heat—as well as two or more secondary neutrons. The secondary neutrons, if moderated in turn, can go on to fission other U235 atoms in natural uranium in an exponential chain: one releases two, two release four, four release eight, eight release sixteen, sixteen release thirty-two—doubling every few millionths of a second until the energy release from the atoms they fission is substantial enough to heat water to steam, which can then drive a turbine to spin a generator to generate electricity.

Fermi used graphite rather than water to moderate CP-1 because graphite absorbs fewer neutrons than water. Natural uranium contains too little U235 to sustain a chain reaction with water moderation. To use water to moderate a reactor, the uranium fuel has to be enriched from .07 percent U235 to at least 3 percent U235. Uranium enrichment, which the United States's Manhattan Project pioneered during World War II, was first used to make nearly pure U235 for the first atomic weapon, the bomb exploded over Hiroshima, Japan, on 6 August 1945. After the war, when the navy developed the first power reactors for nuclear submarines, it used uranium fuel enriched to a level of U235 sufficient to sustain a chain reaction with water moderation.

The laboratory at the University of Chicago where Fermi built CP-1 worked during the early years of World War II to develop the technology of breeding plutonium. An even better bomb fuel than uranium, plutonium was bred in large graphite-moderated, water-cooled production reactors constructed along the Columbia River in eastern Washington State, extracted

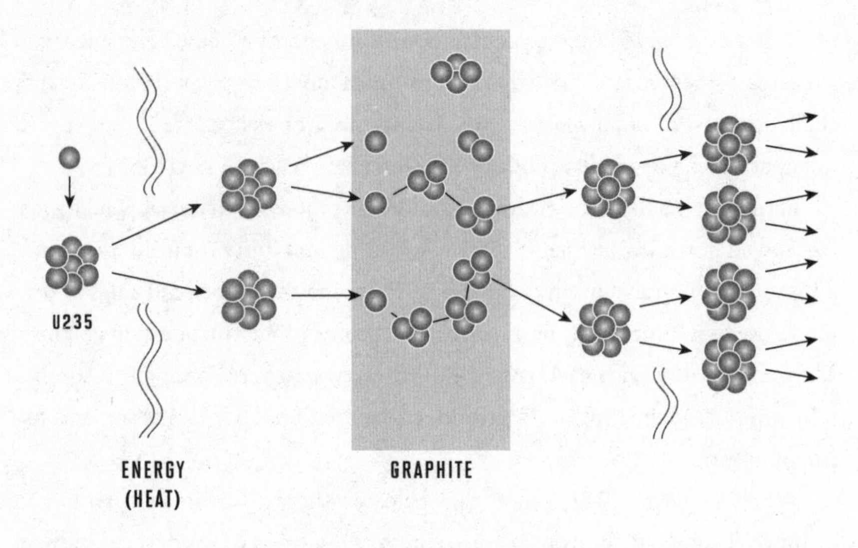

Moderated uranium fission chain reaction: left to right, a neutron enters an atom of U235, causing it to fission (split). This releases energy and two or more secondary neutrons, which slow down bouncing off carbon atoms in graphite moderator and then go on to enter other U235 atoms in a chain-reacting exponential cascade.

chemically from its uranium matrix by remote control, purified and shipped to the bomb laboratory at Los Alamos, New Mexico, to be fashioned into the Nagasaki bomb.

Chicago completed development work by mid-1944, after which a committee of Fermi and his colleagues had time to consider the future of nuclear energy for power. The result was a secret sixty-seven-page report, *Prospectus on Nucleonics.* (*Nucleonics* was a neologism the committee coined as a name for the new field of nuclear technology, by analogy with the word *electronics*—electronics concerning electrons; nucleonics concerning nuclei.)

The *Prospectus* reviews the prewar and wartime history of nucleonics—the early discoveries in physics, that is, and the work of the Manhattan Project—then turns in its final sections to the question of the postwar development of nuclear power. A key issue it raises is the amount of extractable uranium in the world. "It has been estimated," the *Prospectus* reports, "that the total quantity of uranium available in high grade ores on earth is about 20,000 tons, of which about 10,000 tons are on this continent. This latter amount will enable 5 piles of the larger type . . . to run for 75 years."[2] The next paragraph qualifies this estimate, noting that it "is based on the assumption that plutonium [bred in the reactor in the course of its operation from the U238 in natural uranium] is regularly removed and utilized for other purposes."

Those "other purposes" could only be atomic bombs, and the *Prospectus* notes that about one pound of plutonium is bred in an ordinary graphite-moderated reactor for each pound of U235 consumed. If the plutonium were left in the reactor and allowed to fission, the *Prospectus* continues—not likely so long as America remained in the atomic bomb business—then the supply of reactor fuel would last 140 times as long—five reactors for 10,500 years, that is, or, if the plutonium were extracted and used to fuel additional reactors, some multiple of five reactors for a shorter span of time.

Even so, the possibility of large-scale application of nuclear power looked dim: counting only proven reserves, the *Prospectus* estimates, "the energy available in 10,000 tons of uranium will not be sufficient to permit this material to replace coal and other combustible materials, or falling water, as an energy source."[3] The *Prospectus* estimates that US coal consumption at the time amounted to about a billion tons per year. And since uranium fission was at least a million times as energetic as coal burning, ten thousand tons of uranium—seemingly the entire US supply known to the scientists as of 1944—would be equivalent to ten billion tons of coal, meaning, at most, ten years of coal-equivalent atomic energy.

Qualifying that gloomy prediction, the *Prospectus* judges it "quite likely" that methods would be found to extract uranium from lower-grade ores. It mentions thorium as an alternative and much more abundant reactor fuel. (It still is, having not yet been used commercially as reactor fuel.) The report even speculates about controlled thermonuclear fusion of hydrogen for power as a development later on, "thus making available the water of all the oceans as a fuel or power source." But it thought that possibility lay "in an as yet dimly perceived (but by no means fantastic or necessarily remote) future." (More than eighty years later, it still does. Controlled thermonuclear fusion—harnessing the fusion of light elements rather than the fission of heavy elements—is a hard problem, since it requires generating temperatures in the millions of degrees and confining them within a closed vessel. As of 2018, no experimental fusion reactor had yet advanced beyond breakeven—meaning produced any net energy beyond the energy supplied to run the reactor.)

The *Prospectus on Nucleonics* isn't footnoted. Nowhere does it explain who provided its low and inaccurate estimate of the amount of uranium ore on earth. Even as it was being written, General Leslie R. Groves, who ran the Manhattan Project, was directing a secret organization code-named the Murray Hill Area Project (continuing the Manhattan theme) to identify all the world's proven reserves of uranium ore and buy up the mineral rights. The investment would extend to more than fifty countries. The Manhattan Project itself had drawn heavily on two sources for its uranium: the Shinkolobwe mine in what was then the Belgian Congo in central Africa, and the Eldorado mine on Great Bear Lake in northwestern Canada. But Groves was well aware that there was uranium on the Colorado Plateau, a component of the vanadium ores being mined there, because he used some sixty-four thousand pounds of uranium oxide from Arizona during the war as part of the feedstock for the first atomic bombs.[4]

Since the scientists at the US government's national laboratories at Oak Ridge, Tennessee, and outside Chicago in the Argonne National Forest believed mineable uranium to be scarce, they assumed the only sustainable route to commercial nuclear power was via breeder reactors. The breeder in its first conception consisted of a power reactor with a "blanket" of natural uranium wrapped around it to absorb excess fission neutrons to breed plutonium, which could then be separated chemically and used as fuel. This additional function—breeding plutonium as well as generating power—complicated research, design, and development, and necessarily postponed the time when nuclear energy might become available for commercial electricity generation. As one director of the US Geologic Survey explained, developing the breeder would require "an R&D program involving such an enormous outlay of public capital that it would be unwise to make the investment until absolutely necessary."[5] After the end of the war, through the late 1940s, Oak Ridge and Argonne focused on basic research in nuclear technology rather than reactor development.

When the Soviet Union tested its first atomic bomb in August 1949, however, the three-year-old US Atomic Energy Commission responded with panic.

Besides chasing an ill-advised program to build a hydrogen bomb (which no one yet knew how to do, or even if it could be done), the AEC undertook to identify and develop domestic sources of uranium in case foreign sources were somehow cut off.

To that end, in March 1951, the AEC more than doubled its previous offering price for high-grade uranium ore and offered as well a $10,000 bonus ($100,000 today) should the prospector identify a productive new mine. (Uranium as yellow carnotite or gray-black uraninite tends to occur in lens-shaped formations in hundred-million-year-old streambeds because it's easily dissolved in sulfated water. Small deposits often turn up incorporated into or encrusted onto trunks of petrified wood. A prospector whom journalist Tom Zoellner interviewed about the 1950s uranium rush the AEC spurred told him, "I saw a man once find a whole tree that was just high-grade uranium. It came off like black powder, really soft, like pepper. It was a [petrified] tree about two feet in thickness, and the branches went off maybe ten feet in each direction. He blasted that thing out of there and got sixty-five hundred dollars. When the tree was depleted, that was the end of his uranium ore. There was no more."[6]

Prospectors, some two thousand in all, came running from every direction at the AEC's enticing call. Magazines—*Life, Popular Mechanics, National Geographic*—featured the uranium rush on their covers. Geiger counter sales boomed. A few miners got rich. Most earned a little, or nothing.

By the mid-1950s the AEC had as much uranium ore as it needed for bombs, from South Africa as well as the American West. Finding more had turned out to be a matter of pricing. "Tens of millions of tons come into prospect in the price range of $30–$100 per pound," a Geologic Survey geologist wrote in 1972.[7] The AEC continued to reward prospectors for another ten years to accumulate reserves for the anticipated development of commercial nuclear power.

One source of uranium left unexploited was coal. Lignite or "brown coal," which contains about 60 percent to 70 percent carbon, has a chemical affinity for uranium, a fact first noted by the Swiss American mining engineer Edward L. Berthoud in 1875.[8] "Peat, lignite, and subbituminous coal," two scientists reported in 1954, "can extract more than 98 percent of the uranium in a liquid solution of uranium sulfate." That was why ancient trees decaying in streambeds accumulated uranium.[9] There are extensive beds of uranium-bearing lignite

1950s uranium mining on the Colorado Plateau made the Geiger counter a familiar instrument. The US Atomic Energy Commission offered the reward.

coal in the Dakotas, although the uranium content is low grade. One suggestion for refining it has been to burn the lignite in coal power plants to generate electricity and then extract the uranium from the fly ash, where it concentrates. In 2007 China began exploring such extraction, drawing on a pile of some 5.3 million metric tons of lignite fly ash at Xiaolongtang in Yunnan Province. The Chinese ash averages about 0.4 pounds of U_3O_8 per metric ton. Central Europe and South Africa are also exploring uranium extraction from coal fly ash.[10] Coal, with its ubiquitous content of uranium and thorium, releases more radioactivity into the environment when it is burned than any other fuel.

US Navy engineering officer Hyman Rickover's relentless pursuit of a submarine power plant opened the way to the development of nuclear power. Rickover was a small, quick, intelligent first-generation Jewish immigrant from the Russian Pale of Settlement who escaped the czar's pogroms with his parents in 1908, when he was eight years old. His father, a tailor, settled the family in Chicago, where Rickover grew up. Working as a Western Union telegraph delivery boy connected him with Congressman Adolph Sabath, a fellow Jewish immigrant, who nominated him for appointment to the US Naval Academy when he was eighteen. Rickover graduated in 1922, took up duty on a new destroyer, and within a year had won appointment as the ship's engineering officer, the youngest in the squadron.

After five years of sea duty, Rickover qualified for further education. He used the opportunity to earn a master's degree in electrical engineering at the Naval Postgraduate School and to study at Columbia University, after which he looked around and decided that the navy's submarine service offered the fastest route to advancement. It turned out not to be. An intense workaholic with a Chicago accent and a nasal, high-pitched voice, Rickover didn't look or act like a standard-issue naval officer, nor did his Jewish heritage improve his prospects in what was then an anti-Semitic branch of service. (Years later, choosing officers for his enlarging fleet of nuclear submarines, he deliberately picked tall, rugged, Nordic-looking men as potential commanders, provided that they were smart and ambitious as well.[11]) He forged ahead anyway, serving for three years as engineering officer on an outdated submarine, the USS *S-48,* learning enough to redesign and rebuild its cantankerous electric propulsion motors.

Through the rest of the 1930s and the years of World War II, Rickover worked at engineering duty on ships and in shipyards. His one command, possibly a cruel joke, was what one of his protégés, Theodore Rockwell, calls "a wretched rust-bucket, an aging minesweeper named the *Finch.*"[12] With typical brio, Rickover drove his crew to renovate from bilge to binnacle. From the *Finch,* in 1937, he applied for engineering-duty-only status—EDO—received it, and spent the war commanding the electrical section of the Bureau of Ships. He oversaw the development of an infrared communications system for nighttime talk ship-to-ship, replacing the visible lights that made American ships targets for enemy bombs and torpedoes. Late in the war, he reorganized the vast navy supply depot at Mechanicsburg, Pennsylvania, reducing delivery time for spare parts from months to days.

Rickover was one of five naval officers assigned to Oak Ridge in 1946 to learn about atomic energy and nuclear power. Since he outranked the other four, he took command and directed their study and collective research. Navy scientists had investigated nuclear propulsion for ships during the war, but the technology had not yet won serious commitment. With the Manhattan Project itself an army operation, the navy brass worried more about army domination of the new atomic weapons than about nuclear energy for power.

Rickover, after studying the subject, concluded that the ultimate naval fighting platform would be a nuclear-powered submarine, its boiler a nuclear reactor that required no oxygen, capable of sailing underwater for weeks on end, quiet and immensely lethal. As atomic weapons had finally given the US Air Force the bombs it needed to fulfill its fantasy of winning wars from the air, so would atomic torpedoes—and, more to the point, nuclear missiles, when those came along—give the navy the ultimate power of nuclear deterrence.

The navy didn't know that yet, but Rickover did. Across the next three years, against immense resistance, he applied what he called his "orthodontic approach"—the quiet, unrelenting pressure that moves teeth—to the challenge of moving the navy to commit to building nuclear submarines. "To many in high places, however," writes a later member of Rickover's team, "the proposal sounded like a trip to the Moon."[13] A key strategy was arranging his dual appointment to the Atomic Energy Commission as well as to the navy, which allowed him to fight resistant bureaucrats from two corners at once. "If the Navy doesn't like what we're doing," Rockwell quotes him as saying, "we'll do it with our AEC hat."[14] He needed AEC authority anyway, since the commission alone was authorized to sign contracts involving atomic materials and atomic secrets.

Along the way, Rickover made the historic decision to moderate his submarine and large-ship reactors with water rather than a less familiar but more efficient coolant such as liquid sodium. Sodium, he argued, had its disadvantages as well as its advantages. It requires heavy shielding because it gives off dangerous high-energy gamma rays on exposure to radiation. Complicating maintenance, it burns in air, explodes in water, and needs a week after being exposed in a reactor to shed enough radiation to allow access for maintenance. Crucially, ordinary water was a conservative medium familiar to naval engineers, long used in steam boilers and for heat transfer. Rickover had enough troubles building the first power reactor and somehow packing it into the hull of a submarine without pioneering exotic moderators as well. But the choice, locking in a technology, would reverberate through the years. Other countries made other choices: heavy water, helium, sodium, lead, or, as at Chernobyl, graphite moderation with water cooling.

The Atomic Energy Act of 1946 made atomic energy in all its manifestations a monopoly of the US government. All discoveries concerning atomic energy were "born" secret—treated as secret until declassified—and the penalty for divulging atomic secrets was life imprisonment or death. All fissionable materials became the property of the US government, as beached whales once became the property of kings. No one might build or operate a reactor except under government license, nor might such devices be owned privately. Authority over atomic energy was vested in a commission of civilians, the Atomic Energy Commission, responsible to the president—"the most totalitarian governmental commission in the history of the country," one historian has called it.[15]

Proponents of the bill that became the Atomic Energy Act had argued that atomic energy was too important to be left to the military. Apparently, it was also too important to be conveyed to the people. Few outside the Manhattan Project and even fewer in Congress knew much about atomic energy. Almost everyone believed that atomic bombs were original inventions rather than straightforward applications of nuclear physics and explosives engineering.

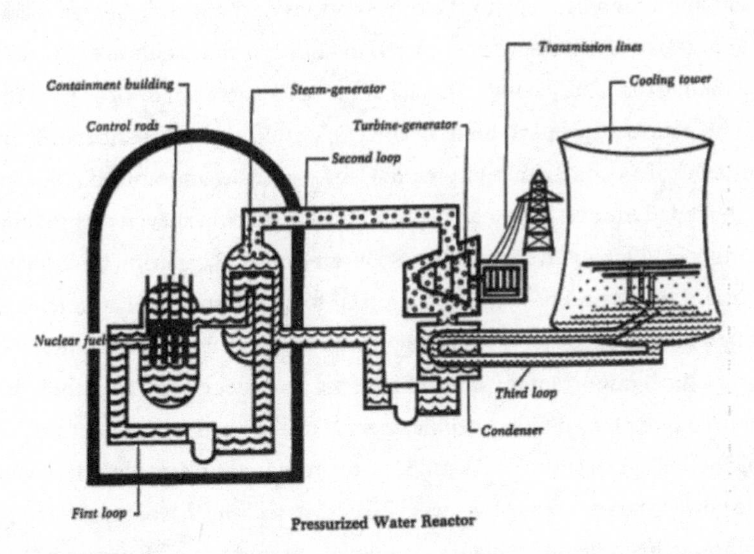

A pressurized-water nuclear power plant. Water carries reactor heat to a heat exchanger to isolate radioactivity within the reactor pressure vessel. From the secondary loop outward, the operation is identical to a conventional power plant. The steam turbine drives a generator to make electricity.

Keep the "secrets" of their design, then, and the United States could maintain a monopoly for years, perhaps for decades, to come.

The Soviet test of an atomic bomb in August 1949 disabused American leaders of such misunderstandings. The AEC answered the Soviet test by expanding its manufacture of atomic bombs well beyond the 170 it held in stockpile in 1949. When Dwight Eisenhower took office as president in January 1953, less than three months after the United States had tested its first hydrogen device, he inherited a nuclear arsenal containing 841 atomic bombs.[16] A believer in balanced budgets, he was determined to reduce government expenditures. The former general would do so by cutting conventional military expenses while enlarging the US stockpile of nuclear weapons, a policy that his secretary of state, John Foster Dulles, would describe as the "deterrent of massive retaliatory power": massive retaliation.[17] In a secret October 1953 National Security Council report, NSC 162/2, Eisenhower had confirmed the policy with a chillingly simple finding: "In the event of hostilities, the United States will consider nuclear weapons to be as available for use as other munitions."[18]

The Cold War that Eisenhower inherited was in part a competition with the Soviet Union for alliance with the other nations of the world. A large increase in the number of US nuclear weapons, in addition to Eisenhower's threat of using them to escalate a conventional conflict and a new policy of sharing them with the nation's European allies in NATO,* could make America look like a warmonger. To counter that impression, the president announced a new program he called Atoms for Peace. He proposed it in part in a speech to the General Assembly of the United Nations on 8 December 1953, offering to contribute "normal uranium and fissionable materials" to an international atomic energy agency and inviting others—meaning the Soviet Union—to match the US effort. The most important responsibility of the new agency, he said, would be to devise methods for applying atomic energy "to the needs of agriculture [and] medicine." A "special purpose" would be "to provide abundant electrical energy in the power-starved areas of the world."[19] ("Our technical experts assured me," Eisenhower, ever the poker player, revealed in his memoirs, "that even if Russia

*The North Atlantic Treaty Organization, a group of US and western European nations allied against the Soviet Union and its eastern European satellites.

agreed to cooperate in such a plan solely for propaganda purposes, the United States could afford to reduce its atomic stockpile by two or three times the amount the Russians might contribute, and still improve our relative position." The United States eventually contributed more than 40,000 kilograms of fuel uranium to the peaceful stockpile, most of which was dispersed to universities around the world in small, inherently safe research reactors.[20])

The First Lady, Mamie Eisenhower, launched the navy's first nuclear submarine, the USS *Nautilus,* on 21 January 1954. A month later, Eisenhower sent a message to Congress proposing amendments to the 1946 Atomic Energy Act that would release nuclear materials and technology for civilian development—and, incidentally, allow the president to fold NATO into an alliance of nuclear defenses.[21]

By then, the utility industry was beginning to worry about commercial competition for nuclear power. It was not yet economical in the United States, but it was already competitive with non-nuclear power in Western Europe and Japan, areas with few native fossil-fuel resources. Canada was building a pressurized heavy-water reactor fueled with natural uranium with a 200-megawatt design capacity, a better configuration for countries without enrichment capabilities. There was a lucrative foreign market opening. "If we are outdistanced by Russia in this race," worried influential senator John O. Pastore of Rhode Island, "it would be catastrophic. If we are outdistanced either by the United Kingdom or France, there would be a tremendous economic tragedy to our commerce."[22] Firms such as Westinghouse and General Electric were eager to compete, but they were barred from doing so by the terms of the Atomic Energy Act.

The Soviets startled the world on 26 June 1954 by announcing that they had begun generating grid-connected electricity with a power reactor, a world first. The 5-megawatt-electric Soviet reactor was operating at V Laboratory Nuclear Center in Obninsk, about sixty miles southwest of Moscow. (It was graphite moderated and water cooled, a configuration with an inherent design flaw of which the Soviet scientists were aware: because water absorbs neutrons, a loss-of-coolant accident—the water draining from the fuel channels or boiling away—would increase the reaction rate. "In the case of five liters of water penetrating the graphite," one of them wrote later, "and its homogeneous distribution in the core, nuclear runaway would occur."[23] They studied the problem at the time and concluded that the water wouldn't disperse through the core if a fuel channel leaked. The graphite-water system continued to be a preferred configuration for Soviet reactors because it made them dual-use, capable of breeding plutonium for weapons while generating power. The reactors at Chernobyl would be graphite-water.)

Heavy infighting in Congress over whether the first US commercial nuclear power plant would be built by the government or by private enterprise complicated making law of Eisenhower's proposals. The AEC went ahead with preparations, drawing on Rickover's expertise. In 1950 the chief of naval operations had proposed that the navy develop a large ship reactor prototype to power an aircraft carrier, which became an on-again, off-again project across the next three years. By July 1953, the AEC had assigned the civilian reactor project to Rickover and his team.[24] The large ship reactor plans would serve as a beginning for Rickover's civilian design.

At the same time, Rickover made a crucial decision to change the form of the fuel from uranium metal to uranium dioxide, a ceramic. "This was a totally different design concept from the naval reactors," writes Theodore Rockwell, "and required the development of an entirely new technology on a crash basis."[25] Rockwell told me that Rickover made the decision, despite the fact that it complicated their work, to reduce the risk of nuclear proliferation: it's straightforward to turn highly enriched uranium metal into a bomb, while uranium dioxide, which has a melting point of 5,189° F (2,865° C) requires technically difficult reprocessing to convert it back into a metal.

By October 1953, the AEC was ready to announce the commissioning of an AEC-owned demonstration power plant of 60 megawatts. It would be built on the Ohio River at Shippingport, Pennsylvania, about thirty-eight miles northwest of Pittsburgh, jointly by Westinghouse and Pittsburgh's Duquesne Light Company. Rickover's Naval Reactor Group would direct the construction, Rickover wearing his reversible AEC hat.

Philip A. Fleger, the chairman of Duquesne's board of directors, told me that the basic reason his company went nuclear was "pollution control."[26] The first US commercial nuclear-power plant, that is, was offered and welcomed as "green" technology. Polluted Pittsburgh had begun urban redevelopment in the late 1940s, instituting strict smoke control. By the time the AEC solicited bids from private industry for the Shippingport project, sulfur-oxide controls were also pending in the Pittsburgh area. Duquesne had been petitioning to build a coal-fired power plant on the Allegheny River, but area residents were resisting. "We encountered a great deal of harassment and delay from objectors," Fleger said. "It began to look as if we wouldn't be able to complete the plant in time to meet the power demands we were facing." The AEC's PWR project was a godsend: no expensive precipitators for smoke control, no expensive scrubbers for sulfur-oxide control, 60 megawatts of peak-load power, and a leg up on nuclear-power technology.

The economics of the project also looked promising. "I realized that at this stage of the art, it would be very difficult for any company to foresee the ultimate cost of the project," Fleger recalled. "But here we could negotiate a contract and know there'd be a ceiling on cost."

Duquesne already owned 271 acres of flatland beside the Ohio River at the village of Shippingport, near the Pennsylvania-Ohio border. The company bought 237 acres more, creating a relatively isolated site. It proposed to build the necessary structures for a nuclear power plant, to install a 100-megawatt turbogenerator to produce electricity from reactor-heated steam, and to put up the equivalent in manpower and services of $5 million toward the reactor's cost. The $5 million was roughly equivalent to the cost of the boiler plant that Duquesne would have had to buy if the power system had been conventional instead of nuclear.

The AEC liked Duquesne's bid. It liked Duquesne's location in the same city as the Westinghouse facility where the reactor would be designed and built. It received ten bids. Duquesne's was by far the best, and the AEC accepted it.

Waving a magic wand—a neutron source—over a transmitter in Denver, where he was recovering from a heart attack, Eisenhower activated a robot bulldozer in Shippingport to turn the first dirt for the new power plant on Labor Day, 6 September 1954. Excavation and building began in earnest the following spring. Work progressed smoothly. "We are able to do our work with few letters and no fuss," Rickover told Congress. "There has never been a single letter written between the Commission and the Duquesne Light Company since the contract was signed with them. It has never been necessary."

A hands-on director, Rickover would arrive at the site from Washington in the evening or late on a Friday afternoon, to keep his managers worrying nights and weekends. "There was a motto down here," Duquesne president Stanley Schaffer remembered, "that some of us who were the doers learned to hate: the admiral's motto, 'Full Power in Fifty-Seven.' " Not everyone loved the admiral, but he got the job done.

The pressure vessel that would hold the reactor's hot, radioactive core, thirty-three feet long and nine feet in diameter, with walls a half foot thick, required two and a half years to fabricate. Westinghouse, Duquesne, the navy, the AEC, and all their several contractors had to coordinate their efforts on and off the site. The uranium-oxide fuel in its first use had to be fabricated and clad with zirconium. Rickover's group had stimulated the creation of a new industry to bring that exotic metal's production up to the new nuclear industry's demands. A by-product of zirconium production was the excellent neutron absorber hafnium, element 72, which would serve for the Shippingport reactor's control rods.

Then there was the midget welder. "In building an atomic plant, you spend a lot of time just looking for weak and leaky joints," the Westinghouse project manager for Shippingport told a journalist. "One day an X-ray revealed a defect inside a bend of fifteen-inch pipe. It was a hard place to get at. We considered dismantling the pipe, but that would have been costly as hell in time and money. Then we learned of a firm in Georgia that hires out midget welders for just such jobs. They sent us one who was just thirty-nine inches tall, and he crawled into the pipe and made a good, solid repair."

The plant needed to be solid. Water pressurized to 2,000 pounds per square inch would be pumped through the reactor core at 45,000 gallons per minute. The core was a hybrid: 115 pounds of bomb-grade U235 metal as "seed" in plates at the center ($1 million worth of U235 at $8,700 per pound) and 12 tons of natural uranium oxide in rods blanketed around.

Rickover sounded testy toward the end, early in 1957, when the congressional committee overseeing the project came out to Pittsburgh to take a look. "I think we have babied a lot of people in this country too long with the glamour of atomic energy," he told the congressmen, "and I think as soon as possible, we have got to get down to do it like any other business." Someone made the mistake of asking him about the new, larger power reactors then being designed. They were supposed to be more efficient.

Rickover sneered. "Any plant you haven't built yet is always more efficient than the one you have built. This is obvious. They are all efficient when you haven't done anything on them, in the talking stage. Then they are all efficient. They are all cheap. They are all easy to build, and none have any problems."

He was candid about Shippingport's problems. Costs had increased by at least 50 percent. People, Rickover said, had the idea that reactors were "much further advanced than they are." Their designers and builders lacked much of the necessary basic technology. The "reactor game" hung "on a much more slender thread than most people are aware. There are a lot of things that can go wrong, and it requires eternal vigilance. All we have to have is one good accident in the United States, and it might set the whole game back for a generation."

Shippingport was completed in good time by any standard less rigorous than Rickover's. "A little over two and a half years," Duquesne's president recalled, "by comparison with today, which may be twelve to fourteen years from the time of a plant's inception. I think it moved very expeditiously."

Fifteen years to the day after Enrico Fermi first operated his Chicago Pile, on 2 December 1957, Shippingport went cold critical, meaning its operators ran the reactor for testing but not for power production. They cut in power on 18 December at 12:39 a.m. By 3:00 a.m., the plant was producing more electricity than the 8 megawatts it consumed. By 7:00 in the morning, it was generating 12 megawatts. It would generate 20 megawatts by that night and 60 within a few days. Budgeted at $47.7 million, it cost $84 million. Another $36 million went to reactor research and development. Shippingport electricity came to 55 to 60 mills per kilowatt hour, although the AEC sold it to Duquesne at 8 mills—eight-tenths of one cent—a figure that stood in for the equivalent charge Duquesne would have paid for conventional fuel.

Democrats in Congress, the *New York Times* reported, "have been urging a government program for building atomic plants," so the AEC's announcement of Shippingport coming online emphasized its service to domestic power. It missed its chance to score a "psychological triumph" to offset "the Soviet satellite achievement," the *Times* observed: *Sputnik,* the first artificial earth satellite, had achieved orbit on 4 October 1957, two months ahead of the Shippingport startup.

The AEC made the best of being outshone. Shippingport, it announced, was "the world's first full-scale atomic electric plant devoted exclusively to peacetime uses." The qualified superlative exempted the Soviet reactor at Obninsk and Britain's 70-megawatt, air-cooled power reactor at Calder Hall in Cumberland, England, both of which made plutonium for weapons as well as domestic electricity.

"Between 1953 and 1957," a historian writes of Eisenhower's atomic legerdemain, "he engaged in the greatest nuclear arms buildup in the history of the world—at precisely the same time that he was promoting Atoms for Peace."[27] From the 841 atomic bombs of 1952, Eisenhower, by the end of his second term in January 1961, had sponsored the production of an arsenal of some 19,000 nuclear weapons totaling 30,000 megatons of explosive force—the equivalent of ten tons of TNT for every person on earth.[28] Behind that martial shield, commercial nuclear power had found a modest beginning.

Shippingport was an experimental reactor intended to test new fuel configurations as well as to generate commercial power. In 1977 it was modified into a light-water breeder by replacing its uranium core with a thorium core and blanket. Thorium bred uranium 233, another fissile isotope, which chain-reacts as U235 and plutonium do. The reactor operated on thorium for five years thereafter, until it was decommissioned on 1 October 1982.[29]

But neither the Shippingport reactor nor even Fermi's CP-1 were the first uranium-fission reactors assembled on earth. Two billion years earlier, in a rich bed of uranium ore in what is now the West African nation of Gabon, the first in a series of natural reactors went critical when water seeping through the ore bed created the conditions for a nuclear slow-neutron chain reaction to occur. It could do so because the uranium in the ore beds of Gabon contained a higher percentage of U235 than uranium in ores today, in Gabon or anywhere else on earth. The element, which is about as common in the earth's crust as tin, is mildly radioactive. It gives off alpha particles, each of which consists of two protons and two neutrons. Ejecting an alpha particle changes a uranium atom. Taking away two protons from a uranium nucleus transforms it from element 92 to element 90, which is thorium. Thorium is also mildly radioactive and decays in its turn. Through a progression of such decays, the uranium on earth is slowly converting itself into lead.

The two main isotopes of natural uranium, U238 and U235, decay at different rates. U238 takes about four and a half billion years for half the atoms to decay to thorium. U235, however, has a half-life of only seven hundred million years. Which means that two billion years ago, there was more U235 in natural uranium than there is today—not 0.7 percent, but about 3.5 percent. That was enough to enable the Gabon uranium ore to sustain a chain reaction with water moderation.

The Gabon reactors were intermittent. As water flowed through the ore bed, it moderated neutrons spontaneously released from the ore's content of U235. Fission began, a chain reaction followed, the energy release heated up the ore bed, the water boiled away, the reaction shut down, the ore bed cooled, water began flowing again, and the cycle repeated, probably on a scale of hours or days. The natural reactors were like the geysers of Yellowstone National Park, which cycle similarly because water seeps into underground magma chambers where the hot magma heats it to steam, which spouts up through natural channels in the overburden and sprays into the air, emptying the underground chambers and shutting down the geysering. Then water seeps into the magma chambers again, and the cycle repeats.

In May 1972 a staff member at the Eurodif uranium-enrichment plant in Pierrelatte, France, first noticed the discrepancy that led to the discovery of the Gabon reactors: analyzing a standard sample prepared from uranium ore, he was surprised to find it slightly depleted in U235. Checking other samples, Eurodif found U235 similarly missing in ore shipments from Gabon going back to 1970. The ore shipments had yielded some 700 tons of natural uranium, so about 200 kilograms of U235 were missing, enough to make at least ten atomic bombs.[30]

At that point, the Commissariat à l'énergie atomique—CEA, the French atomic energy commission—got involved. In the course of the subsequent investigation, a CEA scientist recalled discussions in the mid-1950s, when reactor science first went public through Atoms for Peace, about the possibility of natural reactors forming earlier in the history of the earth. "It is also interesting to extrapolate back two thousand million years," two Chicago scientists had speculated at a 1953 conference, "where the U^{235} abundance was 6 percent [*sic*; 3.5] instead of 0.7. Certainly, such a deposit would be closer to being an

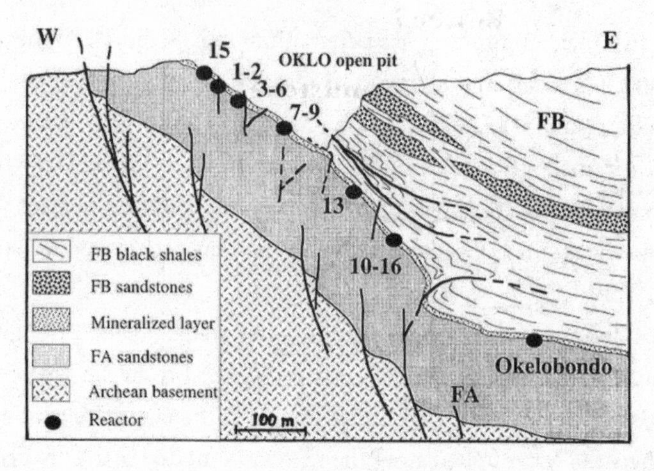

A cross section of the Gabon ore bed showing the location of the natural reactor remains.

operating pile. Certainly, to date we have no evidence for such a thing occurring."[31] A Japanese American nuclear chemist, Paul Kuroda, speculated further in a 1956 journal article that South African uranium ore "plus water were nuclear-physically 'unstable' 2,100 million years ago, and the critical uranium chain reactions could have taken place, if the size of the assemblage was greater than, say, a thickness of a few feet. The effect of such an event could have been a sudden elevation of the temperature, followed by a complete destruction of the critical assemblage."[32] The reactions had taken place, but because they were intermittent, they hadn't destroyed the assembly as Kuroda speculated they might.

In one regard, the natural reactors of Gabon proved to be more than a curiosity. "Without any containment structure whatsoever," writes Theodore Rockwell, the nuclear engineer who built Shippingport, "the plutonium and radioactive fission products created by these natural reactors have stayed immobilized in the soil around the reaction, and have not created any problems for the plants and animals in the area. That's how Nature answered the question, 'What can be done with all that nuclear waste?' "[33]

EIGHTEEN

AFFECTION FROM THE SMOG

America discovered air pollution in Pennsylvania, in a small town on the Monongahela River twenty-eight miles due south of Pittsburgh, on Halloween weekend in October 1948. Before then, smoke had been a nuisance in many cities and towns throughout the world, but in only one reported instance had complex smoke, fog, and air pollutants killed. That previous disaster, when more than sixty people died, had occurred under similar circumstances in Liège, Belgium, a town on the River Meuse, between 1 and 5 December 1930.[1]

In 1948, across the Great Depression and World War II, few outside Belgium remembered the Meuse Valley disaster. The fog that killed twenty people and sickened some six thousand more in Donora, Pennsylvania, in 1948 compelled public attention because it led the United States Public Health Service to begin investigating air pollution.[2]

The Monongahela makes a sharp horseshoe bend around Donora, with bluffs across the river from the town that rise 450 feet to confine it on three sides and hills behind it that rise even higher, so that it sits in a bedpan-shaped trough. In 1948 Donora was a town of 12,300 people, about 3,000 of whom worked in one of the steel or zinc mills that lined the riverbank.

"The river is densely industrialized," the *New Yorker* journalist Berton

Roueché catalogues the setting, noting that Monongahela shipping "exceeds in tonnage that of the Panama Canal." Roueché found "trucking highways along its narrow banks and interurban lines and branches of the Pennsylvania Railroad and the New York Central and smelters and steel plants and chemical works and glass factories and foundries and coke plants and machine shops and zinc mills, and its hills and bluffs are scaled by numerous blackened mill towns."[3]

The blackest of the mill towns, Roueché notes, was Donora, with streets of concrete or cobble but many of dirt and crushed coal. The town was treeless and almost grassless, probably because of the pollutants the mills wheezed: a steel plant, a wire plant (which made the cable for San Francisco's Golden Gate Bridge when that grand harp was strung in 1937), a zinc and sulfuric-acid plant, "huge mills . . . two blocks long . . . five or six stories high, and all of them bristle with hundred-foot stacks perpetually plumed with black or red or sulphurous yellow smoke."[4]

Onto this grim town, across that 1948 weekend, a fog settled—"greasy, gagging," Roueché calls it.[5] One of the public health officials who interviewed Donora residents later heard of "streamers of carbon [in the fog that] appeared to hang motionless in the air."[6] Visibility was so poor that even natives lost their way. People turned out Friday evening for the annual Halloween parade, but "everybody was talking about the fog," a medical secretary told Roueché. "As far as the parade was concerned, it was a waste of time. You really couldn't see a thing. . . . Everybody was coughing."[7] The town mortician said the fog "had the smell of poison."[8]

The first death eased its victim's gagging, retching cough at two in the morning on Saturday. More deaths followed as Donora's few doctors endured a two-day marathon of house calls. The fire department helped, soon exhausting its supply of oxygen and sending off to towns downriver for more. The Red Cross arrived and opened an emergency center. By midnight seventeen people were dead. Two more died on Sunday, when the rain began that would dissipate the fog. (The last of the twenty died a week later.)

The US Public Health Service postmortem took a year. Nine engineers, seven physicians, six nurses, five chemists, three statisticians, two meteorologists, two dentists, and a veterinarian studied, examined, and interviewed the townspeople and investigated the environment.[9] Besides coughing, choking, tearing, headache, difficulty breathing, chest pain, and sore throat, they heard of nausea, vomiting, and diarrhea. The number of reported symptoms increased with age: small children did well, but of people over sixty-five, almost half were severely affected, and all those who died were fifty-two or older. Of the total area population, 42.7 percent—5,910 persons—reported being affected. Or as the heath service termed it, "reported some affection from the smog."[10]

Despite the ongoing disaster, the general superintendent of the steel and wire works and the head of the zinc plant continued operating their mills, pumping poisonous smoke into the fog until Sunday morning, when U.S. Steel, the parent company, ordered them to bank their fires and shut down. Nevertheless, they told the mayor that they "were sure the mills had nothing to do with the trouble."

The Public Health Service investigators concluded that no single pollutant burdening the Donora air was present in sufficient volume to cause the town's gust of deaths and severe illnesses. The investigators found only trace amounts of hydrogen and zinc chlorides, nitrogen oxides, hydrogen sulfide, and cadmium oxide. "Sulfur dioxide and total sulfur," they wrote, had been "significant constituents of the over-all atmospheric pollution load"—that caustic yellow fog—but said they had no way post-disaster to measure what the chemicals' concentrations had been. "It seems reasonable to state," they concluded, "that while no single substance was responsible for the October 1948 episode, the syndrome could have been produced by a combination . . . of two or more of the contaminants." In his introduction to the report, as if in partial apology, the surgeon general claimed that "it was not until the tragic impact of Donora that the Nation as a whole became aware that there might be a serious danger to health from air contaminants."[11]

A consultant named Philip Sadtler would claim to the contrary that the Public Health Service report was a whitewash, and the real cause of sickness and death was exposure to toxic concentrations of fluorine. Sadtler, the president of Sadtler Research Laboratories in Philadelphia, had experience with fluorine contamination from previous work as an investigator in a lawsuit brought against DuPont, which manufactured hydrogen fluoride for the Manhattan Project during World War II. Peach orchardists in Salem County, New Jersey, whose 1944 crops had been destroyed by fluorine contamination, brought the lawsuit, seeking $400,000 in damages on their behalf.*

A Manhattan Project officer, Lieutenant Colonel Cooper B. Rhodes, had attended a meeting at the US Food and Drug Administration (FDA) concerning the so-called Peach Crop Cases. The Department of Agriculture had called the meeting to discuss condemning "vegetable produce" grown in the part of New Jersey affected by the DuPont fluorine plant. A DuPont lawyer, C. E. Geuther, had requested the meeting. According to Rhodes, Geuther opened the meeting by stating that any FDA action condemning New Jersey "vegetable produce" would create "a bad public relations situation" for DuPont. Geuther said DuPont didn't admit that the fluorine content of the vegetables "was in any way attributable to activities or operations of the DuPont Company," but that the "company was taking steps to eliminate entirely the present small amount of [hydrogen fluoride] which is now being discharged into the atmosphere by the Chambers Works."[12] In plain English: we don't admit doing it, but we're cleaning it up. The USDA proposal followed from its investigation of fluorine contamination of New Jersey crops, and Rhodes learned after the meeting with Geuther that information from Philip Sadtler had triggered the investigation.

It was against this background that Sadtler looked into the Donora disaster. He visited the town soon after the event: a brief news story in the trade magazine *Chemical and Engineering News* on 13 December 1948 reports his findings. "Circumstantial and actual proof has been found," the story begins, "of acute fluorine poisoning by the smog in the Monongahela River Valley to persons who already had chronic fluorine intoxication"—intoxication in this case meaning chronic exposure among factory workers.

Sadtler had questioned the physicians who treated the victims and autopsied the dead. He told the *News* that "analysis of blood of deceased and hospitalized victims showed 12 to 25 times the normal quantity of fluorine." He had found as well that "corn crops, very sensitive to fluorine, were severely damaged, and all of the vegetation north of the town [downwind from the factories] was

*Samuel Philip Sadtler, who founded the firm in 1901 with his son Samuel Schmucker Sadtler, died in 1923; Philip Sadtler, who took over the business from his father in 1947, was the founder's grandson.

killed." Tellingly, all of those who died had shown earlier symptoms of chronic fluorine poisoning.[13]

Interviewed in 1996, after a long and distinguished career, Sadtler told a journalist bluntly, "It was murder. The directors of U.S. Steel should have gone to jail for killing people."[14] After the *Chemical and Engineering News* story appeared, Sadtler said, though he had been a frequent contributor, the editor told him not to send the journal any more stories. The American Chemical Society published the *News*; a former president of the society, Dr. Edward R. Weidlein, an industrial chemist with ties to Union Carbide, had blacklisted him. In a subsequent 1950 class-action lawsuit brought by the families of Donora victims, faced with the possibility that the judge might release analyses of several air samples that U.S. Steel had taken at the height of the Donora disaster, the steel company settled out of court. Nor did the US Public Health Service ever release a final report.[15]

Knowledge of fog contaminants had not improved by 1952, when a poisonous fog that smothered London between 5 and 9 December resulted in about three thousand excess deaths, more than three times the norm.[16] Like the "presumptuous smoke" of John Evelyn's day, this fog invaded the indoors. A Londoner remembered going to work: "By two o'clock in the afternoon, you couldn't see across the corridor. It was thick, yellow, and somebody knocked on the door of my office and said, 'Go home, we're closing.' . . . I'll never forget it because the smog was so thick you really felt like you were walking into a war.' "[17] Perversely, the Ministry of Health attributed the excess deaths to an influenza epidemic, even though unexpected deaths in the first three months of 1953 were seven times the number of flu deaths.[18] The London press knew better: it called the event the "Great Killer Fog."[19] Its main ingredient was sulfur dioxide coating black particles of coal smoke.

Parliament resisted acting on the evidence of deadly air pollution, fearing the cost to British industry, but the committee it appointed to investigate proved serious about its work and returned a report recommending an 80 percent reduction in London's smoke across the next decade. A Clean Air Act in 1956 prohibited the emission of "dark smoke . . . from a chimney of any building," the burning of waste, or the installation of any furnace or boiler that emitted smoke. Local authorities could designate smoke-control areas, while the

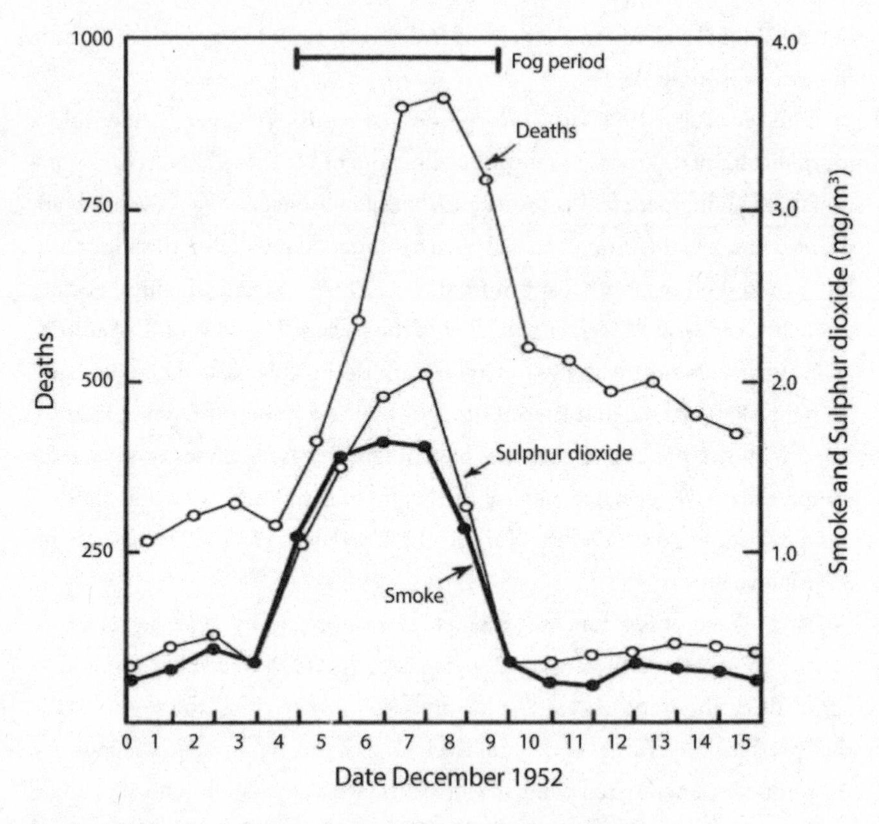

London's "Killer Fog" of December 1952. Deaths correlated closely with a peak increase in sulfur dioxide and smoke during a winter temperature inversion.

government could exempt well-designed fireplaces from these restrictions.[20] The act essentially switched heating from solid fuels such as coal to gas: town gas; imported liquid propane, beginning in 1959; and then, after the discovery in 1965 of large gas fields off the Yorkshire coast, natural gas. Since natural gas burns hotter than propane or town gas, some twenty million gas appliances had to be adapted or replaced. That took ten years from 1967 forward. Another killer fog recurred in London in 1962, but year by year, the United Kingdom slowly cleared its air.[21]

A different kind of "fog" had begun affecting the air over Los Angeles in the decade of World War II and its aftermath, 1940 to 1950, and beyond. People called the substance by its present name, smog, but experts still attached the older meaning to the word: "smoky fog." That usage dates back to 1905, when the honorary treasurer of the British Coal Smoke Abatement Society coined it to describe London's smoke-soused glooms.[22] US motor vehicle registrations more than doubled between 1945 and 1955, from about twenty-six million to more than fifty-two million, a hefty share of those unfiltered engine exhausts discharging into the natural basin of Los Angeles County. Inversions were frequent, the air turning hazy and brown and painful to breathe, and residents complained with increasing bitterness, voting out one or another of the mayors of the forty-five crazy-quilt townships that patched the valley. Engine exhaust escaped blame, however. The blame fell on the basin's factories, oil refineries, and even backyard trash burning. "Some people have blamed smog on the Mexican volcano Paricutin," an oil industry periodical noted cavalierly; "some have suggested the atom bomb; and still others have considered smog to be a manifestation of the occult."[23]

"At that time," the Caltech chemist and instrument inventor Arnold Beckman recalled, referring to the late 1940s, "there was great confusion and uncertainty over the causes of smog and what should be done to eliminate it, much legal and political snarling, frenzied exhortations by innumerable citizens' groups demanding an immediate cure for smog." Cities called in experts to diagnose the cause, among them Dr. Edward Weidlein, of all people, the industry-friendly director of Pittsburgh's Mellon Institute of Industrial Research and the man who would blackball Philip Sadtler. With staff from the US Bureau of Mines, Weidlein spent two weeks studying California smog. "They were of no help," Beckman said. "According to their own tests, which measured chiefly soot and SO_2 [sulfur dioxide], Los Angeles's polluted air was cleaner than Pittsburgh's purified air! They were baffled."[24]

In 1947, to facilitate action across township lines, California authorized the creation of air pollution control districts in every county in the state. The oil industry had opposed this state legislation and managed for a time to bottle it up in committee. The executive vice president of the Union Oil Company of California, however, W. L. Stewart Jr., had informed a meeting of industry executives that his company's future depended on Los Angeles's welfare. (The company was one of the first oil companies in California, founded by his father in 1890.) Union Oil, Stewart told the executives, would support the state bill. The other companies saw Stewart's point: the bill passed unanimously. The

Los Angeles County Air Pollution Control District was the first established in the state and in the nation. In October 1947 a University of Illinois classmate of Beckman's, Dr. Lewis C. McCabe, was appointed director.

Everyone assumed that the primary cause of Los Angeles smog was sulfur dioxide. "They took it for granted," Beckman said, "SO_2 was the chemical villain responsible for haze and for eye irritation. There was some justification for this view, for in Pittsburgh and St. Louis, the air pollution villains were indeed soot and SO_2. As there was no soot in the Los Angeles area, obviously SO_2 must be the obnoxious pollutant." SO_2 does irritate the eyes and the lungs. "Amateur scientists," as Beckman calls them, also condemned SO_2, noting that it could oxidize to SO_3, sulfur trioxide, the main ingredient in acid rain. Nylon stockings, a new hosiery product at the time, developed holes and runs in the smoggy air, another indication to Beckman's "amateur scientists" that SO_3—which dissolves in water to form strongly corrosive sulfuric acid—was at work.[25]

Beckman thought otherwise, for a simple reason: as a chemical engineer, he knew that SO_2 had a characteristically pungent smell, like rotten eggs. "I did not smell SO_2 in the air," he said, "and was therefore reluctant to believe that it was responsible for Los Angeles smog." With a chemist colleague, Beckman met with McCabe. "The poor chap was being harassed from all sides," Beckman recalled. He urged the new director of the air pollution district "to do a little research," including analyzing the air. "There's no time for research," the harried McCabe countered. Eventually the two chemists convinced him.[26]

To analyze an air sample, they needed a skilled microchemist. Beckman had a candidate in mind, a colleague and personal friend whom he admired. Arie Haagen-Smit was a Dutchman, an organic chemist, and a professor of chemistry at the California Institute of Technology in Pasadena, eleven miles northeast of downtown Los Angeles. Born in Utrecht in 1900, Haagen-Smit specialized in the biosynthesis of essential oils. He was working at that time on isolating the flavor of pineapple by laboriously condensing pineapple fumes from a workroom full of ripe fruit.

The equipment set up to analyze pineapple essence would work with smog. Haagen-Smit, a big, bluff man, drew thirty thousand cubic feet of Pasadena smog through an open window and passed it through a trap chilled with liquid air.[27] Along with frozen water vapor, his trap caught what Beckman calls "a couple of drops of dark-brown, vile-smelling liquid."[28] When the Dutch chemist analyzed it, he reported it to be saturated and unsaturated hydrocarbons from petroleum products and industrial solvents, adding, "This includes all the material lost at the oil fields, refineries, filling stations, automobiles, etc." What was unusual about this newly identified source of air pollution was its two-stage formation. It entered the air invisibly from all the sources Haagen-Smit lists, but then it oxidized—rusted—into brownish haze "under the influence of sunlight and ozone, and possibly other air contaminants such as nitrogen oxide. . . . In these reactions, aerosols are formed which have eye-irritating properties and which, because of the small size of their particles, are able to decrease the visibility."[29] Los Angeles smog wasn't smoke and fog; it was oxidized hydrocarbons. Since *smog* was too common a word to abandon, it acquired a qualifying adjective characterizing its two-stage creation: photochemical smog. Haagen-Smit's students at Caltech, riffing on his name, called it Haagen-Smog.

The problem of smog would not be so simple, then, as banning home trash incineration or switching from coal. Haagen-Smit estimated that the quantity of ozone in the Los Angeles basin during severe smog attacks was some 500 tons, and the actual weight of the 25-by-25-mile, 1,000-foot-thick smog layer no less than 650 million tons. "This enormous tonnage," the Dutch chemist wrote, "is often overlooked in 'quick cures' for smog, when proposals are made to place fans on the mountains or drill tunnels through the mountains to drive away the smog." He indicted the petroleum industry, which allowed its gasoline products to evaporate from lake-scale open ponds and stored gasoline in roofed tanks with space between the roof and the fuel surface that allowed for continual evaporation. (Eventually he convinced them to cover their ponds and install floating roofs on their storage tanks, saving them money as well as saving the air.) But the half million cars driving around Los Angeles, he wrote, burned "approximately 12,000 gallons of gasoline daily," and "even if the combustion were 99 percent complete, which it certainly is not, 120 tons of unburned gasoline would be released" daily into the California air.[30]

"This conclusion produced impassioned outbursts of protests," Beckman reports, "especially from the oil companies and the auto manufacturers. Haagen-Smit was all wet, they said." The industry-supported Stanford Research Institute (SRI) ran tests of its own, got different results, and claimed the

Dutch chemist had made a serious mistake. (In fact, it was SRI that had erred, despite access to expensive and sophisticated instruments. It used an unrealistically high concentration of gases in its measurements, and the gases quenched the smog reaction. Haagen-Smit bent strips of old tire inner tubes and measured their degree of cracking with increasing ozone exposure.)

Stung and furious, Haagen-Smit determined to prove SRI wrong. He set aside his pineapple research and proceeded over the next six months to tease out the complex of gases and particulates in photochemical smog. "He introduced a wholly new concept of air pollution," Beckman concludes, "that brought about a revolution in efforts to obtain clean air. He identified and courageously named the major sources of air pollution: the automobile, oil refineries, power plants, and steel factories." Haagen-Smit's work prepared the way for the long, fiercely resisted national evolution toward pollution-control systems on automobiles and stricter standards for industry. Los Angeles, and beyond that pleasure palace the world, owe their cleaner air in part to the scientific skill and moral indignation of a stubborn Dutchman.[31]

Across the next decades, industry resistance gave way to reluctant compliance. In July 1970, by executive order, President Richard Nixon created the Environmental Protection Agency (EPA), the real beginning of effective federal air pollution control. Five months later, by large majorities, Congress passed the Clean Air Act.[32] Power shifted then to federal regulators, who imposed increasingly strict standards nationwide. Unleaded gasoline was mandated for new cars in 1975, not to reduce the volume of atmospheric lead pollution but to protect the catalytic converters required for the exhaust systems of all new models. The phasedown of leaded gasoline followed in the 1980s as research revealed the neurological damage that lead caused in children.[33] Much of the soot pollution in New York City, researchers found, came not from automobiles but from unfiltered waste incineration, which was eventually brought under control.[34] (In 1959–60, when I lived in Manhattan, greasy soot quickly rendered any typing paper I left exposed on my apartment desk too dirty to use.)

In 1991, while studying the possible environmental impacts of the North American Free Trade Agreement (NAFTA) then being negotiated by the

Smog obscuring the George Washington Bridge, New York City, 1973.

United States, Canada, and Mexico, two Princeton University economists discovered a curious trend. When they examined the relationship between the level of air pollution and increasing income in a cross section of urban areas in forty-two countries, they found that for two pollutants—sulfur dioxide and "smoke"—concentrations increased with per capita gross domestic product (GDP) at low national income levels but decreased with GDP growth at higher levels of income.[35] The graph of the SO_2 finding in their influential 1991 paper looks like this:

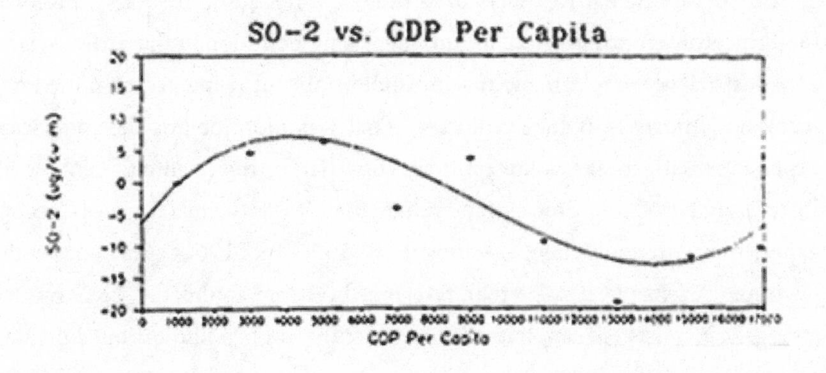

The curve on the Princeton economists' graph happened to resemble a Kuznets curve, a visualization of a controversial economic theory named after the twentieth-century Belarussian American economist Simon Kuznets. Kuznets had related increasing income and income inequality, a different relationship entirely. The Princeton economists' version thus came to be called an environmental Kuznets curve (EKC). In its standard form, it looks like this:

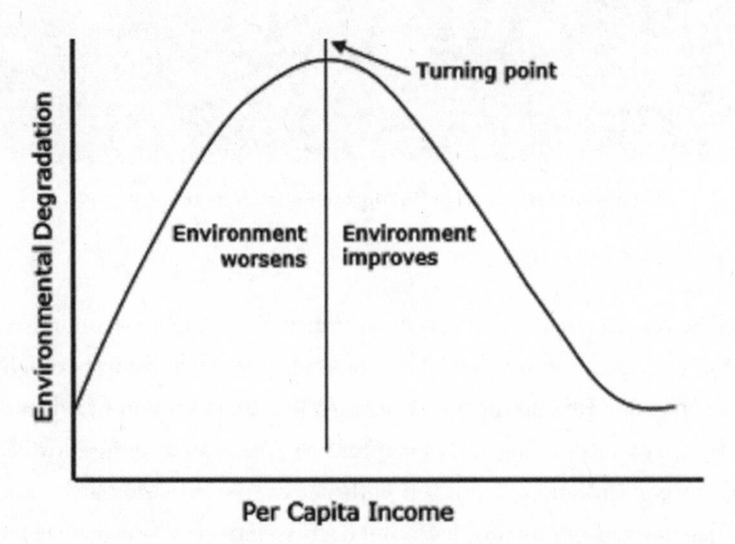

Environmental Kuznets curve.

An environmental Kuznets curve models a relationship such as the one the Princeton economists had found: increasing pollution in the earlier stages of industrialization and then, at a threshold point of rising personal income, increasing efforts to reduce pollution. That was more or less the American experience with smoke pollution in the first half of the twentieth century in Pittsburgh, New York, Chicago, and other cities. It was the Los Angeles County experience with photochemical smog in the 1950s and 1960s, and nationwide beginning in the 1970s. Further research, however, found the EKC model strangely inconsistent, applying most evidently to air pollution but not, say, to water quality or increasing levels of carbon dioxide, the greenhouse gas.[36]

Despite its inconsistencies, the EKC model has been the darling of capitalist conservatives, who like to claim it proves that economic growth unfettered by government regulation will sweep away pollution with a broom worked by invisible hands.

A partly overlapping model of environmental improvement conceives such improvement to be a luxury good: elites buying cleaner air and water and less crowded spaces ("environmental amenities" in economist-speak). In this model, necessities remain affordable as production becomes more efficient and material substitutions replace depleting natural resources, while environmental amenities become more expensive: electricity stays cheap, but there's only one Grand Canyon, where more people means less access. Economists haven't liked this model, and claim to have refuted it.[37]

Somewhere between, perhaps, lies the province of post–World War II longing for peace, quiet, family, and a fresh start—for a clean new world. Veterans came home weary from war, sick of sweat, stink, boredom, and slaughter. The nation met them with providence: vast savings that had been encouraged during the war to limit inflation; pent-up demand for goods rationed or unavailable during wartime and during the Depression that preceded it; a GI Bill that paid for college or vocational training; support for home ownership. "The Great Leap Forward of the American level of labor productivity that occurred in the middle decades of the twentieth century," writes the economist Robert J. Gordon, "is one of the greatest achievements in all of economic history. Had the economy continued to grow at the average annual growth rate that prevailed during 1870 to 1928, by 1950, output per hour would have been 52 percent higher than it had been in 1928. Instead, the reality was a 1928–50 increase of 99 percent."[38]

Economic growth put more money in Americans' pockets. It stimulated a psychological transformation as well. For most people, the Great Depression had been gloom and doom and even horror. The suicide rate jumped from 12 per 100,000 in the 1920s to almost 19 per 100,000 in 1929. It stayed high until World War II. By 1933, a quarter of the population was unemployed. The gross domestic product declined by almost half. "I see one-third of a nation ill-housed, ill-clad, ill-nourished," Franklin Roosevelt declared in 1936 in his Second Inaugural Address. World War II redeemed the nation's economic promise and reduced the gap between rich and poor. Looking back from today, when wealth inequality has once again emerged to depress the American working class, the middle decades of the twentieth century glow like an egalitarian promised land.

The economist Robert Higgs specifies the change in mood:

> The war economy . . . broke the back of the pessimistic expectations almost everybody had come to hold during the seemingly endless Depression. In the long decade of the 1930s, especially its latter half, many people had come to believe that the economic machine was irreparably broken. The frenetic activity of war production . . . dispelled the hopelessness. People began to think: if we can produce all these planes, ships, and bombs, we can also turn out prodigious quantities of cars and refrigerators.[39]

In the decades after World War II, Americans went to work improving their material lives and improving the environment in which they lived those lives as well. If the environmental Kuznets curve is a valid model, the United States and other advanced industrial societies crossed the income threshold toward environmental remediation in the mid-1950s. Yet even as they prospered, they began to question if growth was good.

NINETEEN

THE DARK AGE TO COME

Rachel Carson was mortally ill with metastasizing breast cancer when she wrote *Silent Spring,* the 1962 book credited with founding the environmental movement. She was suffering the toxic effects of chemotherapy when she condemned the toxic effects of pesticides. Radiation treatments sickened her as she denounced radiation's deadly risks.

Was her body the paradisiacal small town of her book's first chapter, "where all life seemed to live in harmony with its surroundings"? Where, then, "a strange blight crept over the area and . . . everywhere was a shadow of death"?[1] Writers write best when the passion they bring to their work is personal. Pesticides, Carson warned in a commencement speech in the spring before her book's September song, "are being introduced into our environment at a rapid rate. . . . There simply is no time for living protoplasm to adjust to them."[2] She died less than two years later, in April 1964. US vice president Al Gore would say that without her book "the environmental movement might have been long delayed or never have developed at all."[3]

Yet other books at other times have invoked apocalypse without resonance, without lighting national and international movements. Why did *Silent Spring* resonate? Who or what prepared the way? How did the optimism of rising expectations in the United States in the decade after World War II canker to a demoralized pessimism about the human future and the destruction of the natural world?

One anticipation of *Silent Spring* was the work of a personable nuclear geochemist named Harrison Brown, the son of a Wyoming rancher and a church organist, a veteran of the Manhattan Project, and a professor at Caltech. Carson never referenced Brown's work, a book titled *The Challenge of Man's Future,* published in 1954. The conditions it explored, however, related to those that troubled her. As scientists, they worked in related fields. Both were deeply apprehensive of the future, warning of world-scale disaster if their prescriptions were ignored.

After two atomic bombs that Brown had participated in creating marked the close of the most destructive war in human history, the young scientist had worked to educate the public about the new weapons' unique dangers. Carson herself located her motivation for investigating humankind's "tampering" with nature in the disillusion the bombs brought. In a letter to a friend, she wrote, "I suppose my thinking began to be affected soon after atomic science was firmly established. . . . [T]he old ideas die hard, especially when they are emotionally as well as intellectually dear to one. It was pleasant to believe, for example, that much of Nature was forever beyond the tampering reach of man."[4]

It's difficult at this distance to appreciate the immediacy of concern that thoughtful people felt for the human future in the aftermath of two world wars that together had accounted for the violent deaths of at least eighty million human beings. Years later, interviewing the physicist Richard Feynman, I asked him about that time. (I was not yet aware of his beloved wife Arline's death from tuberculosis in June 1945, two months before the war ended, or of the grief that consumed him afterward—that part of his life story had not been revealed yet.) Feynman told me he'd been sitting in a Times Square bar one day in the summer of 1946, looking out through the darkened windows onto the passing crowd. "And I thought to myself," he said, " 'You poor fools. Here you are, busy with your lives, and you have no idea that within a few more years you'll all be dead.' "

Of course his wife's death cast a pall over Feynman's view of the world. But like many others in that time of postwar transition, he believed the Soviet Union would develop a nuclear arsenal just as the United States was doing, at which point an arms race would ensue that would culminate in a world-destroying nuclear war. After he mourned, Feynman was able to set aside his fatalism, begin again to play, and in play revived his enthusiasm for life and for creative work.

Yet fatalism seemingly underlay the busy postwar transformation as veterans came home, settled down, and began building their lives. Remarkably few of them admitted to having experienced psychological trauma in their experiences of war, though most had been exposed to such trauma for years. Instead, they sucked it up and allayed their symptoms with smoking, drinking, and compulsive overwork. Many of them sought help only in retirement, decades later, when they could no longer sustain their tough, soldierly facades.[5]

In such a hooded cast of mind, it wasn't difficult to believe that the world itself was infected. For Rachel Carson, radiation and physical poisons caused the infection. For Harrison Brown and the neo-Malthusians who endorsed his views, other human beings threatened the world. Brown's protégé John Holdren, President Barack Obama's science adviser, would eulogize Brown as a "warm and witty man, cheerful, always a twinkle in his eye, and surprisingly modest."[6] He may have been, but his vision of the future was darkened with antihumanism.

In *The Challenge of Man's Future,* Brown wrote that much of humanity was behaving as if it "would not rest content until the earth is covered completely and to a considerable depth with a writhing mass of human beings, much as a dead cow is covered with a pulsating mass of maggots."[7] Given the supposedly selective forces of overpopulation, Brown claimed, to prevent what he called "the long-range degeneration of human stock," it would be necessary to prevent "breeding in persons who present glaring deficiencies clearly dangerous to society and which are known to be of a hereditary nature." For example, he continues:

> We could sterilize or in other ways discourage the mating of the feeble-minded. We could go further and systematically attempt to prune from society, by prohibiting them from breeding, persons suffering from serious inheritable forms of physical defects, such as congenital deafness, dumbness, blindness, or absence of limbs. . . . Unfortunately, man's knowledge of human genetics is too meager at the present time to permit him to be a really successful pruner.[8]

"Pruning" the species, Brown concludes, could be accomplished in two ways: discouraging "unfit" persons from breeding, and encouraging breeding by people who pass physical and mental testing and whose ancestors were fit. There were more radical methods of pruning undesirables, Brown hinted, only a decade after the Holocaust, but he suspected that such methods "would probably not be palatable to many of us who are alive today."[9]

If this cool invitation to racial selection and maiming echoes with the pseudoscience of Nazi eugenics, it does so for a reason: there is a direct link between the original eugenics movement that inspired Adolf Hitler and the veiled postwar eugenics organizations that endorsed Malthusian nightmares of a "population explosion." Tracing that link in detail is outside the scope of this book, but several scholars have done so, including Robert Zubrin in his 2013 study *Merchants of Despair: Radical Environmentalists, Criminal Pseudo-Scientists, and the Fatal Cult of Antihumanism,* and Pierre Desrochers and Christine Hoffbauer in a lengthy 2009 paper, "The Postwar Intellectual Roots of the Population Bomb."[10]

Thomas Malthus, the eighteenth-century English proto-economist, was himself no piker at human pruning, notoriously proposing:

> Instead of recommending cleanliness to the poor, we should encourage contrary habits. In our towns we should make the streets narrower, crowd more people into the houses, and court the return of the plague. In the country, we should build our villages near stagnant pools, and particularly encourage settlements in all marshy and unwholesome situations. But above all, we should reprobate [i.e., reject, repudiate] specific remedies for ravaging diseases, and those benevolent, but much mistaken men, who have thought they were doing a service to mankind by projecting schemes for the total extirpation of particular disorders.[11]

An insolent contempt for humanity channels through all these high-minded works. The most widely read popular formulation of the neo-Malthusian argument, entomologist Paul R. Ehrlich's *The Population Bomb,* commissioned by the Sierra Club in 1967, opens with the Ehrlich family culture-shocked into panic on "a stinking hot night in Delhi a couple of years ago":

> My wife and daughter and I were returning to our hotel in an ancient taxi. The seats were hopping with fleas. . . . The temperature was well over 100, and the air was a haze of dust and smoke. The streets seemed alive with people. People eating, people washing, people sleeping. People visiting, arguing, and screaming. People thrusting their hands through the taxi window, begging. People defecating and urinating. . . . As we moved slowly through the mob, hand horn squawking, the dust, noise, heat, and cooking fires gave the scene a hellish aspect. Would we ever get to our hotel? All three of us were, frankly, frightened. It seemed that anything could happen—but, of course, nothing did. Old India hands will laugh at our reaction. We were just some overprivileged tourists, unaccustomed to the sights and sounds of India. Perhaps, but since that night, I've known the feel of overpopulation.[12]

It seemed that anything could happen—but, of course, nothing did. How Delhi street life might have felt to those who lived there—the "streets alive with people"—evidently didn't concern Ehrlich. In *The Population Bomb,* discussing "the last tragic category"—"those countries that are so far behind in the population-food game that there is no hope that [American] food aid will see them through to self-sufficiency"—he recommended that India "should receive no more food"—that is, should be left to starve, an outcome that he, at least, believed inevitable.[13]

The Brazilian physician and geographer Josué de Castro, writing a decade earlier about South American conditions, took umbrage at such murderous indifference. He condemned condescension such as Ehrlich's in his classic *The Geography of Hunger.* "The neo-Malthusian doctrine of a dehumanized economy," de Castro wrote, "which preaches that the weak and the sick should be left to die, which would help the starving to die more quickly, and which even goes to the extreme of suggesting that medical and sanitary resources should not be made available to the more miserable populations—such policies merely reflect the mean and egotistical sentiments of people living well, terrified by the disquieting presence of those who are living badly."[14]

The same street scenes as those the Ehrlichs recoiled from in the early 1960s can be experienced in Delhi today. The difference is that today, on the other side of the Green Revolution, India, China and the other countries Ehrlich wrote off

The "population bomb" combines crawling-mass-of-humanity imagery with the destructive associations of a characteristic nuclear-explosion mushroom cloud.

are producing a surplus of food despite their much larger populations, though food insecurity continues to be a problem. Like most false prophets, Ehrlich's answer to his failed predictions of catastrophe has been to move the date of the end of the world a few more decades along the calendar. By now, he's reached the 2050s. The end is still not in sight, but Ehrlich, eighty-six years old in 2018, is still certain it's coming.

The small-world, zero-population-growth, soft-energy-path faction of the environmental movement that emerged across the 1960s and 1970s knowingly or unknowingly incorporated the antihumanist ideology of the neo-Malthusians into its arguments. That ideology supported, even determined, its contradictory stance on nuclear power. David Brower, the influential president of California's Sierra Club, was blunt about his hostility to expanded energy production. "More power plants create more industry," he complained at a Sierra Club board meeting in 1966; "that in turn invites greater population density. If a doubling of the state's population in the next twenty years is encouraged by providing the power resources for this growth, the state's scenic character will be destroyed."[15] In a contest between human beings and landscape, Brower was clear on where he stood. He did not include himself in his head count of undesirables.

Government encouragement alone wasn't sufficient to persuade utilities to go nuclear in the years after President Eisenhower's Atoms for Peace initiative. Shippingport came online in 1957. Several other reactor demonstrations followed in other places. In 1958 the Atomic Energy Commission was supporting the development of no fewer than eleven different types of reactors, with pressurized water (PWR) and boiling water designs the farthest advanced.[16] But utilities weren't buying: nuclear reactors still cost too much to compete with coal-fired plants.

General Electric forced the breakthrough. The company had been developing a boiling water reactor (BWR) in parallel with Westinghouse's pressurized water reactor. A BWR operates at normal temperature and pressure, simplifying its design and making it much less expensive. Late in 1963 GE signed a fixed-price contract with Jersey Central Power and Light to deliver a 515-megawatt turnkey plant to a site at Oyster Creek, New Jersey, about eighty-five miles south of New York City on the New Jersey shore. It was the first of twelve such turnkey units that GE and Westinghouse would sell in the next three years and on which they would collectively lose nearly $1 billion.

"Only extreme confidence in future technological progress," writes the sociologist James M. Jasper, "could allow this [loss] to happen."[17] That confidence stemmed in part from inexperience. Both companies hired new CEOs in 1963, one a former light-bulb salesman, the other a financial analyst. Both companies had been competitors since the days of Thomas Edison and George Westinghouse. Both CEOs wanted to demonstrate their leadership abilities. From 1963 onward, they offered loss-leader reactors, hoping declining costs would eventually make them profitable. What utilities executive Philip Sporn christened a "bandwagon" market ensued. The "bandwagon effect," Sporn told Congress in 1967, resulted from "many utilities rushing ahead to order nuclear power plants, often on the basis of only nebulous analysis and frequently because of a desire to get started in the nuclear business."[18] Utilities executives sometimes decided to go nuclear after bantering with executives of other utilities over games of golf. The Atomic Energy Commission had not focused on reactor safety during the years of design development in the 1950s. No more did the utilities in the 1960s. Between 1965 and 1970, US utility companies placed orders for some one hundred reactors. The bandwagon rolled.

Nuclear-power advocates responded to the neo-Malthusianism of the 1950s and 1960s by arguing that atomic energy could resolve the apparent dilemma of population growth outpacing natural resources. The most prominent spokesperson for this advocacy view was Alvin Weinberg, a Manhattan Project nuclear physicist who directed Tennessee's Oak Ridge National Laboratory for twenty-five years after the war.

Weinberg, a man of messianic spirit, spoke and testified widely across his long career. His large vision for nuclear power opened him to the criticism that he and his fellow technologists, in Weinberg's own words, "overestimate the benefits and underestimate the risks of their new inventions." Weinberg argued to the contrary that critics of technology made an error that was the mirror image of the error they accused technologists of making: *underestimating* the indirect results, the broader human benefits, that society reaps from new technologies. The critics imagined that there were few large-scale benefits from nuclear energy; that its main benefits were higher dividends for the utilities and profits for reactor manufacturers. The actual case, Weinberg insisted, was far otherwise.[19]

Weinberg made that claim in 1970, when the challenge under discussion across America was the neo-Malthusian fear of overpopulation with resource depletion. He presented his argument for nuclear power's nonmarket benefits in that context. In doing so, he took it for granted that the challenge was real. The United Nations's medium estimate for world population growth by 2030, he noted, was ten billion (a number that by 2017 the United Nations had revised, extending it seventy years forward to 2100).[20] In 1970, going Malthusian, Weinberg postulated a world population of *eighteen* billion by 2050.[21] No credible authority projects such an increase today—almost two and a half times the 2017 world population of seven and a half billion—because the demographic transition from high birth rates to lower birth rates that follows from increasing prosperity is advancing today throughout the world. The United Nations's projection of ten billion people by 2100 also projects that the growth *rate* will decline by then to 0 percent, and world population growth will cease.[22]

But whatever the world's population, Weinberg argued in 1970 that nuclear energy could help prevent Malthusian catastrophe. It could do so not only by generating cheap, clean electricity. It could do so as well by desalting seawater and by splitting water to generate hydrogen as a fuel for transportation and industry. "If we have hydrogen," he wrote, "we can reduce metals from their ores; we can hydrogenate coal [to make liquid fuels]; we can manufacture ammonia for agriculture."[23] The first stage in the process Weinberg envisioned of fending off Malthusian disaster with nuclear power would use what the Oak Ridge director called "catalytic nuclear burners," or what we today call breeder reactors. Those would fission uranium or thorium for energy while breeding plutonium, uranium 233, or tritium. Those fuels would then produce energy via either nuclear fission or thermonuclear fusion, if and when that difficult technology is finally mastered.

These optimistic visions depended on how much it would cost to produce the energy, as Weinberg well understood. He estimated that the cost would be competitive with other energy sources, but he was estimating at the beginning of the nuclear-power era. He had not yet encountered the opposition of the environmental activists who were just then turning their attention to nuclear power. He had not yet had occasion to observe the one-way ratchet effect of tightening government regulations driven by fear of both realistic and exaggerated and hypothetical nuclear accidents, an effect that increased capital costs without necessarily improving safety. President Jimmy Carter had not yet officially foreclosed fuel reprocessing, as he would do in 1977, swerving nuclear policy away from breeder-reactor development.

By Carter's own account, his poor opinion of nuclear power originated in personal experience. In 1952 the future president was a US Navy lieutenant with submarine experience stationed at General Electric in Schenectady, New York, training in nuclear engineering under Hyman Rickover. That December, an experimental Canadian 30-megawatt heavy-water moderated, light-water cooled reactor at Chalk River, Ontario, experienced a runaway reaction, surging to 100 megawatts, exploding and partly melting down. It was the world's first reactor accident, a consequence of a fundamental design flaw of the kind that would destroy a Soviet reactor at Chernobyl three decades later. Since Carter had clearance to work with nuclear reactors, which were still classified as military secrets, he and twenty-two other cleared navy personnel went to Ontario early in 1953 to help dismantle the ruined machine. Because it was radioactive, the calculated maximum exposure time around the damaged structure itself was only ninety seconds. That exposure would be the equivalent of a worker's defined annual maximum dose of radiation—in those days, 15 rem (roentgen equivalent man). More than a thousand men and two women, most of them Chalk River staff, would participate in the cleanup.[24]

The Canadians had constructed a reactor mock-up on the laboratory tennis court where the cleanup teams could train. "A team of three of us practiced several times on the mock-up," Carter writes, "to be sure we had the correct tools and knew exactly how to use them. Finally, outfitted with white protective clothes, we descended into the reactor and worked frantically for our allotted time." At eighty-nine seconds, they climbed out and dashed outside, and another team took their place.

"Each time our men managed to remove a bolt or fitting from the core," Carter writes, "the equivalent piece was removed on the mock-up." For several months afterward, the future president recalled, he and his men were required to submit urine and stool samples to monitor their radiation exposure. "There were no apparent aftereffects from this exposure," Carter concludes—"just a lot of doubtful jokes among ourselves about death versus sterility."[25]

Had he known the long-term outcome of the Chalk River radiation exposures, Carter might have felt friendlier to nuclear power. A thirty-year outcome study, published in 1982, found that lab personnel exposed during the reactor cleanup were "on average living a year or so longer than expected by comparison with the general population of Ontario." None died of leukemia, a classic disease of serious radiation overexposure. Cancer deaths were below comparable averages among the general population. In the dubious tradition of radiation-dose science, the outcome study assigned this reduced mortality following low-level radiation exposure to a healthy-worker effect. That effect, a form of selection bias, results from workers being healthier on average than the general population—healthy enough to work. In the Chalk River study, however, deaths from lung cancer (probably from smoking) and from cardiovascular disease were higher among the exposed than among the general Ontario population. If the healthy-worker effect applied, then the radiation workers' reduced mortality was all the more remarkable.[26]

Fear of radiation and misunderstanding of its effects were powerful drivers of antinuclear sentiment. Activists encouraged this response over the years with claims that a meltdown would destroy an area "the size of Pennsylvania" (Ralph Nader) or that "nearly a million" had died from Chernobyl fallout (Helen Caldicott, the Australian physician).[27] Bolstering the claims of extremists, the United States in the 1950s adopted a flawed model of radiation exposure that made radiation safety seemingly impossible to guarantee. How that standard emerged is a sorry tale of scientific ineptitude if not actual misconduct and of good intentions gone bad.

In the alarm that followed the first Soviet test of an atomic bomb in August 1949, America accelerated efforts to accumulate a nuclear arsenal. Between 1946 and 1958, the United States tested no fewer than 193 atomic and hydrogen weapons, all exploded in the atmosphere on Bikini Atoll and other locations in the western Pacific northeast of Papua New Guinea or north of Las Vegas at Yucca Flats, Nevada. During the same period, the Soviet Union tested 86 such weapons on its proving grounds at Semipalatinsk, Kazakh SSR, or on the island of Novaya Zemlya in the Arctic Ocean north of European Russia.

Radioactive material from these many tests fell out across the world. One fallout product in particular, strontium-90, which the body absorbs like calcium and deposits in bone, frightened mothers when it began showing up in their breast milk.

The United States conducted the most notorious of its weapons tests, Castle Bravo, at Bikini on 1 March 1954. Bravo was the first test of a new "dry" hydrogen bomb, Shrimp, fueled with solid lithium deuteride rather than the cryogenically cooled liquid deuterium used in the first test of a thermonuclear device, Mike I, on 1 November 1952. The dry bomb was much lighter and would therefore be an easier lift for US bombers. (Mike, an experiment rather than a weapon, weighed 82 tons; Shrimp, only 10.5 tons.) The lithium deuteride in the Shrimp device would make its own hydrogen by transmuting its lithium to a hydrogen isotope, tritium, an instant after detonation.

But as one of the Castle physicists, Harold Agnew, told me later, the Shrimp designers had not realized that one of the two isotopes of lithium in the fuel, $lithium^7$, would be reactive. They had believed that the Li^7, representing 60 percent of Bravo's lithium content, would essentially be inert under the conditions of the experiment. It actually reacted as energetically as the 40 percent content of Li^6, tripling the device's expected yield, from a projected 5 megatons to 15 megatons—that is, from 5 to 15 *million* tons of TNT equivalent. Bravo

Shrimp in its shot cab, ready for detonation. The device was fifteen feet long and four and a half feet in diameter. The pipes extending from its body are light pipes for diagnostics.

exploded into a 4.5-mile-diameter fireball, vaporized a crater 250 feet deep and more than a mile wide on the ocean floor, and spread dangerous fallout across more than 7,000 square miles of the Pacific Ocean. Besides exposing Pacific islanders, who had to be evacuated, twenty-three crew members of the Japanese fishing vessel *Lucky Dragon*, which was operating outside the designated test exclusion zone, were contaminated with fallout and experienced acute radiation sickness. One crew member eventually died. There was an international outcry as the radiation cloud, lofted into the jet stream, fell out around the world.

The chairman of the US Atomic Energy Commission at the time of the Bravo accident was a wealthy Wall Street financier and Naval Reserve admiral named Lewis Strauss, an arrogant and vengeful Eisenhower political appointee. Strauss's immediate response was to issue a cold disclaimer of responsibility. Bravo, he announced at a Washington press conference on 31 March, "was a very large blast, but at no time was the testing out of control." The *Lucky Dragon*, he insisted, "must have been well within the danger area."[28] (Privately, he told Eisenhower's press secretary that the fishing vessel was probably a "Red spy ship."[29]) Strauss went on to claim that the fallout—he mistakenly called it "natural"—was less than the fallout measured from some previous tests and "far below the levels which could be harmful in any way to human

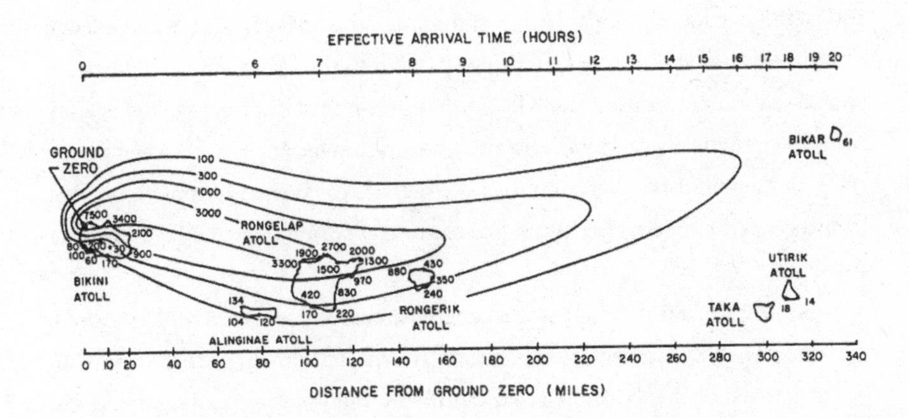

Radioactivity from the Castle Bravo test fell out locally far from the test site as well as around the world. Number contours indicate rems of radiation. (Receiving 500 or more rem in a short period of time without medical intervention is usually fatal.)

beings." Which clearly wasn't true for the people of Bikini and the *Lucky Dragon* fishermen.

One of the pioneers of modern genetics, Alfred H. Sturtevant, a professor at Caltech, took public issue with Strauss's disingenuous claims. The son of a university mathematician who had given up teaching for farming in southern Alabama, Sturtevant discovered gene mapping while still a Columbia University undergraduate. That work, begun in 1911 and published in 1913, won him a desk in the famous "fly room" at Columbia where fruit flies—*Drosophila melanogaster*—served as model organisms for genetic research. Sturtevant moved to Caltech in 1928 and taught and did research there for the rest of his career, studying the genetics of *Drosophila*, horses, fowl, mice, moths, snails, iris, and evening primrose.[30] He issued his challenge to the AEC chairman in Washington in June 1954 during his annual address as president of the Pacific Division of the American Association for the Advancement of Science.

Sturtevant was disturbed, he said, that a figure of authority such as Chairman Strauss should have claimed that low levels of radiation were harmless. To the contrary, he argued, every competent geneticist knew that "any level whatever" of radiation exposure was certain to be "at least genetically harmful to human beings." Low doses of high-energy radiation, he emphasized, were a "biological hazard."[31]

The 1954 fallout debate was the origin point of the AEC decision at the end of the decade to base radiation standards on a linear model, with damage postulated even at low doses, all the way down to the lower limit of detection. Such a seeming reversal of the AEC's position came about because the agency admitted geneticists onto its committees, resulting in a debate between physicians and geneticists—between health concerns for present populations, that is, and genetic concerns for future populations—which the geneticists eventually won.

A key figure in this debate was a geneticist named Hermann Muller, a small dynamo of a man, five feet two inches tall, born in Manhattan in 1890 and another denizen of the fly room as a Columbia undergraduate and graduate student. In 1926, now a professor of zoology at the University of Texas in Austin, Muller reported discovering artificial mutation—demonstrating for the first time that irradiating *Drosophila* with X-rays increased its mutation rate. That discovery, giving researchers an artificial method of inducing mutation, made it possible to shape experiments to study the genetic basis of life. Muller received the 1946 Nobel Prize in Physiology or Medicine for his discovery.

An iconoclastic socialist, Muller had a deep interest in eugenics. Since he believed eugenic improvement would be ethical only in a classless society—he denounced the American eugenics movement for its racism and elitism—he moved to the Soviet Union from 1933 to 1937 to do genetics research. When the rise of Lysenkoism* put him at political risk, he escaped purging by volunteering to serve with the International Brigades in the Spanish Civil War. Working in Madrid with the Canadian surgeon Norman Bethune, Muller investigated using fresh cadaver blood for blood transfusion and helped develop blood preservation and storage, pioneering a lifesaving new technology.[32]

From Spain, Muller moved to Edinburgh to work at the university there, finally returning to the United States in 1940. After teaching at Amherst College under temporary appointments during World War II, he found a home in 1945 at the University of Indiana, where he would teach and work for the rest of his career. His life to that point had been filled with conflict and financial strain, making him defensive about his reputation. James Watson, in 1953 one of the codiscoverers of the structure of DNA, took Muller's course in advanced genetics at Indiana in the late 1940s and recalled of Muller that "the component of past hurt was never deep below the surface, and over and over he would tell how he was the first to see the consequences of a given experiment. But . . . it always seemed natural that he had been the first."[33]

Muller believed in human perfectibility through voluntary "germinal choice," as he called it. He also knew from his genetics research that mutations are random and mostly deleterious: they damage far more often than they improve.[34] Artificial mutation, he said in his Nobel lecture, allowed him "to obtain, by a half-hour's treatment, over a hundred times as many mutations in a group of treated germ cells as would have occurred in them spontaneously in the course of a whole generation." Under the conditions of his experiments

*Trofim Denisovich Lysenko was a Soviet biologist who promoted a pseudoscientific theory of the inheritance of acquired characteristics. Under Stalin's patronage, he purged many reputable Soviet geneticists.

with fruit flies, the frequency of such mutations was directly proportional to the dose of radiation applied. It followed, he argued, that there was "no escape from the conclusion that there is no threshold dose": every hit caused a mutation, most mutations were damaging or lethal, and they were irreversible and permanent as well.[35]

For Muller, that meant research and medicine, and even more so industry, especially with "the coming increasing use of atomic energy," labored under an urgent obligation: to take "simple precautions" to shield the reproductive organs from radiation. He knew "with certainty," he said, from experiments on lower organisms—primarily fruit flies—that all high-energy radiation would produce such mutations in human beings.[36]

When Muller, in Stockholm in December 1946 to receive his Nobel Prize, asserted confidently that any radiation exposure, no matter how limited, was genetically contaminating, he deliberately avoided mentioning that he had just seen evidence to the contrary. In mid-November, shortly before Muller left for Sweden, a colleague with whom he had worked on Manhattan Project genetic testing, Curt Stern, had sent him a draft research paper by another Manhattan Project scientist, an insect behaviorist named Ernst Caspari. Muller's work, and that of others across the years, had explored the effects of high and medium doses of radiation. Caspari had extended that research into the low-dose range and had asked in particular whether the effect would be the same when the dose was spread over a period of time ("chronic") rather than delivered all at once ("acute"). Caspari's predecessors had found with acute irradiation that mutations increased in proportion to dose, from 4,000 r* all the way down to 25 r. Caspari's startling new finding was that fruit flies exposed to a daily dose of 2.5 r for twenty-one days, for a total of 52.5 r, showed *no increase* in their mutation rate. "This result," he wrote, "seems to be in direct contradiction to [a previous] finding that acute irradiation with 50 r and even 25 r X-rays causes a significant increase in mutation rate."[37]

Caspari's new results presented Muller with a dilemma. He was about to receive a Nobel Prize for his discovery of artificial mutation, which he claimed

*R for "roentgen," a measure of radiation, named after Wilhelm Roentgen, the discoverer of X-rays.

extended throughout the range of doses all the way down to zero. Caspari's new results contradicted that claim, or at least raised doubts about it. What should he do?[38] What he should have done was qualify his Nobel lecture. He understood that Caspari's evidence indicated a possible low-dose threshold to radiation damage, because he wrote Stern on 12 November acknowledging having received the draft, apologizing for being busy with work before leaving for Stockholm in early December but promising to "do all I can to go through it in a reasonable time" because "it is of such paramount importance, and the results seem so diametrically opposed to those which you and the others have obtained."[39] In Stockholm, however, Muller accepted his Nobel Prize and then deliberately ignored Caspari's findings in his lecture.

Back in Indiana in January 1947, Muller reviewed Caspari's paper more carefully. He found very little to criticize—he told Stern he had "so little to suggest in regard to the manuscript"—although he thought someone should redo the experiment to see if they could replicate the results. He then served as the paper's prepublication reviewer for the journal *Genetics*. When the paper appeared there, in January 1948, it reflected two changes: Muller's name now appeared among the acknowledgments, and one crucial sentence had been deleted. The deleted sentence was the sentence that questioned Muller's theory.[40]

Having suppressed an evidence-based challenge to his "linear no-threshold" (LNT) model of radiation effects, Nobel laureate Muller thereafter continued to promote and defend the LNT model whenever and wherever the question arose. He was not alone. As the AEC pursued its effort to establish radiation standards, first for radiation workers and then for the general public in the matter of weapons-test fallout, AEC and independent scientists participated in various institutional committee meetings throughout the 1950s. Apparently the LNT model was accepted as a standard throughout.

Edward Calabrese, the University of Massachusetts professor of toxicology who recently unearthed Hermann Muller's efforts to suppress opposition to his research findings, has charged that the LNT model and its acceptance were the result, in his words, of "untruths, artful dodges, and blind faith."[41] Jan Beyea, a physicist and public health consultant, has argued to the contrary that many other studies and committees across the past fifty years have found in favor of an LNT standard, which remains in place. As a Canadian scientist writes,

"It was the leading physicists responsible for inventing nuclear weapons who instilled a fear of small doses in the general population. In their highly ethical endeavor to stop preparations for atomic war, they were soon joined by many scientists from other fields. Eventually, this developed politically into opposition against atomic power stations and all things nuclear."[42]

The LNT debate was crucial to the long-term decline of nuclear power in the United States. Celebrated at first as a clean and potentially inexhaustible new form of energy, nuclear power lost popular support in large part because of a general public fear of radiation. That fear was exacerbated by the Damoclean sword of nuclear war that hung over the world in the long years of the Cold War, as well as by the three accidents that have occurred at nuclear power plants since the introduction of commercial nuclear power: Three Mile Island, Chernobyl, and Fukushima Daiichi. The AEC and its successors, as well as the nuclear-power industry, have attempted to assuage public fears by asserting minimal or no damage from low-level radiation—even sometimes arguing for a positive, hormetic effect. An antinuclear movement that originated in hostility to population growth in a supposedly Malthusian world promoted in turn the LNT model, exaggerating its effects.

One of Jan Beyea's conclusions seems relevant to this ongoing, contentious, and destructive debate. "What is often lost in disputes over radiation dose response models," he writes, "is that risks at the low dose levels that are being debated are also low."[43] Low doses of radiation are not only low risk; they're also lost in the noise of other sources of environmental insult.

Hermann Muller, offering society a rule to follow in judging whether or not to expose its citizens to radiation, proposed that "every definite risk to an outsider must therefore be measured against the likelihood and amount of good to be conferred, and the risk should be resorted to only if it is clearly less than the good to be derived by using this particular method [of treatment, of research, of power production] rather than some modified or alternative one."[44] Is there good, beyond the great good of carbon-free energy, that has come from nuclear power?

By 2013, according to scientists at the NASA Goddard Institute for Space Studies and the Columbia University Earth Institute, "global nuclear power [had] prevented an average of 1.84 million air pollution-related deaths and 64 gigatonnes* of CO_2-equivalent greenhouse gas emissions that would have resulted from fossil-fuel burning." They estimated as well, "on the basis of global projection data that take into account the effects of the Fukushima accident," that "nuclear power could additionally prevent an average of 420,000–7.04 million deaths and 80–240 [gigatons of CO_2-equivalent] emissions due to fossil fuels by midcentury [2050], depending on which fuel it replaces."[45]

This projection, needless to say, assumes that nuclear power will continue to find political support as one component of the largest energy transition in human history, the ultimate transition, the one the world faces today as it confronts global climate change. Or can renewables save the day?

*A gigaton is 1 billion metric tons.

TWENTY

ALL ABOARD

The great era of wind energy was the eighteenth and nineteenth centuries, when sails drove ships of war, exploration, and commerce across the oceans of the world. As a source of electricity, wind energy dates from 1887, when a Scottish physicist, James Blyth, built a horizontal wind turbine to generate electricity to light his holiday cottage at Marykirk, in eastern Scotland. Modeled on the anemometer, that unit featured cloth sails, but a later and sturdier turbine that Blyth had built for the nearby Montrose Lunatic Asylum used wooden sails and stood by as an emergency power source for twenty-seven years.[1]

A few months after Blyth built his cottage unit in Scotland, a wealthy American inventor, Charles F. Brush, independently designed and built a much larger wind turbine in his five-acre backyard in Cleveland. Modeled on a standard farm windmill for pumping water but much larger, it stood sixty feet tall, weighed eighty thousand pounds, and charged a basement full of batteries to light and power Brush's home.

Two brothers, Marcellus and Joseph Jacobs, manufactured the first commercial wind turbines in the 1920s. They modeled their product on a wind turbine they designed and built for their parents' ranch in Montana, to substitute prairie wind for the gasoline they had been fetching laboriously by horse

James Blyth's lunatic asylum windmill, thirty-three feet tall, with wooden sails, operated for twenty-seven years.

Charles Brush's home wind turbine. For scale, note the man mowing lawn at right.

and wagon from forty miles away to power a farm generator. (Only one-tenth of American farms had central-station electricity by 1930.) The Jacobs brothers' first system used a Ford truck rear axle and gearbox to power a generator, with three propeller blades and a fly-ball governor to protect it from damage in gusty wind conditions. Sales by word-of-mouth of the 1-kilowatt units led the brothers to incorporate in 1928; in 1931 they moved to Minnesota to open a factory closer to their suppliers, with a new direct-drive generator mounted atop the tower. Across the next thirty years, the Jacobs Wind Electric Company of Minneapolis, Minnesota, sold about twenty thousand 1- to 3-kilowatt wind-powered generators. Other small wind-electric systems by the hundreds of thousands delivered electricity to rural America until the 1950s, when connection approached 100 percent.[2]

Like wind, sunlight can be harnessed for energy directly as heat and light or through conversion to electricity. Worldwide, solar energy serves primarily to grow green plants for food and raw materials. Solar electricity, however, awaited the development of the silicon photovoltaic (PV) cell at Bell Telephone Laboratories in Murray Hill, New Jersey, in the 1950s. Earlier light-sensitive materials such as selenium operated at extremely low efficiencies, not above 1 percent. Bell's first silicon PV cells managed 6 percent, but efficiencies went up with improved boron doping and production techniques, to 11 percent by 1955. Bell scientists estimated that the maximum efficiency of their PV cells was about 22 percent. Reflection, surface and contact resistance, and other factors halved that efficiency, suggesting that 11 percent was probably a maximum.[3]

The real problem with early PV cells was cost: after one six-month trial application in a Georgia telephone system, Bell made no further effort to use

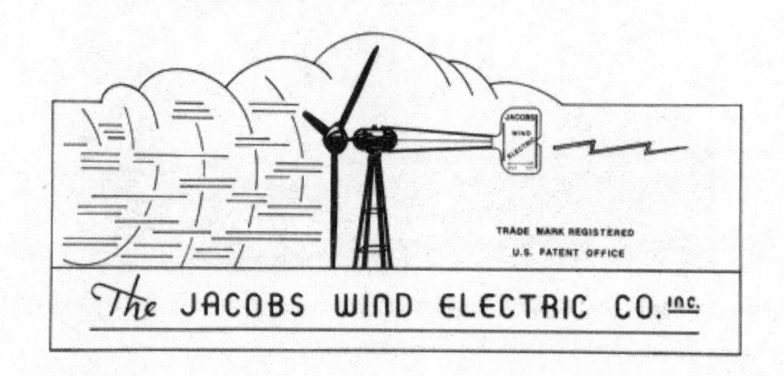

them.[4] One of the German scientists who came to the United States after World War II, working for the army at its Signal Corps laboratory at Fort Monmouth, New Jersey, was impressed with Bell's PV cells. Hans K. Ziegler championed installing them on the first US communications satellite, *Vanguard I,* a project funded by the Department of Defense and the US Navy. The two funders fought over the power source: the Navy preferred batteries. In a compromise that became a seminal experiment, both power sources were installed on the 6.4-inch, 3.5-pound *Vanguard I,* which was launched on 17 March 1958 atop a three-stage Vanguard rocket. By June, the batteries had run down; the PV cells continued supplying 1 watt of power for six years.[5]

Early PV cells were expensive because they were sawn from large single crystals grown for use making transistors, even though PV didn't need silicon of transistor purity. Two Hungarian refugee scientists at the COMSAT satellite research center in Clarksburg, Maryland, Joseph Lindmayer and Peter Varadi, decided to start their own company, Solarex, to make PV cells for terrestrial applications. They first tried marketing cells made from satellite silicon rejects.

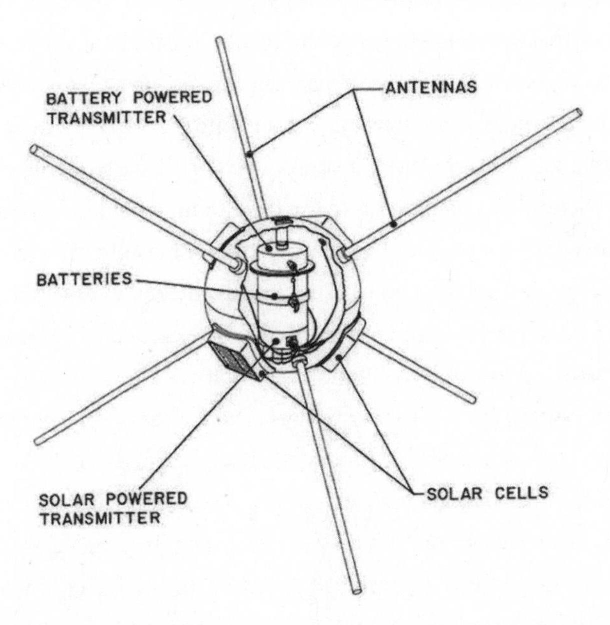

Vanguard I, the first US communications satellite, launched in 1958, only 6.4 inches in diameter, was powered by both batteries and early Bell Laboratories PV cells. The satellite went silent in 1964 but logged fifty-nine years in orbit in 2017.

Those worked well enough, but supply was intermittent, and they were still expensive. Solarex found a market for them, Varadi recalled, in the "vast areas where nobody was living but communication was still needed": federal lands maintained by the US Forest Service, the Bureau of Land Management, the National Weather Service, "and in many states, e.g., Arizona, the police. PV was tried and was performing very well, and these organizations used it."[6] The 1973–74 Arab oil embargo, stimulating demand for solar power, put Solarex on the map, and in the early 1980s the company developed a multicrystalline precursor by simply melting silicon, pouring it into a mold, and then cutting wafers from the resulting block. Large rooftop units followed and then solar farms. Southern California Edison turned on the first megawatt solar field in 1982 near the town of Hesperia in the high Mojave Desert northeast of Los Angeles.

Solar energy capacity worldwide was minimal in 2000 and had increased to only 50 gigawatts* by 2010. By 2017, solar energy was delivering a small but increasing share of world electricity: 305 gigawatts out of total world installed capacity of some 25 million gigawatts—much less than 1 percent. "Solar is still a relative minnow in the electricity mix of most countries," the *Guardian* reported that year. "Even where the technology has been embraced most enthusiastically, such as in Europe, solar on average provides 4% of electricity demand." Improved photovoltaic systems are under development using flexible and printable polymer thin films, and growth is brisk, particularly in China and the United States.[7] For many areas of the developing world lacking an electrical grid, improved solar technology with lowered cost may offer the same possibility of leapfrogging over the grid barrier that advantaged mobile phones.

In 2016 total installed wind electrical capacity reached 487 gigawatts. That's much less than 1 percent of world total electricity. Numbers for these intermittent energy sources are misleading, however, since they represent installed capacity rather than actual energy generated. Their "capacity factor"—how much of the time they actually generate electricity—is a problem for all intermittent energy sources. The sun doesn't always shine, nor the wind always blow, nor water always fall through the turbines of a dam. In the United States in 2016, nuclear power plants, which generated almost 20 percent of US electricity, had

*A gigawatt is 1 million kilowatts.

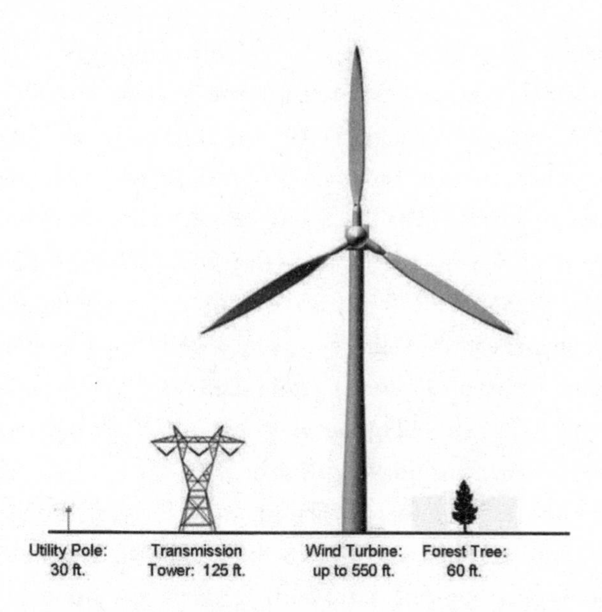

A modern wind turbine is an industrial-scale machine, up to 550 feet tall. Its generator and controls are housed in a streamlined nacelle behind the rotating blades.

an average capacity factor of 92.1 percent, meaning they operated at full power on 336 out of 365 days per year. (The other 29 days they were taken off the grid for maintenance—not all at the same time, of course.) In contrast, US hydroelectric systems delivered power 38 percent of the time (138 days per year); wind turbines, 34.7 percent of the time (127 days per year); and solar PV farms, only 27.2 percent of the time (99 days per year). Even plants powered with coal or natural gas generate electricity only about half the time.[8]

Electric charge flows nearly instantaneously from generators to outlets, which means it has to be generated in real time to meet demand. It can be stored temporarily in batteries or by using it to move masses of material, typically water, to higher ground, to be released later to generate power. In 2018 it was hardly stored anywhere except on a local and extremely limited scale. The United States had only 225 megawatts of grid battery storage in place in 2016, most of it installed by one Mid-Atlantic interconnection grid. When intermittent sources aren't generating electricity, or when their supply is ramping up and down because of blowing clouds or gusty wind conditions, they have to be backed up with a load-following source, typically natural gas. Unfortunately, natural gas is about 75 percent methane, a potent greenhouse gas some eighty-four times as effective at trapping heat in the atmosphere as CO_2.[9] Burned, natural gas produces about half as much carbon dioxide as coal burning, as well as smog-generating nitrogen oxides, sulfur, mercury, and particulates at lower levels than gasoline and much lower levels than coal.

The great challenge of the twenty-first century will be limiting global warming while simultaneously providing energy for a world population not only increasing in number but also advancing from subsistence to prosperity. A world population in 2100 of ten billion people is two and a half billion more—25 percent more—than the world population of 2017.

Another way to say "limiting global warming" is to speak, as energy experts do, of "decarbonizing" the energy sources the world uses. Switching from coal to natural gas is decarbonizing, since burning natural gas produces about half the carbon dioxide of burning coal. Switching from coal to nuclear power is radically decarbonizing, since nuclear power produces greenhouse gases (GHG) only during construction, mining, fuel processing, maintenance, and decommissioning—about as much as solar power does.[10] Both nuclear and solar generate only about 2 percent to 4 percent as much CO_2 as a coal-fired power plant and about 4 percent to 5 percent as much as a natural-gas-fired power plant.[11]

Yet nuclear power development was slowed in 2017 in the wake of the third nuclear accident worldwide in more than forty years of development. The accident in Pennsylvania at Three Mile Island, in 1979, destroyed the reactor but not its reinforced steel-and-concrete confinement structure and released minimal radioactivity into the atmosphere. The accident at Chernobyl, in what was then Soviet Ukraine in 1986, destroyed the reactor (which lacked a confinement structure, a design illegal in the United States) and released substantial airborne radioactivity as it burned out of control for fourteen days.[12]

The accident in Japan at Fukushima Daiichi in March 2011 followed a major earthquake and tsunami. The tsunami flooded out the power supply and cooling systems of three power reactors, causing them to melt down and explode, breaching their confinement. Although 154,000 Japanese citizens were evacuated from a twenty-kilometer (12.4-mile) exclusion zone around the power station, radiation exposure beyond the station grounds was limited.

According to the report submitted to the International Atomic Energy Agency (IAEA) in June 2011:

> No harmful health effects were found in 195,345 residents living in the vicinity of the plant who were screened by the end of May 2011. All the 1,080 children tested for thyroid gland exposure showed results within safe limits. By December, government health checks of some 1,700 residents who were evacuated from three municipalities showed that two-thirds received an external radiation dose within the normal international limit of 1 mSv/year, 98% were below 5 mSv/year, and ten people were exposed to more than 10 mSv. So while there was no major public exposure, let alone deaths from radiation, there were reportedly 761 victims of "disaster-related death," especially old people uprooted from homes and hospitals because of forced evacuation and other nuclear-related measures. The psychological trauma of evacuation was a bigger health risk for most than any likely exposure from early return to homes, according to some local authorities.[13]

These accidents need not have happened. Three Mile Island was a financial disaster for its owners, but its minor releases of radiation, well within accepted limits of radiation exposure, harmed no one.[14] Chernobyl resulted from a deeply flawed dual-use reactor design maintained as a military secret and an ill-advised experiment in shutdown control by poorly trained operators which required disarming all the reactor's safety systems. An overly slow reinsertion of the reactor's graphite-tipped control rods allowed it to runaway to one hundred times normal power. The resulting steam and hydrogen explosions destroyed the reactor core and blew pieces of its heavy biological shield through the roof of the reactor hall high into the air. Fire followed, releasing most of the core's radioactivity into the environment.

The Belarussian nuclear physicist Stanislav Shushkevich, in 1992 the first head of state of independent Belarus, told me about his experience of Chernobyl. Shushkevich directed a physics laboratory in a forested grove on the outskirts of Minsk, 285 miles north of Chernobyl. On the morning of 26 April 1986 a radiation alarm sounded in his laboratory. He and his colleagues assumed someone in the lab had spilled something radioactive. The safety officer began

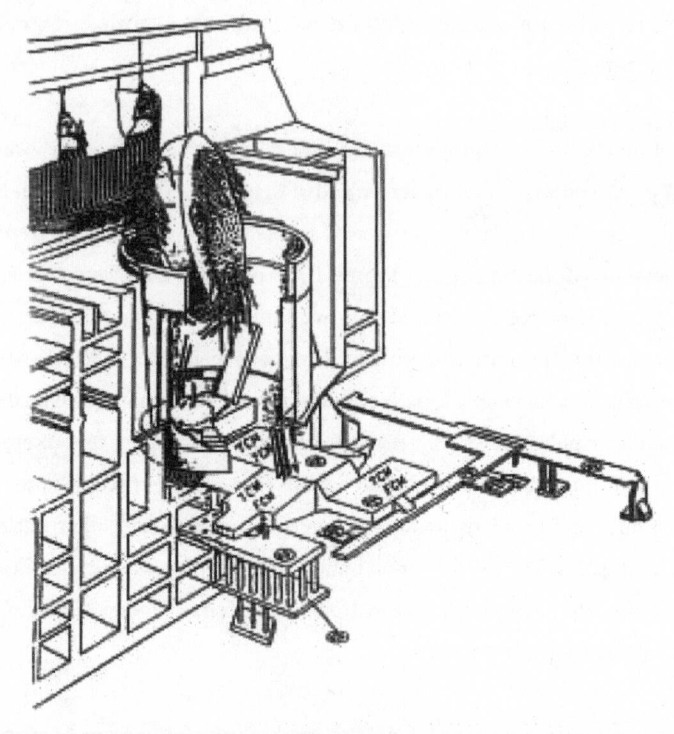

Chernobyl Reactor No. 4 after steam and hydrogen explosions. The bowl-shaped object on top of the reactor vessel is its massive lid, blown up and tilted. Objects labeled *TCW* on floor are pieces of reactor core. The reactor burned out of control for two weeks.

checking the lab with a radiation counter. When nothing turned up, he thought to go outside. The radiation level was higher outside than inside. Shushkevich said he then realized that the radiation must be coming from elsewhere. He immediately called the nearest nuclear power plant, at Visaginas, Lithuania, 126 miles north-northwest of Minsk. They reported no problems. Then he tried to call Chernobyl. No one answered the phone. Calling around, he learned of the Chernobyl explosion and fire. The fallout was blowing northward and increasing. It would soon pass over Minsk.

Shushkevich told me he thought immediately about the children in the path of the fallout. Its most threatening component would be iodine 131, a fission product with a short eight-day half-life that radiates energetic beta particles and gamma rays. It's taken up selectively by the thyroid, particularly by the active thyroids of children. The standard prevention for iodine 131 irradiation is a dose of potassium iodide, a salt that saturates the thyroid and temporarily blocks further uptake of iodine. Every fallout shelter in the Soviet Union contained a supply of potassium iodide tablets. To protect the children of Minsk, Shushkevich called his superiors in Moscow to ask permission to break out the tablets and dose the children. "They said," he told me, still angry ten years later, " 'Comrade, why are you making trouble? Do you want to start a riot? Shut up and get back to work.' "

A calm and charismatic man and a Belarussian patriot, Shushkevich decided then that he needed to go into politics. In time he became head of the Belarussian Supreme Soviet. On 18 December 1991 he was one of the three leaders, along with Russian president Boris Yeltsin and Ukrainian president Leonid Kravchuk, who signed the document that dissolved the Soviet Union. Then, as Belarus's first head of state, he transferred all the many nuclear weapons on Belarusian soil back to Russia, wanting no part of them or of the Russian army forces that guarded them.

By Shushkevich's reckoning, the Chernobyl accident was a failure of governance, not of technology. Had the Soviet Union's nuclear-power plants not been dual-use, designed for producing military plutonium as well as civilian power and therefore secret, problems with one reactor might have been shared with managers at other reactor stations, leading to safety improvements such as those introduced into US reactors after the accident at Three Mile Island and into Japanese reactors after Fukushima.

Tokyo Electric Power (Tepco), the company operating Fukushima Daiichi, had management problems as well and a long history of hiding mistakes from public scrutiny. A known engineering blunder made the Fukushima reactors dangerously vulnerable: their backup diesel generators and batteries, intended to supply power to pump cooling water through the reactor cores if the grid power supply failed, had been installed in the reactor hall basement despite the risk of flooding in a tsunami. Tepco had been warned by the Japanese Nuclear and Industrial Safety Agency a decade earlier to prepare for a thousand-year tsunami—one was known to have occurred 1,140 years previously—but had failed to heed the warning. Installation of backup power supplies *above* the reactors rather than below them would have protected the emergency cooling systems from any tsunami short of one massive enough to sweep clean the entire Japanese archipelago.[15]

Every technological system suffers accidents, staged as if by a malevolent god in exactly the crooks and crannies where human operators fail to imagine them occurring. Of all large-scale power technologies, nuclear has experienced the least number of accidents and counts the least number of deaths. A 2007 study in the English medical journal *Lancet* found that nuclear power operations result in "occupational deaths of around 0.019 per TWh*, largely at the mining, milling, and generation stages. These numbers are small in the context of normal operations. For example, a normal reactor of the kind in operation in France would produce 5.7 TWh a year. Hence, more than ten years of operations would be needed before a single occupational death could be attributed to the plant."[16]

Nuclear power's public health record more than compensates for its few occupational accidents. Its limited air pollution combined with its extremely low greenhouse-gas emissions and its 24/7 availability more than 90 percent of the time make it easily the most promising single energy source available to cope with twenty-first-century energy challenges.

Antinuclear activists, whose agendas originated in a misinformed neo-Malthusian foreboding of overpopulation (and a willingness at the margin to condemn millions of their fellow human beings to death from disease and starvation), may fairly be accused of disingenuousness in their successive arguments against the safest, least polluting, least warming, and most reliable energy source humanity has yet devised. Having moved successively through poorly supported arguments about safety and radiation, by the second decade of the twenty-first century they were reduced to two challenges: that nuclear power cost too much and that no safe method had yet been devised to dispose of nuclear waste.

Whether nuclear power costs too much will ultimately be a matter for the markets to decide, but there is no question that a full accounting of the external costs of different energy sources, including their contribution to air pollution and global warming, would find nuclear cheaper than coal or natural gas. The

*Terawatt-hour, a very large number: 1 terawatt-hour equals 1 trillion watt-hours.

disposal of so-called nuclear waste—meaning reactor spent fuel still charged with about 95 percent of its original energy potential—is a political problem in the United States, but it isn't today and has not been for many years an intractable technical problem. The notion that such waste must be successfully protected from exposure for hundreds of thousands of years is counter to how humans handle every other kind of toxic material we produce. We usually bury it, but we also discount its future risk, on the reasonable grounds that we owe concern to one or, at best, two generations beyond our own, since technologies improve over time, and our grandchildren and great-grandchildren will have better ways of dealing with our detritus than we do. (They might even correct our mistake of permanently burying "waste" with major potential for further clean energy production.)

And under what conditions would deeply buried nuclear waste be more than a local problem even if it were to be accidently or deliberately exposed? As I know from personal experience, it's a long, long ride in a narrow elevator cage down through a kilometer of solid rock and rock salt to the waste disposal level at the US Waste Isolation Pilot Plant outside Carlsbad, New Mexico, which I toured in 2015. The 2-kilometer-thick bed of crystalline salt into which WIPP's chambers are carved, the remains of an ancient sea, extends from southern New Mexico all the way northeastward to southwestern Kansas. It could easily accommodate the entire world's nuclear waste for the next thousand years. Finland is even further advanced in carving out a permanent repository in granite bedrock 400 meters (1,300 feet) under Olkiluoto, an island in the Baltic Sea off the nation's west coast. It expects to begin permanent waste storage in 2023.

I've focused on nuclear power in this chapter not because I imagine it to be the only solution to global warming. It's not, any more than renewable energy systems alone are. Every energy system has its advantages and disadvantages, as this excursion through four hundred years of energy developments should have made clear. And given the scale of global warming and human development, we will need them all if we are to finish the centuries-long process of decarbonizing our energy supply—wind, solar, hydro, nuclear, natural gas. As a harbinger of what's coming, the Iranian city of Bandar Mahshar suffered a heat index—a measure of temperature and humidity combined—of 165°F

(74°C) in August 2015. Temperatures in the Middle East in recent years have frequently exceeded 125°F.

Energy: A Human History originated in my encountering the work of an Italian physicist named Cesare Marchetti. Born in 1927, Marchetti has been based for many years at an institute in Laxenburg, Austria, outside Vienna—IIASA, the International Institute for Applied Systems Analysis. IIASA is one productive outcome of the ill-fated Club of Rome, founded in 1968 as a loose organization of European business leaders, scientists, and high-level government officials concerned (again) with overpopulation and resource depletion. With support from the Club of Rome, IIASA was established in 1972 as a think tank that might bridge the increasing digital (and continuing political) divide between the United States and Europe and the Soviet bloc countries.

Marchetti's earlier work had been in nuclear power technology, including reactor design and nuclear waste processing. His interest at IIASA has been energy, especially modeling the regularities of energy transitions. An informative graph based on research he and his colleagues pursued in the late 1970s drew my attention.

In an autobiographical sketch, Marchetti writes that he was asked when he joined IIASA in 1973 to find "a simple and predictive model describing energy markets for the last century or so." His economist colleagues, he jokes, "thought it better to look at their nails"—that is, wanted nothing to do with the assignment—so "with a certain spirit of adventure," Marchetti took up the challenge. Such challenges, "impossible problems," usually lead him to "look at the book of biological systems. Having been around for four billion years in an extremely hostile environment, they are a living library of working solutions."[17] The system he found was species competition for a biological niche. In economic terms, "the basic hypothesis—which has proved very fruitful and powerful—is that *primary energies, secondary energies, and energy distribution systems are just different technologies competing for a market* and should behave accordingly."[18]

The World Primary Energy Substitution graph on the following page is one result of Marchetti's study. A Portuguese political scientist, Luis de Sousa, reviewing the Italian physicist's work in 2007, explains: "This chart showed a very important thing: all of the Industrial Age energy sources follow a similar trend when entering the market. It takes 40 to 50 years for an energy source to

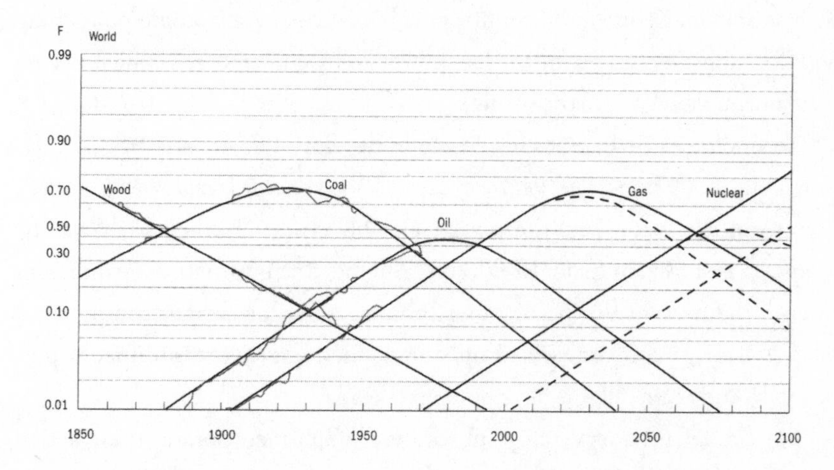

Historical evolution of world primary energy mix. Irregular lines are statistical data as of 1984; smooth lines are computed. Dotted line at right paralleling Nuclear represents a hypothetical new energy source introduced in 2025. Its effect on conventional sources is minimal until after 2050. Renewable sources, having not yet reached even 1 percent penetration, are similarly limited.

go from 1% to 10% of market share, and an energy source that eventually comes to occupy half of the market will take almost a century to do so, from the epoch [when] it reaches 1%."[19]

Why should adoption of a new energy source be so slow? Marchetti's important answer is that society is a learning system. It works by cultural diffusion—the spreading of ideas from one person to another—much like a disease epidemic. Inventing a new technology is only the beginning. Henry Ford's Model Ts needed filling stations. Filling stations needed gasoline, gasoline came from oil, oil had to be discovered, refineries had to process it, pipelines had to deliver the oil to the refineries and the gasoline to the cities where the cars were driven. People had to give up riding horses or horse-drawn buggies and buy cars, learn how to drive—and so on. When zippers began replacing buttons, there were those who resisted the change because they believed zippers were sinful: they made it easier to disrobe.

Coal to many Elizabethans was the Devil's excrement, as nuclear energy is today to many who oppose it. Fossil-fuel companies dislike nuclear and renewables equally: both compete for market space and hurt their bottom lines. As with so much else in American life, energy sources have become politicized, with Republicans embracing nuclear power and Democrats rejecting it, a state of affairs unlikely to save the planet.

Technologies themselves need time to develop. The economist Brian Arthur, in his 2009 book *The Nature of Technology: What It Is and How It Evolves,* points out that a new technology is inevitably crude. "In the early days," he writes, "it is sufficient that it work at all." After its first incarnation, Arthur continues, "the nascent technology must now be based on proper components, made reliable, improved, scaled up, and applied effectively to different purposes."[20] And again: all this development takes time.

Marchetti's primary energy substitution graph incorporates these activities as well as the subtler competition among various forms of energy. In his many papers, he continually expresses his surprise at the deep regularity of energy transitions, which he and his colleagues evidenced in some three thousand examples they researched across the decades. Society is not only a learning system: it is also patterned in waves of technological substitution at regular, more-or-less half-century intervals. This book, among other purposes, is a narrative extension of Marchetti's graph, reaching back in time to Shakespeare's day.

In his 2007 review, Luis de Sousa questioned Marchetti's graphic predictions. De Sousa offered a revised version, reproduced on the next page, that reflects the changes that followed the 1973–74 Arab oil embargo.

"In large measure," de Sousa writes, "the real data moved away from the model of the 1970s. This was probably due to the oil shocks that upset the market, but the prolonged effects are not as easily explainable. What immediately emerges to view is that after the oil crisis was surpassed in the 1980s, the market seems to have frozen, with each energy source maintaining its market share."[21] De Sousa comments in turn on each of the graph's energy components.

Coal: "Since the year 2000, coal has been moving upwards and looks like the best candidate to take oil's dominant place, as soon as the former peaks." Oil: "Although oil has been the most battered energy source since the 1970s, it is the one following closer [to Marchetti's] model." Oil again: "This means that Marchetti probably underestimated oil's trend . . . Today oil is clearly losing ground and will likely follow a downward trend." Natural gas: "[Marchetti's

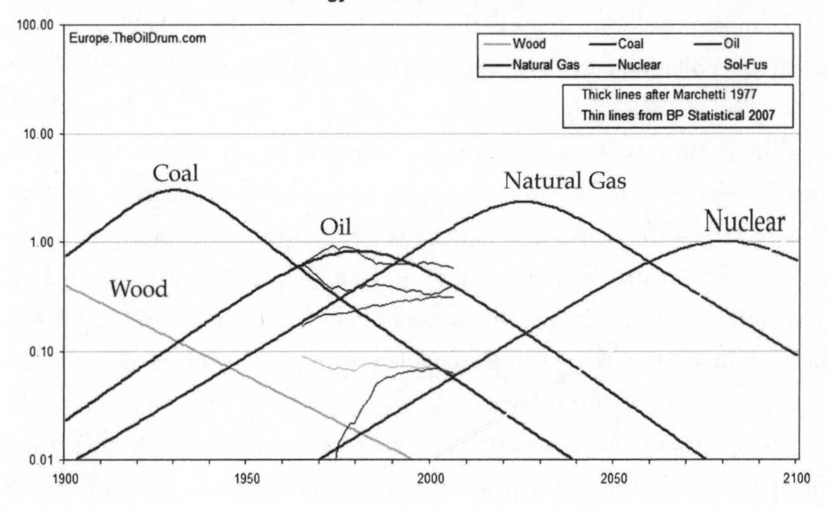

natural gas] model looks highly optimistic. . . . The underestimation of oil is probably reflected here in an overestimation of natural gas."

De Sousa's comments on nuclear power deserve quoting separately:

"Marchetti expected nuclear to enter the 5%–10% interval by the year 2000, but that happened much sooner: nuclear went over 5% in 1987. Up to the 1990s, nuclear energy greatly surpassed his expectations in the wake of the oil crisis (which facilitated market penetration), but as liquid hydrocarbons production started growing again, nuclear's penetration in the market slowed down. By 2000, it hovered around 6.5%, but has fallen below 6% since then."

And, finally, of renewables, de Sousa writes: "There isn't much to say about alternative energy sources, except that they never showed up. As a benchmark, wind energy occupies today 0.2% of the energy market, a point nuclear energy crossed in the 1950s."

Certainly nuclear has faltered against massive political resistance in both Europe and the United States, regions where renewables are heavily subsidized and nuclear heavily regulated. At the same time, a major new pulse of growth is beginning in East and South Asia, particularly India, China, Japan, and South Korea: as of January 2016, 128 operating reactors, 40 more under construction, and firm plans to build another 90. Many more are proposed.[22] These developments reflect the increasing economic expansion toward prosperity of the most populous nations in the world, which have been choking on their fossil-fuel pollution much as Europe and the United States choked on their pollution a hundred years ago.

The prosperous West can—barely—afford to produce all its power with renewables if it decides to do so. The rest of the world doesn't have that option. Cesare Marchetti's graph, assuming its vectors unfreeze themselves as conditions change, predicts a future energy supply dominated by nuclear and natural gas. We will need all that and renewables as well to sustain a world population of ten billion in anything like reasonable prosperity. The boats are lining up. There's room aboard for everyone.

Another book that has been one of my touchstone references since it was first published in 1985 is the scholar and philosopher Elaine Scarry's *The Body in Pain: The Making and Unmaking of the World.* In this remarkable study, Scarry first explores the purpose of the wounding and killing that is war—a contest of belief systems through physical sacrifice. Then, with profound originality, she discovers the purpose of applying the human imagination to the invention of technology:

> The naturally existing external world—whose staggering powers and beauty need not be rehearsed here—is wholly ignorant of the "hurtability" of human beings. Immune, inanimate, inhuman, it indifferently manifests itself in the thunderbolt and hailstorm, rabid bat, smallpox microbe, and ice crystal. The human imagination reconceives the external world, divesting it of its immunity and irresponsibility not by literally putting it in pain or making it animate but by, quite literally, *"making it" as knowledgeable about human pain as if it were itself animate and in pain.*[23]

So shoes protect the feet, a chair relieves the body of gravity's ceaseless burden, a windmill or a nuclear power plant generates electricity to warm or cool and light the way. Ultimately, beyond all arguments about which technology is greener and whether the world is large or small, that is why we reshape the inanimate through human invention. The great human project, Scarry shows, is the progressive alleviation of human suffering.

The population of the earth has increased more than sevenfold since 1850—from one billion to seven and a half billion—primarily because of science and technology, because of improvements in development, public health, nutrition, and medicine. In 1996 two demographers estimated that fully half the population of the United States at that time, 136 million people, were alive because of such improvements in mortality. Without such improvements, a quarter of us—68 million Americans—would have died before reaching the age of reproduction. As a consequence of those early deaths, another 68 million would have never been born.[24] That is more lives saved in the past century in the United States alone than all the man-made deaths from war in the twentieth century throughout the world. In our new century, at the beginning of a new millennium, such improvements in mortality have continued and extended their reach.

Far from threatening civilization, science, technology, and the prosperity they create will sustain us as well in the centuries to come. They are the only institutions human beings have yet devised that consistently learn from their mistakes.

ACKNOWLEDGMENTS

Many people helped me make this book. First and foremost, I thank the officers of the Alfred P. Sloan Foundation, Doron Weber in particular, for voting a grant that made possible the research travel necessary to walk the ground. Michael Keller, Stanford University Librarian, made available his library's outstanding resources.

Ninya Mikhaila of tudortailor.com advised on the winter wear of Elizabethan workmen. Keir Hind at the University of Glasgow Library confirmed current usage of the university's name. Anthony Dawson, author and expert, briefed me on British cavalry horse-buying practices during the Napoleonic Wars. Mike Dunn, manning the replica Newcomen engine at the Black Country Living Museum in Dudley, West Midlands, demonstrated its operation to me.

I corresponded with Ed Calabrese about his investigations of the work of Herman Muller. The late Ted Rockwell shared his experiences working with Hyman Rickover and building the first US commercial nuclear power plant. I benefited from correspondence with Cesare Marchetti and from discussions present or past with Harold Agnew, Hans Bethe, Richard Garwin, Thomas Graham Jr., David Rossin, Michael Shellenberger, Stanislav Shushkevich, Charles Till, Eugene Wigner, and no doubt others whose names escape me. I thank them all.

Anne Sibbald, my agent, represented me with her unfailing intelligence and professionalism. Ben Loehnen, my editor at Simon & Schuster, wielded his keen Occam's razor to make a better (and shorter) book. And Ginger, my steadfast Ginger, was there for me every step of the way: *And now good-morrow to our waking souls.*

BIBLIOGRAPHY

Aaron, Melissa D. "The Globe and Henry V as Business Document." *SEL: Studies in English Literature 1500–1900:* 40.2 (2000) 277–292.

Adams, Edward Dean. *Henry Adams of Somersetshire, England, and Braintree, Mass.: His English Ancestry and Some of His Descendants.* New York: printed privately, 1927.

———. *Niagara Power: History of the Niagara Falls Power Company 1886–1918: Evolution of Its Central Power Station and Alternating Current System.* 2 vols. Niagara Falls, NY: printed privately, 1927.

Adams, Henry. *The United States in 1800.* Ithaca, NY: Cornell University Press, 1955.

Adams, John. *The Works of John Adams.* Vol. 8., edited by Charles Francis Adams. Boston: Little, Brown, 1853.

Adams, Sean Patrick. *Home Fires: How Americans Kept Warm in the Nineteenth Century.* Baltimore: Johns Hopkins University Press, 2014.

Adams, W. Grylls. "The Scientific Principles Involved in Electric Lighting." *Journal of the Franklin Institute* (October 1881): 279–94.

———. "The Scientific Principles Involved in Electric Lighting." *Journal of the Franklin Institute* (November 1881): 364–75.

Adams, Charles Francis, Jr. *Railroads: Their Origin and Problems.* New York: G. P. Putnam's Sons, 1886.

Adas, Michael. *Machines as the Measure of Men: Science, Technology, and Ideologies of Western Dominance.* Ithaca, NY: Cornell University Press, 1989.

Agricola, Georgius. *De re Metallica.* Translated by Herbert Hoover and Lou Henry Hoover. New York: Dover, 1950. First published 1556.

Albion, Robert Greenhalgh. *The Timber Problem of the Royal Navy, 1652–1862.* Cambridge, MA: Harvard University Press, 1926.

Alexander, Thomas G. "Cooperation, Conflict, and Compromise: Women, Men, and the Environment in Salt Lake City, 1890–1930." *Brigham Young University Studies* 35, no. 1 (1995): 6–39.

Alexievich, Svetlana. *Voices from Chernobyl: The Oral History of a Nuclear Disaster.* New York: Picador, 2006.

Allen, John S. "The 1712 and Other Newcomen Engines of the Earls of Dudley." *Transactions of the Newcomen Society.* 37 (1967): 57–87.

———. "The 1715 and Other Newcomen Engines at Whitehaven, Cumberland." *Transactions of the Newcomen Society* 45 (1972): 237–68.

———. "The Introduction of the Newcomen Engine, 1710–1733." *Transactions of the Newcomen Society* 42 (1969): 169–90.

———. "The Introduction of the Newcomen Engine, 1710–1733: Second Addendum." *Transactions of the Newcomen Society* 45 (1972): 223–26.

Allison, Wade. *Nuclear Is for Life: A Cultural Revolution.* Oxford: Wade Allison, 2015.

Anderson, H. R. "Air Pollution and Mortality: A History." *Atmospheric Environment* 43 (2009): 142–52.

Andriesse, C. D. *Huygens: The Man Behind the Principle.* Translated by Sally Miedema. Cambridge: Cambridge University Press, 2005.

Antisell, Thomas. *The Manufacture of Photogenic or Hydro-Carbon Oils, from Coal and other Bituminous Substances, Capable of Supplying Burning Fluids.* New York: D. Appleton, 1860.

Antognazza, Maria Rosa. *Leibniz: An Intellectual Biography.* Cambridge UK: Cambridge University Press, 2009.

Appleby, Joyce. *Inheriting the Revolution: The First Generation of Americans.* Cambridge, MA: Harvard University Press, 2000.

Arago, M. *Historical Eloge of James Watt.* Translated by James Patrick Muirhead. London: John Murray, 1839.

Arthur, W. Brian. "Competing Technologies, Increasing Returns, and Lock-in by Historical Events." *Economic Journal* 99 (March 1989): 116–31.

———. *The Nature of Technology: What It Is and How It Evolves.* New York: Free Press, 2009.

Asbury, Herbert. *The Golden Flood: An Informal History of America's First Oil Field.* New York: Knopf, 1942.

Ashe, W. W. *The Forests, Forest Lands, and Forest Products of Eastern North Carolina. North Carolina Geological Survey Bulletin No. 5.* Raleigh, NC: Josephus Daniels, State Printer and Binder, 1894.

Ashton, T. S. *Iron and Steel in the Industrial Revolution.* Manchester, UK: Manchester University Press, 1951.

Aubrey, John. *Brief Lives.* Edited by John Buchanan Brown. London: Penguin, 2000.

Audubon, John James. *An Audubon Reader.* Edited by Richard Rhodes. New York: Everyman's Library, 2006.

Aurengo, André, Dietrich Averbeck, André Bonnin, Bernard Le Guen, Roland Masse, Roger Monier, Maurice Tubiana (chairman), Alain-Jacques Valleron, and Florent de Vathaire. "Dose-Effect Relationships and Estimation of the Carcinogenic Effects of Low Doses of Ionizing Radiation." Académie des Sciences [Academy of Sciences]—Académie nationale de Médecine [National Academy of Medicine], 2005 (online).

Ausubel, Jesse H. "The Liberation of the Environment." *Daedalus* 125, no. 3 (1996): 1–17.

Ausubel, Jesse H., and H. Dale Langford, ed. *Technological Trajectories and the Human Environment.* Washington, DC: National Academy Press, 1997.

Badash, Lawrence. *A Nuclear Winter's Tale: Science and Politics in the 1980s.* Cambridge, MA: MIT Press, 2009.

Baekeland, Leo H. "The Synthesis, Constitution, and Uses of Bakelite." *Industrial and Engineering Chemistry* 1 (1909): 149–61.

Bagwell, Philip S. *The Transport Revolution from 1770.* London: B. T. Batsford, 1974.

Bailey, Michael R., and John P. Glithero. *The Engineering and History of* Rocket: *A Survey Report.* York, UK: National Railway Museum, 2001.

Bailey, Ronald. *The End of Doom: Environmental Renewal in the Twenty-First Century.* New York: St. Martin's Press, 2015.

Baldwin, John, and Ron Powers. *Last Flag Down: The Epic Journey of the Last Confederate Warship.* New York: Three Rivers Press, 2007.

Barbier, Edward B. "Introduction to the Environmental Kuznets Curve Special Issue." *Environment and Development Economics* 2, no. 4 (November 1997): 369–81.

Bardou, Jean-Pierre, Jean-Jacques Chanaron, Patrick Fridenson, and James M. Laux. *The Automobile Revolution: The Impact of an Industry.* Translated by James M. Laux. Chapel Hill: University of North Carolina Press, 1982.

Barnard, H. E. "Prospects for Industrial Uses of Farm Products." *Journal of Farm Economics* 20 (1938): 119–33.

Bates, David. "A Half Century Later: Recollections of the London Fog." *Environmental Health Perspectives* 110, no. 12 (2002): A735.

Bauer, Martin, ed. *Resistance to New Technology: Nuclear Power, Information Technology and Biotechnology.* Cambridge: Cambridge University Press, 1995.

Baxter, Bertram. "Early Railways in Derbyshire." *Transactions of the Newcomen Society* 26 (1953): 185–97.

Bean, L. H., and P. H. Bollinger. "The Base Period for Parity Prices." *Journal of Farm Economics* 21, no. 1 (1939): 253–57.

Beaton, Kendall. "Dr. Gesner's Kerosene: The Start of American Oil Refining." *Business History Review* 29, no. 1 (1955): 28–53.

Beckerman, Wilfred. *Small Is Stupid: Blowing the Whistle on the Greens.* London: Duckworth, 1995.

———. *Two Cheers for the Affluent Society: A Spirited Defense of Economic Growth.* New York: St. Martin's Press, 1974.

Beddoes, Thomas, and James Watt. *Considerations on the Medicinal Use of Factitious Airs, and on the Manner of Obtaining Them in Large Quantities.* Bristol, UK: J. Johnson, 1794.

Belidor, Bernard Forest de. *Architecture hydraulique, ou L'art de Conduire, d'Élever et de Ménager les Eaux.* Paris: Chez L. Cellot, 1782.

Bell, Charles Henry. *The Bench and Bar of New Hampshire: Including Biographical Notices of Deceased Judges of the Highest Court, and Lawyers of the Province and State, and a List of Names of Those Now Living.* Boston: Houghton Mifflin, 1894.

Bell, Michelle L., and Devra Lee Davis. "Reassessment of the Lethal London Fog of 1952: Novel Indicators of Acute and Chronic Consequences of Acute Exposure to Air Pollution." *Environmental Health Perspectives* 109, no. 3 (2001): 389–94.

Bergin, Mike H., Chinmay Ghoroi, Deppa Dixit, James J. Schauer, and Drew T. Shindell. "Large Reductions in Solar Energy Production Due to Dust and Particulate Air Pollution." *Environmental Science and Technology Letters* 4, no. 8 (2017): 339–44.

Bernton, Hal, William Kovarik, and Scott Sklar. *The Forbidden Fuel: A History of Power Alcohol.* New ed. Lincoln: University of Nebraska Press, 1982.

Berry, Herbert, ed. *The First Public Playhouse: The Theatre in Shoreditch, 1576–1598.* Montreal: McGill-Queen's University Press, 1979.

———. "Shylock, Robert Miles, and Events at the Theatre." *Shakespeare Quarterly* 44, no. 2 (Summer 1993): 183–201.

Berryman, Jack W. "Sport, Health, and the Rural-Urban Conflict: Baltimore and John Stuart Skinner's *American Farmer,* 1819–1829." *Conspectus of History* 1, no. 8 (1982): 43–61.

Berthélemy, Michel, and Lina Escobar Rangel. "Nuclear Reactors' Construction Costs: The Role of Lead-Time, Standardization and Technological Progress." Working Paper 14-ME-01, Interdisciplinary Institute for Innovation, Dallas, October 9, 2013 (online).

Berthoud, E. L. "On the Occurrence of Uranium, Silver, Iron, etc., in the Tertiary Formation of Colorado Territory." *Proceedings of the Academy of Natural Sciences of Philadelphia* 27(2), May–September 1875, 363–66.

Beyea, Jan. "Response to 'On the Origins of the Linear No-Threshold (LNT) Dogma by Means of Untruths, Artful Dodges and Blind Faith.'" *Environmental Research* 148 (2016): 527–34.

Biello, David. "The World Really Could Go Nuclear." *Scientific American* online, last modified September 14, 2015.

Bierck, Harold A., Jr. "Spoils, Soils, and Skinner." *Maryland Historical Magazine* 49, no. 1 (1954): 21–40.

The Big Inch and Little Big Inch Pipelines. Houston: Texas Eastern Transmission Corporation, 2000.

Binder, Frederick Moore. *Coal Age Empire: Pennsylvania Coal and Its Utilization to 1860.* Harrisburg: Pennsylvania Historical and Museum Commission, 1974.

Black, Brian. *Petrolia: The Landscape of America's First Oil Boom.* Baltimore: Johns Hopkins University Press, 2000.

Black, Edwin. *Internal Combustion: How Corporations and Governments Addicted the World to Oil and Derailed the Alternatives.* New York: St. Martin's Press, 2006.

Black, Joseph. *Lectures on the Elements of Chemistry: Delivered at the University of Edinburgh.* 2 vols. Edited by John Robison. Philadelphia: Mathew Carey, 1807.

Blume, David. *Alcohol Can Be a Gas! Fueling an Ethanol Revolution for the 21st Century.* Santa Cruz, CA: International Institute for Ecological Agriculture, 2007.

Bockstoce, John. "From Davis Strait to Bering Strait: The Arrival of the Commercial Whaling Fleet in North America's Western Arctic." *Arctic* 37, no. 4 (1984): 528–32.

Bolles, Albert S. *Industrial History of the United States.* 3rd ed., repr. New York: Augustus M. Kelley, 1966. First published 1881.

Bonner, James. "Arie Jan Haagen-Smit." *Biographical Memoirs of the National Academy of Sciences.* Washington, DC: National Academy of Sciences, 1989.

Bonner, N. *Whales of the World.* New York: Facts on File, 1989.

Boucher, Cyril T. G. *James Brindley, Engineer, 1716–1772.* Norwich, UK: Goose and Son, 1968.

Bowler, Catherine, and Peter Brimblecombe. "Control of Air Pollution in Manchester Prior to the Public Health Act, 1875." *Environment and History* 6, no. 1 (2000): 71–98.

Boyd, T. A. *Professional Amateur: The Biography of Charles Franklin Kettering.* New York: E. P. Dutton, 1957.

Boyle, Robert. *New Experiments Physico-Mechanicall, Touching the Spring of the Air, and Its Effects (Made, for the Most Part, in a New Pneumatical Engine): Written by Way of Letter to the Right Honorable Charles, Lord Vicount of Dungarvan, Eldest Son to the Earl of Corke.* Oxford, UK: Thomas Robinson, 1660.

Bradby, Hannah, ed. *Dirty Words: Writings on the History and Culture of Pollution.* London: Earthscan, 1990.

Braudel, Fernand. *Civilization and Capitalism, 15th–18th Century.* Vol. 1, *The Structures of Everyday Life: The Limits of the Possible.* New York: Harper & Row, 1981.

———. *Civilization and Capitalism, 15th–18th Century.* Vol. 3, *The Perspective of the World.* Berkeley: University of California Press, 1984.

Bray, William, ed. *The Diary of John Evelyn.* 2 vols. London: Everyman Library, 1946.

Brazee, Edward B. *An Index to the* Sierra Club Bulletin, *1950–1976, Volumes 35–61.* Corvallis: Oregon State University Press, 1978.

Brenner, Joel Franklin. "Nuisance Law and the Industrial Revolution." *Journal of Legal Studies* 3, no. 2 (1974): 403–33.

Brice, William R. *Myth Legend Reality: Edwin Laurentine Drake and the Early Oil Industry.* Oil City, PA: Oil City Alliance, 2009.

Brimblecombe, Peter. "Air Pollution in Industrializing England." *Journal of the Air Pollution Control Association* 28, no. 2 (1978): 115–18.

———. *The Big Smoke: A History of Air Pollution in London Since Medieval Times.* London: Methuen, 1987.

Brown, Anthony Cave. *Oil, God, and Gold: The Story of Aramco and the Saudi Kings.* Boston: Houghton Mifflin, 1999.

Brown, A. J. "World Sources of Petroleum." *Bulletin of International News* 17, no. 13 (1940): 769–76.

Brown, F. Hume, ed. *Early Travellers in Scotland.* Edinburgh: Mercat Press, 1973.

Brown, Harrison. *The Challenge of Man's Future: An Inquiry Concerning the Condition of Man During the Years That Lie Ahead.* New York: Viking, 1954.

Brown, William H. *The History of the First Locomotives in America from Original Documents and the Testimony of Living Witnesses.* New York: D. Appleton, 1871.

Browne, D. J. *The Field Book of Manures; or, the American Muck Book.* New York: A. O. Moore, 1858.

Brownlie, David. "The Early History of the Coal Gas Process." *Transactions of the Newcomen Society* 3 (1924): 57–68 (plus plates).

Brox, Jane. *Brilliant: The Evolution of Artificial Light.* Boston: Houghton Mifflin Harcourt, 2010.

Brues, Austin M. "Critique of the Linear Theory of Carcinogenesis: Present Data on Human Leukemogenesis by Radiation Indicate That a Nonlinear Relation Is More Probable." *Science,* September 26, 1958: 693–99.

Bruland, Kristine, and Keith Smith. "Assessing the Role of Steam Power in the First Industrial Revolution: The Early Work of Nick von Tunzelmann." *Research Policy* 42 (2013): 1716–23.

Brundtland, Terje. "From Medicine to Natural Philosophy: Francis Hauksbee's Way to the Air-Pump." *British Journal for the History of Science* 41, no. 2 (2008), 209–40.

Brunskill, R. W. *Timber Building in Britain.* 2nd ed. London: Victor Gollancz, 1994.

Bryan, Ford R. *Friends, Families & Forays: Scenes from the Life and Times of Henry Ford.* Detroit: Wayne State University Press, 2002.

Bryce, Robert. "Guru or Fakir? Amory Lovins Is America's Favorite Green Energy Advocate. Does His Rhetoric Match Reality?" *Energy Tribune* online, November 2007.

———. *Power Hungry: The Myths of "Green" Energy and the Real Fuels of the Future.* New York: Public Affairs, 2010.

Bryson, Chris. "The Donora Fluoride Fog: A Secret History of America's Worst Air Pollution Disaster." ActionPA.org. Last modified December 2, 1998.

Bullard, John M. *The Rotches.* Printed privately, 1947.

Burch, Guy Irving, and Elmer Pendell. *Population Roads to Peace or War.* Washington, DC: Population Reference Bureau, 1945.

Bureau of Mines Bituminous Coal Staff. *Bureau of Mines Synthetic Liquid Fuels Program, 1944–55. Report of Investigations 5506.* Washington, DC: United States Department of the Interior, 1959.

Burn, Robert Scott. *The Steam-Engine, Its History and Mechanism, Being Descriptions and Illustrations of the Stationary, Locomotive, and Marine Engines.* London: H. Ingram, 1854.

Burnett, D. Graham. *Trying Leviathan: The Nineteenth-Century New York Court Case That Put the Whale on Trial and Challenged the Order of Nature.* Princeton, NJ: Princeton University Press, 2007.

Burris, Evadene A. "Keeping House on the Minnesota Frontier." *Minnesota History* 14, no. 3 (1933): 263–82.

Burton, Anthony. *Richard Trevithick: Giant of Steam.* London: Aurum Press, 2000.

Bush, A. L., and H. K. Stager. "Accuracy of Ore-Reserve Estimates for Uranium-Vanadium Deposits on the Colorado Plateau." *Geological Survey Bulletin 1030-D.* Washington, DC: US Government Printing Office, 1956: 137.

Butler, G. M., and M. A. Allen. "Uranium and Radium." *University of Arizona Bulletin Mineral Technology Series No. 27* (December 1, 1921). Tucson: University of Arizona and Arizona Bureau of Mines, 1921.

F. C. *The Compleat Collier: Or, the Whole Art of Sinking, Getting, and Working, Coal-Mines, &c. As Is Now Used in the Northern Parts, especially About Sunderland and New-Castle.* London: G. Conyers, 1708.

Calabrese, Edward J. "Muller's Nobel Lecture on Dose-Response for Ionizing Radiation: Ideology or Science?" *Archives of Toxicology* 85 (2011): 1495–98.

———. "On the Origin of the Linear No-Threshold (LNT) Dogma by Means of Untruths, Artful Dodges and Blind Faith." *Environmental Research* 142 (2015): 432–42.

———. "The Road to Linearity: Why Linearity at Low Doses Became the Basis for Carcinogen Risk Assessment." *Archives of Toxicology* 83 (2009): 203–25.

———. "The Threshold Vs. LNT Showdown: Dose Rate Findings Exposed Flaws in the LNT Model, Part 1: The Russell-Muller Debate." *Environmental Research* 154 (2017): 435–51.

Calabrese, Edward J., and Linda A. Baldwin. "Toxicology Rethinks Its Central Belief: Hormesis Demands a Reappraisal of the Way Risks Are Assessed (Commentary)." *Nature* 421 (February 13, 2003): 691–92.

Caldwell, James. *Report to the United States Shipping Board Emergency Fleet Corporation on Electric Welding and its Application in the United States of America to Ship Construction.* Philadelphia: Emergency Fleet Corporation, 1918.

Camden, William. *Britannia, or a Chorographical Description of the Most Flourishing Kingdoms, England, Scotland, and Ireland, and the lands Adjoining, out of the Depth of Antiquity.* Translated by Philemon Holland. London: George Bishop and Ioannis Norton, 1610.

Campbell, John L. *Collapse of an Industry: Nuclear Power and the Contradictions of U.S. Policy.* Ithaca, NY: Cornell University Press, 1988.

Canby, Edward Tatnall. *A History of Electricity.* New York: Hawthorn Books, 1968.

Cardis, E., E. S. Gilbert, L. Carpenter, G. Howe, I. Kato, BK Armstrong, V. Beral et al. "Effects of Low Doses and Low Dose Rates of External Ionizing Radiation: Cancer Mortality Among Nuclear Industry Workers in Three Countries." *Radiation Research* 142, no. 2 (1995): 117–32.

Carlson, Elof Axel. *Genes, Radiation, and Society: The Life and Work of H. J. Muller.* Ithaca, NY: Cornell University Press, 1981.

Carolan, Michael S. "A Sociological Look at Biofuels: Ethanol in the Early Decades of the Twentieth Century and Lessons for Today." *Rural Sociology* 74, no. 1 (2009): 86–112.

Carson, Rachel. *Silent Spring.* New York: Houghton Mifflin, 1962.

Carter, Jimmy. *Why Not the Best? Jimmy Carter: The First Fifty Years.* Fayetteville: University of Arkansas Press, 1996. First published 1975.

Casey, Robert. *The Model T: A Centennial History.* Baltimore: Johns Hopkins University Press, 2008.

Caspari, Ernest, and Curt Stern. "The Influence of Chronic Irradiation with Gamma-Rays at Low Dosages on the Mutation Rate in *Drosophila Melanogaster.*" *Genetics* 33, no. 1 (1948): 75–95.

Castaneda, Christopher J. *Invisible Fuel: Manufactured and Natural Gas in America, 1800–2000.* New York: Twayne, 1999.

———. *Regulated Enterprise: Natural Gas Pipelines and Northeastern Markets, 1938–1954.* Columbus: Ohio State University Press, 1993.

———. "The Texas-Northeast Connection: The Rise of the Post–World War II Gas Pipeline Industry," *Houston Review* 12, no. 2 (1990).

Castaneda, Christopher J., and Clarance M. Smith (1996). *Gas Pipelines and the Emergence of America's Regulatory State: A History of Panhandle Eastern Corporation, 1926–1993.* Cambridge: Cambridge University Press.

Cerasano, S. P. "The Fortune Contract in Reverse." *Shakespeare Studies* 37 (2009): 79–98.

———. "The Geography of Henslowe's Diary." *Shakespeare Quarterly* 56, no. 3 (2005): 328–53.

———. "Philip Henslowe, Simon Forman, and the Theatrical Community of the 1590s." *Shakespeare Quarterly* 44, no. 2 (1993): 145–58.

Cernansky, Rachel. "State-of-the-Art Soil: A Charcoal-Rich Product Called Biochar." *Nature* 517 (2015): 258–60.

Chalmers, II, Harvey. *The Birth of the Erie Canal.* New York: Bookman, 1960.

Chandler, Charles F. "Address of Acceptance." *Journal of Industrial and Engineering Chemistry* 12, no. 2 (1920): 189–95.

Chandler, John, ed. *John Leland's Itinerary: Travels in Tudor England.* Dover, NH: Alan Sutton, 1993.

Chapelle, Howard I. *The History of American Sailing Ships.* New York: Bonanza, 1988.

Chapin, D. M., C. S. Fuller, and G. L. Pearson. "A New Silicon *P-N* Junction Photocell for Converting Solar Radiation into Electrical Power." *Journal of Applied Physics* 25 (1954): 676–77.

Chapman, Stanley. "British Exports to the U.S.A., 1776–1914: Organisation and Strategy (3) Cottons and Printed Textiles." From *Textiles in Trade: Proceedings of the Textile Society of America Biennial Symposium, September 14–16, 1990, Washington, DC.*

Chard, Jack. *Making Iron & Steel: The Historic Processes, 1700–1900.* Ringwood, NJ: North Jersey Highlands Historical Society, 1995.

Cheney, Margaret. *Tesla: Man out of Time.* New York: Simon & Schuster, 1981.

Chenoweth, William L. *Summary of the Uranium-Vanadium Ore Production, 1947–1969, Monument Valley District, Apache and Navajo Counties, Arizona: Contributed Report CR-14-C.* Tucson: Arizona Geological Survey, 2014.

———. *Uranium Procurement and Geologic Investigations of the Manhattan Project in Arizona: Open-File Report 88-02.* Tucson: Arizona Geological Survey, 1988.

Chesney, Cummings C. "Some Contributions to the Electrical Industry." *Electrical Engineering* 52, no. 12 (1933): 726–30.

Chicago Association of Commerce Committee of Investigation on Smoke Abatement and Electrification of Railway Terminals. *Smoke Abatement and Electrification of Railway Terminals in Chicago.* Chicago: Rand McNally, 1915.

Chillrud, Steven N., Richard F. Bopp, H. James Simpson, James M. Ross, Edward L.

Shuster, Damon A. Chaky, Dan C. Walsh, Cristine Chin Choy, Lael Ruth Tolley, and Allison Yarme. "Twentieth Century Atmospheric Metal Fluxes into Central Park Lake, New York City." *Environmental Science and Technology* 33, no. 5 (1999): 657–62.

Christensen, Leo M., Ralph M. Hixon, and Ellis I. Fulmer. *Power Alcohol and Farm Relief.* Deserted Village 3. New York: Chemical Foundation, 1934: 5–191.

Churchill, Jason Lemoine. "The Limits to Influence: The Club of Rome and Canada, 1968 to 1988." PhD diss., University of Waterloo, Waterloo, Ont., 2006.

Cifuentes, Luis, Victor H. Borja-Aburto, Nelson Gouveia, George Thurston, and Devra Lee Davis. "Hidden Health Benefits of Greenhouse Gas Mitigation." *Science* 293, no. 5533 (August 17, 2001): 1257–59.

Clack, Christopher T. M., Staffan A. Qvist, Jay Apt, Morgan Bazilian, Adam R. Brandt, Ken Caldeira, Steven J. Davis et al. "Evaluation of a Proposal for Reliable Low-Cost Grid Power with 100% Wind, Water, and Solar." *PNAS* 30, no. 20 (2017): 1–6; supporting information, 1–13.

Clark, A. Howard. "The American Whale-Fishery 1877–1886." *Science* ns-9, no. 217S (1887): 321–24.

Clark, Gregory, and David Jacks. "Coal and the Industrial Revolution, 1700–1869." *European Review of Economic History* 11, no. 1 (April 2007): 39–72.

Clark, J. Stanley. *The Oil Century: From the Drake Well to the Conservation Era.* Norman: University of Oklahoma Press, 1958.

Clark, James A. *The Chronological History of the Petroleum and Natural Gas Industries.* Houston: Clark, 1963.

Clark, James Anthony, and Michel T. Halbouty. *The Last Boom.* New York: Random House, 1972.

Clarke, David. *Reflections on the Astronomy of Glasgow: A Story of Some Five Hundred Years.* Edinburgh: Edinburgh University Press, 2013.

Clavering, Robert. *An Essay on the Construction and Building of Chimneys.* London: I. Taylor, 1779.

Clayton, J. C. "The Shippingport Pressurized Water Reactor and Light Water Breeder Reactor." For presentation at 25th Central Regional Meeting, American Chemical Society, Pittsburgh, October 4–6, 1993 (online).

Clegg, Samuel, Jr. *A Practical Treatise on the Manufacture and Distribution of Coal-Gas, Its Introduction and Progressive Improvement.* London: John Weale, 1866.

Clow, Archibald, and Nan L. Clow. "Lord Dundonald." *Economic History Review* 12: nos. 1 and 2 (1942): 47–58.

———. "The Timber Famine and the Development of Technology." *Annals of Science* 12, no. 2 (1956): 85–102.

Cochrane, Thomas. *The Autobiography of a Seaman.* London: Maclaren, n.d.

Cockayne, Emily. *Hubbub: Filth, Noise & Stench in England 1600–1770.* New Haven, CT: Yale University Press, 2007.

Cohen, Aaron J., H. Ross Anderson, Bart Ostra, Kiran Dev Pandey, Michal Kryzanowski, Nino Künzli, Kersten Gutschmidt et al. "The Global Burden of Disease Due to Outdoor Air Pollution." *Journal of Toxicology and Environmental Health, Part A* 68 (2005): 1–7.

Cohen, Bernard L. "High Level Radioactive Waste." *Natural Resources Journal* 21 (1981): 703–21.

Cohen, Michael P. *The History of the Sierra Club 1892–1970.* San Francisco: Sierra Club Books, 1988.

Cole, H. S. D., Christopher Freeman, Marie Jahoda, and K. L. R. Pavitt. *Models of Doom: A Critique of* The Limits to Growth. New York: Universe Books, 1973.

Coleman, D. C. "The Coal Industry: A Rejoinder." *Economic History Review,* n.s. 30, no. 2 (1977): 343–45.

Conca, James. "Pollution Kills More People Than Anything Else." *Forbes* online, last modified November 7, 2017.

———. "Radiation Poses Little Risk to the World." *Forbes* online, last modified June 24, 2016.

Connan, Jacques, Pierre Lombard, Robert Killick, Flemming Høljund, Jean-François Salles, and Anwar Khalaf. "The Archeological Bitumens of Bahrain from the Early Dilmun Period (c. 2200 BC) to the Sixteenth Century AD: A Problem of Sources and Trade." *Arabian Archeology and Epigraphy* 9, no. 2 (November 1998): 141–81.

Corton, Christine L. *London Fog: The Biography.* Cambridge, MA: Harvard University Press, 2015.

Cottrell, F. *Energy and Society: The Relationship Between Energy, Social Change, and Economic Development.* Westport, CT: Greenwood Press, 1955.

Cousteau, J-Y, and P. Diole. *The Whale.* New York: Arrowwood Press, 1972.

Covello, V. T. "The Perception of Technological Risks: A Literature Review." *Technological Forecasting and Social Change* 23 (1983): 285–97.

Cowan, George. "A Natural Fission Reactor." *Scientific American,* July 1976, 36–47.

Cowan, Robin, and Staffan Hultén. "Escaping Lock-In: The Case of the Electric Vehicle." *Technological Forecasting and Social Change* 53, no. 1 (September 1996): 61–79.

Cox, J. Charles. *The Royal Forests of England.* London: Methuen, 1905.

Cox, Louis Anthony (Tony), Jr. "Socioeconomic and Air Pollution Correlates of Adult Asthma, Heart Attack, and Stroke Risks in the United States, 2010–2013." *Environmental Research* 155 (2017): 92–107.

Crevecoeur, J. Hector St. John. *Letters from an American Farmer.* New York: Fox, Duffield, 1904. First published 1782.

Crookes, William. "A New Era in Illumination—Wilde's New Magneto-Electric Machine." *Journal of the Franklin Institute* (December 1866): 400–9.

Cullingford, Benita. *British Chimney Sweeps: Five Centuries of Chimney Sweeping.* Hove, UK: Book Guild, 2000.

Cummings, R. G., and Albert E. Utton. "Managing Nuclear Wastes: An Overview of the Issues." *Natural Resources Journal* 21 (1981): 693–701.

Cummins, C. Lyle, Jr. *Internal Fire: The Internal Combustion Engine, 1673–1900.* Lake Oswego, OR: Carnot Press, 1976.

Curr, John. *The Coal Viewer, and Engine Builder's Practical Companion.* Sheffield, UK: John Northall, 1797.

Cushman, Gregory T. *Guano and the Opening of the Pacific World: A Global Ecological History.* Cambridge: Cambridge University Press, 2013.

Cuttler, Jerry M. "Remedy for Radiation Fear: Discard the Politicized Science." *Dose Response* 12, no. 2 (2014): 170–84.

Daniels, Farrington. "Direct Use of the Sun's Energy." *American Scientist* 55, no. 1 (1967): 15–47.

Darley, Gillian. *John Evelyn: Living for Ingenuity.* New Haven, CT: Yale University Press, 2006.

Darwin, Erasmus. *The Botanic Garden, A Poem, in Two Parts; containing The Economy of Vegetation, and The Loves of the Plants. With Philosophical Notes.* London: Jones & Co., 1824.

Davidson, Cliff I. "Air Pollution in Pittsburgh: A Historical Perspective." *Journal of the Air Pollution Control Association* 29, no. 10 (1979): 1035–41.

Davies, A. Stanley. "The Coalbrookdale Company and the Newcomen Engine, 1717–69." *Transactions of the Newcomen Society* 20 (1941): 45–48.

Davies II, Edward J. *The Anthracite Aristocracy: Leadership and Social Change in the Hard Coal Regions of Northeastern Pennsylvania, 1800–1930.* DeKalb: Northern Illinois University Press, 1985.

Davis, Lance E., Robert E. Gallman, and Karin Gleiter. *In Pursuit of Leviathan: Technology, Institutions, Productivity, and Profits in American Whaling, 1816–1906.* Chicago: University of Chicago Press, 1997.

Davy, Humphry. *The Collected Works.* Vol. 8, *Agricultural Lectures,* pt. 2, edited by John Davy. London: Smith, Elder, 1840.

———. "On the Fire-Damp of Coal Mines, and on Methods of Lighting the Mines So As to Prevent Its Explosion." *Philosophical Transactions of the Royal Society of London* 106 (1816): 1–22.

———. *On the Safety Lamp for Preventing Explosions in Mines, Houses Lighted by Gas,*

Spirit Warehouses, or Magazines in Ships, &c. With Some Researches on Flame. London: R. Hunter, 1825.

Dawson, Frank. *John Wilkinson: King of the Ironmasters.* Edited by David Lake. Stroud, UK: History Press, 2012.

Day, Barry. *This Wooden 'O': Shakespeare's Globe Reborn.* London: Oberon, 1996.

Defoe, Daniel. *A Tour Through the Whole Island of Great Britain.* 3 vols. London: Folio Society, 1983.

Dellapenna, Joseph W. "A Primer on Groundwater Law." *Idaho Law Review* 49 (2012): 265.

———. "The Rise and Demise of the Absolute Dominion Doctrine for Groundwater." *University of Arkansas at Little Rock Law Review* 35, no. 2 (2013): 273.

Dendy Marshall, C. F. "The Rainhill Locomotive Trials of 1829." *Transactions of the Newcomen Society* 9, no. 1 (1928): 78–93.

Department of the Interior. *Hearings Before the Secretary of the Interior on Leasing of Oil Lands and Natural-Gas Wells in Indian Territory and Territory of Oklahoma. May 8, 24, 25, and 29, and June 7 and 10, 1906.* Washington, DC: US Government Printing Office, 1906.

Desrochers, Pierre, and Christine Hoffbauer. "The Postwar Intellectual Roots of the Population Bomb: Fairfield Osborn's 'Our Plundered Planet' and William Vogt's 'Road to Survival' in Retrospect." *Electronic Journal of Sustainable Development* 1, no. 3 (2009): 37–61.

Deutch, John M. *The Crisis in Energy Policy.* Cambridge, MA: Harvard University Press, 2011.

Dibner, Bern. *Oersted and the Discovery of Electromagnetism. 2nd ed.* New York: Blaisdell, 1962.

Dickinson, H. W., and Arthur Titley. *Richard Trevithick: The Engineer and the Man.* Cambridge: Cambridge University Press, 1934.

Diesel, Eugen, Gustav Goldbeck, and Friedrich Schildberger. *From Engines to Autos: Five Pioneers in Engine Development and Their Contributions to the Automotive Industry.* Chicago: Henry Regnery, 1960.

Dillon, Maureen. *Artificial Sunshine: A Social History of Domestic Lighting.* London: National Trust, 2002.

Dimitri, Carolyn, and Anne Effland. "Fueling the Automobile: An Economic Exploration of Early Adoption of Gasoline over Ethanol." *Journal of Agricultural & Food Industrial Organization* 5, no. 2 (2007), 1–21.

Dircks, Henry. *The Life, Times, and Scientific Labours of the Second Marquis of Worcester. To Which is Added, a Reprint of His* Century of Inventions, *1663, with a Commentary Thereon.* London: Bernard Quaritch, 1865.

Dolin, Eric Jay. *Leviathan: The History of Whaling in America.* New York: W. W. Norton, 2007.

Douglas, Ian, Rob Hodgson, and Nigel Lawson. "Industry, Environment and Health Through 200 Years in Manchester." *Ecological Economics* 41 (2002): 235–55.

Douglas, Mary, and Aaron Wildavsky. *Risk and Culture: An Essay on the Selection of Technological and Environmental Dangers.* Berkeley: University of California Press, 1982.

Downer, John. "Disowning Fukushima: Managing the Credibility of Nuclear Reliability Assessment in the Wake of Disaster." *Regulation & Governance* 8, no. 3 (September 2014): 287–309.

Downey, Morgan. *Oil 101.* N.p.: Wooden Table Press, 2009.

Dudley, Dud. *Mettallum Martis, or, Iron made with pit-coale, sea-coale, &c.: and with the same fuell to melt and fine imperfect metals, and refine perfect metals.* Reprint of London: printed by T. M. for the author, 1665. Eindhoven, Nederland: De Archaeologische Pers. (1988).

Duffy, David C. "The Guano Islands of Peru: The Once and Future Management of a Renewable Resource." *Bird Life Conservation Series,* no. 1 (1994): 68–76.

Dunlap, Riley E. "Trends in Public Opinion Toward Environmental Issues: 1965–1990." *Society and Natural Resources* 4, no. 3 (1991): 285–312.

DuPont, Robert L. (1981). "Perspectives of Nuclear Risk: The Role of the Media." Presented at the Annual Meeting of the Canadian Nuclear Association, Ottawa, Canada, June 9, 1981.

Dyni, John R. *Geology and Resources of Some World Oil-Shale Deposits: Scientific Investigations Report 2005–5294.* Reston, VA: US Geological Survey, US Department of the Interior, 2006.

Eaton, S. J. M. *Petroleum: A History of the Oil Region of Venango County, Pennsylvania.* Philadelphia: J. P. Skelly, 1886.

Eavenson, Howard N. *The First Century and a Quarter of American Coal Industry.* Pittsburgh: printed privately, 1942.

Eberhart, Mark E. *Feeding the Fire: The Lost History and Uncertain Future of Mankind's Energy Addiction.* New York: Harmony Books, 2007.

Ehrlich, Paul R. *The Population Bomb.* San Francisco: Sierra Club, 1969.

Ekirch, A. Roger. *At Day's Close: Night in Times Past.* New York: W. W. Norton, 2005.

Eliot, Charles W., ed. *The Harvard Classics.* Vol. 35, *Chronicle and Romance: Froissart, Malory, Holinshed.* New York: P. F. Collier & Son, 1938.

Epstein, Alex. *The Moral Case for Fossil Fuels.* New York: Portfolio/Penguin, 2014.

Eskew, Garnett Laidlaw. *Salt: The Fifth Element: The Story of a Basic American Industry.* Chicago: J. G. Ferguson, 1948.

Espinasse, Francis. *Lancashire Worthies.* London: Simpkin, Marshall, 1874.

Esty, William Suddards. *Dynamos and Motors: A Text Book for Colleges and Technical Schools.* New York: Macmillan, 1909.

Evans, Brock. "Sierra Club Involvement in Nuclear Power: An Evolution of Awareness." *Oregon Law Review* 54 (1975): 607–21.

Evans, Oliver. *The Abortion of the Young Steam Engineer's Guide.* Philadelphia: printed for the author by Fry and Kammerer, 1805.

Evelyn, John. *A Character of England.* London: Joseph Crooke, 1659. Early English Books Online.

———. *Fumifugium: or, the Inconvenience of the Aer, and Smoake of London Dissipated.* London: printed by W. Godbid, for Gabriel Bedel, and Thomas Collins, 1661. Reprinted for B. White, 1672.

———. *Sylva, or a Discourse of Forest-Trees and the Propagation of Timber in His Majesty's Dominions.* London: Robert Scott et al., 1664.

Eyles, Joan M. "William Smith, Richard Trevithick, and Samuel Homfray: Their Correspondence on Steam Engines, 1804–1806." *Transactions of the Newcomen Society* 43, no. 1 (1970): 137–61.

Fanning, Leonard M. *The Rise of American Oil.* New York: Harper & Brothers, 1948.

Farey, John. *A Treatise on the Steam Engine, Historical, Practical, and Descriptive.* London: Longman, Rees, Orme, Brown, and Green, 1827.

Fenger, Jes, O. Hertel, and F. Palmgren, eds. *Urban Air Pollution—European Aspects.* Dordrecht, Neth.: Springer, 1999.

Fermi, Enrico. "Atomic Energy for Power." In *The Future of Atomic Energy: The George Westinghouse Centennial Forum, May 16, 17, and 18, 1946.* Vol. 1. Pittsburgh: Westinghouse Educational Foundation.

———. "Experimental Production of a Divergent Chain Reaction." *American Journal of Physics* 20, 536–58, 1952.

Fischer, R. P., and L. S. Hilpert. "Geology of the Uravan Mineral Belt. Contributions to the Geology of Uranium." *US Geological Survey Bulletin* 988-A, 1952.

Fisher, Howard J. *Faraday's Experimental Researches in Electricity: Guide to a First Reading.* Santa Fe, NM: Green Lion Press, 2014.

Fletcher, William. *English and American Steam Carriages and Traction Engines.* Repr. Devon, UK: David & Charles, 1973. First published 1904.

Flink, James J. *America Adopts the Automobile, 1895–1910.* Cambridge, MA: MIT Press, 1970.

Flinn, Michael W. "Timber and the Advance of Technology: A Reconsideration," *Annals of Science,* 15:2 (1959): 109–20.

Ford, Alice. *The 1826 Journal of John James Audubon.* Norman: University of Oklahoma Press, 1967.

Forrester, Jay W. *World Dynamics.* Cambridge, MA: Wright-Allen Press, 1971.

Foster, Abram John. *The Coming of the Electrical Age to the United States.* New York: Arno Press, 1979.

Fouquet, Roger, and Peter J. G. Pearson. "A Thousand Years of Energy Use in the United Kingdom." *Energy Journal* 19, no. 4 (1998): 1–41.

Fox, Stephen. *Wolf of the Deep: Raphael Semmes and the Notorious Confederate Raider* CSS Alabama. New York: Vintage, 2007.

Franklin, Benjamin. *Experiments and Observations on Electricity Made at Philadelphia in America.* London: E. Cave, 1751.

———. *The Papers of Benjamin Franklin.* Vol. 5, *July 1, 1753, Through March 31, 1755,* edited by Leonard W. Labaree. New Haven, CT: Yale University Press, 1962.

Franklin, William Studdards, and William Esty. *Dynamos and Motors.* New York: Macmillan, 1909.

Franks, Angela. *Margaret Sanger's Eugenic Legacy: The Control of Female Fertility.* Jefferson, NC: McFarland, 2005.

Franks, Kenny A., Paul F. Lambert, and Carl N. Tyson. *Early Oklahoma Oil: A Photographic History, 1859–1936.* College Station: Texas A&M University Press, 1981.

Friedel, Robert, and Paul Israel. *Edison's Electric Light: Biography of an Invention.* New Brunswick, NJ: Rutgers University Press, 1986.

Frye, Northrop. *Northrop Frye on Shakespeare.* Edited by Robert Sandler. Markham, Ont.: Fitzhenry & Whiteside, 1986.

Fthenakis, Vasilis M., and Hyung Chul Kim. "Greenhouse-Gas Emissions from Solar Electric- and Nuclear Power: A Life-Cycle Study." *Energy Policy* 35 (2007): 2549–57.

Fulton, John F., and Elizabeth H. Thomson. *Benjamin Silliman 1779–1864: Pathfinder in American Science.* New York: Henry Schuman, 1947.

Funigiello, Philip J. *Toward a National Power Policy: The New Deal and the Electric Utility Industry, 1933–1941.* Pittsburgh: University of Pittsburgh Press, 1973.

Galbraith, John Kenneth. *American Capitalism: The Concept of Countervailing Power.* Boston: Houghton Mifflin, 1952.

Gallopin, Gilberto C. "Branching Futures and Energy Projections." *Renewable Energy for Development* 10, no. 3 (1997) (online).

Galloway, Robert L. *Annals of Coal Mining and the Coal Trade: The Invention of the Steam Engine and the Origin of the Railway.* London: Colliery Guardian, 1898.

———. *A History of Coal Mining in Great Britain.* London: Macmillan, 1882.

Galvani, Luigi. *Commentary on the Effect of Electricity on Muscular Motion (De Viribus Electricitatis in Motu Musculari Commentarius).* Translated by Robert Montraville Green. Cambridge, MA: Elizabeth Licht, 1953. First published 1791.

Gannon, Michael. *Operation Drumbeat: The Dramatic True Story of Germany's First*

U-Boat Attacks Along the American Coast in World War II. New York: Harper & Row, 1990.

Gauthier-Lefaye, François. "2 Billion Year Old Natural Analogs for Nuclear Waste Disposal: The Natural Nuclear Fission Reactors in Gabon (Africa)." *Comptes Rendus R. Physique* 3, nos. 7–8 (September/October 2002): 839–49.

Gelber, Steven M., and Martin L. Cook. *Saving the Earth: The History of a Middle-Class Millenarian Movement*. Berkeley: University of California Press, 1990.

Gesner, Abraham. *A Practical Treatise on Coal, Petroleum, and Other Distilled Oils*. 2nd ed., rev. and enl. by George Weltden Gesner. New York: Bailliere Brothers, 1865.

Ghobadian, B., and H. Rahimi. "Biofuels—Past, Present and Future Perspective." *Proceedings of the Fourth International Iran & Russia Conference in Agriculture and Natural Resources*. Shahrekord, Iran: University of Shahrekord, 2004.

Gibbon, Richard. *Stephenson's Rocket and the Rainhill Trials*. Oxford: Shire, 2010.

Gibbs, Ken. *The Steam Locomotive: An Engineering History*. Stroud, UK: Amberley, 2012.

Gibney, Elizabeth. "Why Finland Now Leads the World in Nuclear Waste Storage." *Nature News*, December 2, 2015.

Giddens, Paul H. *The Birth of the Oil Industry*. New York: Macmillan, 1938.

———. *Early Days of Oil: A Pictorial History of the Beginnings of the Industry in Pennsylvania*. Princeton, NJ: Princeton University Press, 1948.

———, ed. *Pennsylvania Petroleum, 1750–1872: A Documentary History*. Titusville, PA: Pennsylvania Historical and Museum Commission, 1947.

Giebelhaus, August W. "Farming for Fuel: The Alcohol Motor Fuel Movement of the 1930s." *Agricultural History* 54, no. 1 (1980): 173–84.

Gies, Frances and Joseph. *Cathedral, Forge, and Waterwheel: Technology and Invention in the Middle Ages*. New York: HarperPerennial, 1994.

Gilmer, Robert W. "The History of Natural Gas Pipelines in the Southwest." *Texas Business Review* (May/June 1981): 129–35.

Goddard, Stephen B. *Getting There: The Epic Struggle Between Road and Rail in the American Century*. Chicago: University of Chicago Press, 1994.

Goettemoeller, Jeffrey, and Adrian Goettemoeller. *Sustainable Ethanol: Biofuels, Biorefineries, Cellulosic Biomass, Flex-Fuel Vehicles, and Sustainable Farming for Energy Independence*. Maryville, MO: Prairie Oak, 2007.

Goklany, Indur. *Clearing the Air: The Real Story of the War on Air Pollution*. Washington, DC: Cato Institute, 1999.

Goldstein, Eli (2012). "CO_2 Emissions from Nuclear Plants." Submitted as coursework for PH241, Introduction to Nuclear Energy, Stanford University, Winter 2012 (online).

Gordon, Robert J. *The Rise and Fall of American Growth: The U.S. Standard of Living Since the Civil War.* Princeton, NJ: Princeton University Press, 2016.

Graham, Gerald S. "The Migrations of the Nantucket Whale Fishery: An Episode in British Colonial Policy." *New England Quarterly* 8, no. 2 (1935): 179–202.

Graham, John W. *The Destruction of Daylight: A Study of the Smoke Problem.* London: George Allen, 1907.

Granqvist, Claes G. "Transparent Conductors as Solar Energy Materials: A Panoramic Review." *Solar Energy Materials & Solar Cells* 91 (2007): 1529–98.

Gray, Earle. "Gesner, Williams and the Birth of the Oil Industry." *Oil-Industry History* 9 (1) 2008: 12–23.

Gray, Thomas. *Observations on a General Iron Rail-way, or Land Steam-Conveyance; to Supersede the Necessity of Horses in all Public Vehicles; Showing Its Vast Superiority in Every Respect, Over all the Present Pitiful Methods of Conveyance by Turnpike Roads, Canals, and Coasting-Traders, Containing Every Species of Information Relative to Railroads and Loco-motive Engines.* 5th ed. London: Baldwin, Cradock, and Joy, 1825.

Gray, William. *Chorographia, or, A Survey of Newcastle upon Tine.* Newcastle, UK: Printed by S. B., 1649. Early English Books Online.

Great Britain. *Proceedings of the Committee of the House of Commons on the Liverpool and Manchester Railroad Bill: Sessions, 1825.*

Green, Constance McLaughlin, and Milton Lomask. *Vanguard, A History.* Washington, DC: National Aeronautics and Space Administration, 1970 (online).

Greene, Ann Norton. *Horses at Work: Harnessing Power in Industrial America.* Cambridge, MA: Harvard University Press, 2008.

Gresley, William Stukeley. *A Glossary of Terms Used in Coal Mining.* London: E. and F. N. Spon, 1883.

Griffiths, John. *The Third Man: The Life and Times of William Murdoch, 1754–1839, the Inventor of Gas Lighting.* London: Andre Deutsch, 1992.

Grodzins, Morton, and Eugene Rabinowitch, eds. *The Atomic Age: Scientists in National and World Affairs.* New York: Basic Books, 1963.

Grossman, Peter Z. *US Energy Policy and the Pursuit of Failure.* New York: Cambridge University Press, 2013.

Grossman, Gene M., and Alan B. Krueger. "Environmental Impacts of a North American Free Trade Agreement." Working Paper 3914. Cambridge, MA: National Bureau of Economic Research, 1991.

Grosso, Michael. *The Millennium Myth: Love and Death at the End of Time.* Wheaton, IL: Quest Books, 1997.

Grübler, Arnulf. "Diffusion: Long-Term Patterns and Discontinuities." *Technological Forecasting and Social Change* 39 (1991): 159–80.

———. *Technology and Global Change.* Cambridge: Cambridge University Press, 1998.

Grübler, Arnulf, and Nebojsa Nakicenovic. "Decarboning the Global Energy System." *Technological Forecasting and Social Change* 53 (1996): 97–110.

Grübler, Arnulf, Nebojsa Nakicenovic, and David G. Victor. "Dynamics of Energy Technologies and Global Change." *Energy Policy* 27 (1999): 247–80.

Gugliotta, Angela. "Class, Gender, and Coal Smoke: Gender Ideology and Environmental Justice in the City: A Theme for Urban Environmental History." *Environmental History* 5, no. 2 (2000): 165–93.

Gunter, Pete A. Y. "Whitehead's Contribution to Ecological Thought: Some Unrealized Possibilities." *Interchange* 31, nos. 2 and 3 (2000): 211–33.

Guroff, Margare. *The Mechanical Horse: How the Bicycle Reshaped American Life.* Austin: University of Texas Press, 2015.

Guy, Andy. *Steam and Speed: Railways of Tyne and Wear from the Earliest Days.* Newcastle, UK: Tyne Bridge, 2003.

Guy, Andy, and Jim Rees. *Early Railways 1569–1830.* Oxford: Shire, 2011.

Haagen-Smit, A. J. "The Air Pollution Problem in Los Angeles." *Engineering and Science* 14 (December 1950): 7–13.

———. "The Control of Air Pollution." *Scientific American* 210, no. 1 (1964): 25–31.

———. "The Control of Air Pollution in Los Angeles." *Engineering and Science* 18, no. 3 (December 1954): 11–16.

———. "A Lesson from the Smog Capital of the World." *Proceedings of the National Academy of Sciences* 67, no. 2 (1970): 887–97.

———. "Smog Research Pays Off." *Engineering and Science* 15, no. 8 (May 1952): 11–16.

Haagen-Smit, A. J., and M. M. Fox. "Photochemical Ozone Formation with Hydrocarbons and Automobile Exhaust." *Air Repair* 4, no. 3 (1954): 105–36.

Haagen-Smit, Zus (Maria) Interview. Shirley K. Cohen, interviewer, 16, 20 March 2000, Archives, California Institute of Technology, Pasadena, California (online).

Hadfield, Charles. *The Canal Age.* London: Pan Books, 1968.

Hamilton, Alice. *Exploring the Dangerous Trades: The Autobiography of Alice Hamilton, M. D.* Boston: Little, Brown, 1943.

Hamilton, Alice, Paul Reznikoff, and Grace M. Burnham. "Tetra-ethyl Lead." *Journal of the American Medical Association* 84, no. 20 (1925): 1481–86.

Hammersley, G. "The Charcoal Iron Industry and Its Fuel, 1540–1750." *Economic History Review* 24 (1973): 593–613.

Handler, Philip. "Some Comments on Risk Assessment." In *The National Research Council in 1979: Current Issues and Studies.* Washington, DC: National Academy of Sciences, 1979.

Hardin, Garrett. "The Tragedy of the Commons." *Science* 162, no. 3859 (December 13, 1968): 1243–48.

Harkness, Deborah E. *The Jewel House: Elizabethan London and the Scientific Revolution.* New Haven, CT: Yale University Press, 2007.

Harris, Kenneth. *The Wildcatter: A Portrait of Robert O. Anderson.* New York: Weidenfeld & Nicolson, 1987.

Hart, Cyril E. *Royal Forest: A History of Dean's Woods as Producers of Timber.* Oxford: Clarendon Press, 1966.

Hartley, Janet M., Paul Keenan, and Dominic Lieven, eds. *Russia and the Napoleonic Wars (War, Culture and Society, 1750–1850).* London: Palgrave Macmillan, 2015.

Hatcher, John. *The History of the British Coal Industry.* Vol. 1, *Before 1700: Towards the Age of Coal.* Oxford: Clarendon Press, 1993.

Haupt, Lewis M. "The Road Movement." *Journal of the Franklin Institute* 135, no. 1 (1893): 1–16.

Hawken, Paul, ed. *Drawdown: The Most Comprehensive Plan Ever Proposed to Reverse Global Warming.* New York: Penguin, 2017.

Hawkins, Laurence A. *William Stanley (1858–1916)—His Life and Work.* New York: Newcomen Society of North America, 1951.

Hawley, Ellis W. *The New Deal and the Problem of Monopoly: A Study in Economic Ambivalence.* New York: Fordham University Press, 1995. First published 1966.

Hays, Samuel P. *Beauty, Health, and Permanence: Environmental Politics in the United States, 1955–1985.* Cambridge: Cambridge University Press, 1987.

Health Effects Institute. *State of Global Air/2017.* Boston: HEI, 2017 (online).

Heard, B. P., B. W. Brook, T. M. L. Wigley, and J. C. A. Bradshaw. "Burden of Proof: A Comprehensive Review of the Feasibility of 100% Renewable-Electricity Systems." *Renewable and Sustainable Energy Reviews* 76 (2017): 1122–33.

Heflin, Wilson, Mary K. Bercaw Edwards, and Thomas Farel Heffernan, eds. *Herman Melville's Whaling Years.* Nashville: Vanderbilt University Press, 2004.

Heilbron, J. L. "The Contributions of Bologna to Galvanism." *Historical Studies in the Physical and Biological Sciences* 22, no. 1 (1991): 57–85.

Heilbroner, Robert L. "Ecological Armageddon." *New York Review of Books* online. April 23, 1970.

Heinrich, Thomas R. *Ships for the Seven Seas: Philadelphia Shipbuilding in the Age of Industrial Capitalism.* Baltimore: Johns Hopkins University Press, 1997.

Henderson, W. O. "Wolverhampton as the Site of the First Newcomen Engine." *Transactions of the Newcomen Society* 26 (1953): 155–59.

Henry, J. T. *The Early and Later History of Petroleum, with Authentic Facts in Regard to Its Development in Western Pennsylvania.* Philadelphia: Jas. B. Rodgers, 1873.

Herrick, Rufus Frost. *Denatured or Industrial Alcohol: A Treatise on the History, Manufacture, Composition, Uses, and Possibilities of Industrial Alcohol in the Various Countries Permitting its Use, and the Laws and Regulations Governing the Same, Including the United States.* New York: John Wiley & Sons, 1907.

Hewlett, Richard G., and Jack M. Holl. *Atoms for Peace and War, 1953–1961.* Berkeley: University of California Press, 1989.

Hibbert, Harold. "The Role of the Chemist in Relation to the Future Supply of Liquid Fuel." *Journal of Industrial and Engineering Chemistry* 13 (1921): 841–43.

Hickam, Homer H., Jr. *Torpedo Junction: U-Boat War Off America's East Coast, 1942.* Annapolis: Naval Institute Press, 1989.

Hidalgo, César. *Why Information Grows: The Evolution of Order, from Atoms to Economies.* New York: Basic Books, 2015.

Hill, Colin K. "The Low-Dose Phenomenon: How Bystander Effects, Genomic Instability, and Adaptive Responses Could Transform Cancer-Risk Models." *Bulletin of the Atomic Scientists* 68, no. 3 (2012): 51–58.

Hills, Richard L. "The Origins of James Watt's Perfect Engine." *Transactions of the Newcomen Society* 68 (1997): 85–107.

———. *Power from Steam: A History of the Stationary Steam Engine.* Cambridge: Cambridge University Press, 1989.

———. *James Watt: Volume 1: His Time in Scotland, 1736–1774.* London: Landmark.

Himmelfarb, Gertrude. *The Idea of Poverty: England in the Early Industrial Age.* New York: Vintage, 1983.

Hinchman, Lydia S. *Early Settlers of Nantucket: Their Associates and Descendants.* 2nd ed. and enl. ed. Philadelphia: Ferris & Leach, 1901.

Holdren, John, and Philip Herrera. *Energy: A Crisis in Power.* San Francisco: Sierra Club, 1971.

Holland, John. *The History and Description of Fossil Fuel, the Collieries, and Coal Trade of Great Britain.* 2nd ed. London: Whittaker, 1841.

Houghton-Alico, Doann. *Alcohol Fuels: Policies, Production, and Potential.* Boulder, CO: Westview Press, 1982.

House of Commons [H. C.]. *Proceedings of the Committee on the Liverpool and Manchester Railroad Bill,* 1825.

Howe, Henry. *Memoirs of the Most Eminent American Mechanics: Also, Lives of Distinguished European Mechanics; Together With a Collection of Anecdotes, Descriptions, &c., &c. Relating to the Mechanic Arts.* New York: Alexander V. Blake, 1841.

Howsley, R. "The IAEA and the Future of Nuclear Power: A View from the Industry." *Journal of Nuclear Materials Management* 30, no. 2 (2002): 21–23.

Hubbert, M. King. "Nuclear Energy and the Fossil Fuels." Publication no. 95, Shell

Development Company, Exploration and Production Research Division, Houston, June 1956 (online).

Hunt, Bruce J. *Pursuing Power and Light: Technology and Physics from James Watt to Albert Einstein.* Baltimore: Johns Hopkins University Press, 2010.

Hunt, Charles. *A History of the Introduction of Gas Lighting.* London: Walter King, 1907.

Hunt, Gaillard. *Life in America One Hundred Years Ago.* Williamstown, MA: Corner House, 1914.

Hunter, John P. *A Brief History of Natural Gas: Its Advantages, Use, Supply, and Economy as a Fuel to Manufacturers.* Verona, PA: Dexter Spring, 1886.

Hurley, Andrew. "Creating Ecological Wastelands: Oil Pollution in New York City, 1870–1900." *Journal of Urban History* 20, no. 3 (2004): 340–63.

Hutchins, Teresa Dunn. "The American Whale Fishery, 1815–1900: An Economic Analysis." PhD diss., Department of Economics, University of North Carolina at Chapel Hill, 1988.

Hutchinson, G. Evelyn. "The Biogeochemistry of Vertebrate Excretion (Survey of Contemporary Knowledge of Biogeochemistry)." *Bulletin of the American Museum of Natural History* 96: 1950. New York: By Order of the Trustees.

Hyde, Charles K. *Technological Change and the British Iron Industry, 1700–1870.* Princeton, NJ: Princeton University Press, 1977.

"Industrial News: Fluorine Gases in Atmosphere as Industrial Waste Blamed for Death and Chronic Poisoning of Donora and Webster, PA, Inhabitants." *Chemical and Engineering News* 26, no. 50 (December 13, 1948): 3692.

Inglis, David Rittenhouse. "Nuclear Energy and the Malthusian Dilemma." *Bulletin of the Atomic Scientists* 27, no. 2 (1971): 14–18.

Inhaber, Herbert. "Risk Analysis Applied to Energy Systems." *Encyclopedia of Energy,* vol. 5: 1–14. (2004).

Inman, Mason. *The Oracle of Oil: A Maverick Geologist's Quest for a Sustainable Future.* New York: W. W. Norton, 2016.

IPCC. *Climate Change 2014 Synthesis Report.* Intergovernmental Panel on Climate Change, 2014 (online).

Irwin, Paul G. "Overview: The State of Animals in 2001." In *The State of the Animals 2001,* edited by D. J. Salem and A. N. Rowan. Washington, DC: Humane Society Press, 2001: 1–19.

Jacobs, Meg. *Panic at the Pump: The Energy Crisis and the Transformation of American Politics in the 1970s.* New York: Hill and Wang, 2016.

Jacobs, Chip, and William J. Kelly. *Smogtown: The Lung-Burning History of Pollution in Los Angeles.* New York: Overlook Press, 2013.

Jacobson, Mark Z., and Mark A. Delucchi. "A Path to Sustainable Energy by 2030." *Scientific American*, November 2009, 58–64.

James I. "Speech of 1609." In *The Political Works of James I.* Reprinted from the edition of 1616. Cambridge, MA: Harvard University Press, 1918.

Jaworowski, Zbigniew. "Observations on Chernobyl After 25 Years of Radiophobia." *21st Century Science & Technology*, Summer 2010, 30–44.

———. "Observations on the Chernobyl Disaster and LNT." *Dose-Response* 8, no. 2 (2010): 148–71.

Jay, Mike. *The Atmosphere of Heaven: The Unnatural Experiments of Dr. Beddoes and His Sons of Genius.* New Haven, CT: Yale University Press, 2009.

Jeaffreson, J. C. *The Life of Robert Stephenson, F. R. S.* 2 vols. London: Longman, Green, Longman, Roberts, & Green, 1864.

Jedicke, Peter. "The NRX Incident." Canadian Nuclear Society online. Last modified 1989. www.cns-snc.ca/media/history/nrx.html.

Jefferson, Thomas. "Observations on the Whale Fishery," 1791. Jefferson Papers, Avalon Project, Yale Law School Lillian Goldman Law Library online.

Jeffries, Zay, Enrico Fermi et al. (1944). *Prospectus on Nucleonics.* Chicago: Metallurgical Laboratory MUC-RSM-234 (online).

Jenkins, Rhys. "Coke: A Note on Its Production and Use, 1587–1650." *Transactions of the Newcomen Society* 12 (1933): 104–7.

———. "The Heat Engine Idea in the Seventeenth Century: A Contribution to the History of the Steam Engine." *Transactions of the Newcomen Society* 17 (1937): 1–11.

———. "Savery, Newcomen and the Early History of the Steam Engine," pt. 1. *Transactions of the Newcomen Society* 3, no. 1 (1922): 96–118.

———. "Savery, Newcomen and the Early History of the Steam Engine," pt. 2. *Transactions of the Newcomen Society* 4, no. 1 (1923): 113–31.

———. "A Sketch of the Industrial History of the Coalbrookdale District." *Transactions of the Newcomen Society* 4 (1925): 102–7.

Jenner, Mark. "The Politics of London Air: John Evelyn's *Fumifugium* and the Restoration." *Historical Journal* 38, no. 3 (September 1995): 535–51.

Johnson, Arthur M. *The Development of American Petroleum Pipelines: A Study of Private Enterprise and Public Policy, 1862–1906.* Westport, CT: Greenwood Press, 1982.

———. *Petroleum Pipelines and Public Policy, 1906–1959.* Cambridge, MA: Harvard University Press, 1967.

Johnston, Fay H., Shannon Melody, and David M. J. S. Bowman. "The Pyrohealth Transition: How Combustion Emissions Have Shaped Health Through Human History." *Philosophical Transactions of the Royal Society B* 371 (2016): 1–10.

Joint Secretariat: "One Decade After Chernobyl: Summary of Conference Results, Joint Secretariat of the Conference," Vienna, Austria, 1996. European Commission, International Atomic Energy Agency, and World Health Organization.

Jones, Christopher F. *Routes of Power: Energy and Modern America*. Cambridge, MA: Harvard University Press, 2014.

Joskow, Paul L. "The Economic Future of Nuclear Power." *Daedalus* (Fall 2009): 45–59.

———. "Electricity from Uranium, Pt. 2: The Prospects for Nuclear Power in the United States." *Milken Institute Review* (4Q 2007): 32–43.

———. "The Future of Nuclear Power After Fukushima." Massachusetts Institute of Technology Center for Energy and Environmental Research Working Paper 2012-001 (online).

———. "Natural Gas: From Shortages to Abundance in the US." Massachusetts Institute of Technology Center for Energy and Environmental Research Working Paper 2012-001 (online).

Jungers, Frank. *The Caravan Goes On: How Aramco and Saudi Arabia Grew Up Together*. Surbiton, UK: Medina, 2013.

Jungk, Robert. *The New Tyranny: How Nuclear Power Enslaves Us*. New York: Grosset & Dunlap, 1979.

Kanefsky, John, and John Robey. "Steam Engines in 18th-Century Britain: A Quantitative Assessment." *Technology and Culture* 21, no. 2 (1980): 161–86.

Kasun, Jacqueline. *The War Against Population: The Economics and Ideology of World Population Control*. San Francisco: Ignatius Press, 1999.

Kean, Sam. "The Flavor of Smog." *Distillations*, Fall 2016, Chemical Heritage Foundation online, www.chemheritage.org/distillations/magazine/the-flavor-of-smog.

Kelley, Brooks Mather. *Yale: A History*. New Haven, CT: Yale University Press, 1974.

Kemble, Frances Ann. *Record of a Girlhood*. Vol. 2. London: Richard Bentley and Son, 1878.

Kerker, Milton. "Science and the Steam Engine." *Technology and Culture* 2, no. 4 (Autumn 1961): 381–90.

Kerridge, Eric. "The Coal Industry in Tudor and Stuart England: A Comment." *Economic History Review*, n.s. 30, no. 2 (1977): 340–42.

Kettering, Charles F. "More Efficient Utilization of Fuel." SAE technical paper 190010, 1919.

Keuchel, Edward F. "Coal-Burning Locomotives: A Technological Development of the 1850s." *Pennsylvania Magazine of History and Biography* 94, no. 4 (1970): 484–95.

Kharecha, Pushker A., and James E. Hansen. "Prevented Mortality and Greenhouse

Gas Emissions from Historical and Projected Nuclear Power." *Environmental Science and Technology* 47 (2013): 4889–95.

Kidder, Tracy. "The Nonviolent War Against Nuclear Power." *Atlantic Monthly* 242, no. 3 (1978): 70–76.

Kiefner, John F., and Cheryl J. Trench. "Oil Pipeline Characteristics and Risk Factors: Illustrations from the Decade of Construction." *American Petroleum Pipeline Committee Publication.* Washington, DC: American Petroleum Institute, 2001.

Kiester, Edwin, Jr. "A Darkness in Donora," *Smithsonian* online. November 1999.

King-Hele, Desmond. *Erasmus Darwin: A Life of Unequalled Achievement.* London: DLM, 1999.

Kintner, C. J. "History of the Electrical Art in the United States Patent Office." *Journal of the Franklin Institute* 121, no. 5 (1886): 377–96.

Kintner, E. E. "Admiral Rickover's Gamble: The Landlocked Submarine." *Atlantic Monthly* online, January 1959.

Kirsch, David A. "The Electric Car and the Burden of History: Studies in Automotive Systems Rivalry in America, 1890–1996." PhD diss., Stanford University, 1996.

Kitsikopoulos, Harry. "The Diffusion of Newcomen Engines, 1706–73: A Reassessment." Economic History Association online, 2013.

Kolbert, Elizabeth. *Field Notes from a Catastrophe: Man, Nature, and Climate Change.* New York: Bloomsbury, 2006.

———. "Mr. Green: Environmentalism's Optimistic Guru Amory Lovins." *New Yorker* online, January 22, 2007.

Kotchetkov, L. A. "Obninsk: Number One." *Nuclear Engineering International:* July 13, 2004, www.neimagazine.com/features.

Köteles, G. J. "The Low Dose Dilemma." *Central European Journal of Occupational and Environmental Medicine* 4, no. 2 (1998): 103–13.

Kovarik, William. "Charles F. Kettering and the Development of Tetraethyl Lead in the Context of Alternative Fuel Technologies." SAE Technical Paper 941942, 1994, n.p.

———. "Henry Ford, Charles F. Kettering, and the Fuel of the Future," *Automotive History Review* 32 (Spring 1998): 727.

———. "Environmental Conflict over Leaded Gasoline and Alternative Fuels." Paper to the American Society for Environmental History Annual Conference, March 26–30, 2003 (online).

———. "History of Biofuels." In *Biofuel Crops: Production, Physiology and Genetics.* Edited by B. P. Singh. Wallingford, UK: CABI, 2013.

Krebs, Frederik C. "Fabrication and Processing of Polymer Solar Cells: A Review of Printing and Coating Techniques." *Solar Energy Materials & Solar Cells* 93 (2009): 394–412.

Krehl, Peter O. K. *History of Shock Waves, Explosions and Impact: A Chronological and Biographical Reference.* Berlin: Springer, 2009.

Kubler, George. "Towards Absolute Time: Guano Archaeology." *Memoirs of the Society for American Archaeology 4: A Reappraisal of Peruvian Archeology* (1948): 29–50.

Kuroda, P. K. "On the Nuclear Physical Stability of the Uranium Minerals." *Journal of Chemical Physics* 25, no. 4 (1956): 781–82.

Kutz, Charles W., and American Members of the International Waterways Commission. *Reports on the Existing Water-Power Situation at Niagara Falls, So Far as Concerns the Diversion of Water on the American Side.* Washington, DC: US Government Printing Office, 1906.

Kuznets, Simon. "Economic Growth and Income Inequality." *American Economic Review* 45, no. 1 (1955): 1–27.

Kyvig, David E. *Daily Life in the United States, 1920–1940: How Americans Lived Through the "Roaring Twenties" and the Great Depression.* Chicago: Ivan R. Dee, 2002.

Labouchere, Rachel. *Abiah Darby 1716–1793 of Coalbrookdale, Wife of Abraham Darby II.* York, UK: William Sessions, 1988.

Ladd, Brian. *Autophobia: Love and Hate in the Automotive Age.* Chicago: University of Chicago Press, 2008.

Lafitte, Jacques. *Reflections on the Science of Machines.* Unpublished first draft, translation by J. F. Hart. London, Ont.: University of Western Ontario Computer Science Department, 1969.

Lambert, Jeremiah D. *The Power Brokers: The Struggle to Shape and Control the Electric Power Industry.* Cambridge, MA: MIT Press, 2015.

La Mettrie, Julien Offray de. *Man a Machine and Man a Plant.* Translated by Justin Leiber. Indianapolis: Hackett, 1994. First published 1751.

Landes, David S. *The Unbound Prometheus: Technological Change and Industrial Development in Western Europe From 1750 to the Present.* Cambridge UK: Cambridge University Press, 1969.

Larsen, Ralph I. "Air Pollution from Motor Vehicles." *Annals of the New York Academy of Sciences* 136 (1966): 277–301.

Latimer, L. H., C. J. Field, and John W. Howell. *Incandescent Electric Lighting: A Practical Description of the Edison System.* New York: D. van Nostrand, 1890.

Laughlin, Robert B. *Powering the Future: How We Will (Eventually) Solve the Energy Crisis and Fuel the Civilization of Tomorrow.* New York: Basic Books, 2011.

Law, R. J. *James Watt and the Separate Condenser: An Account of the Invention.* London: Her Majesty's Stationery Office, 1969.

Layton, Walter T. *The Discoverer of Gas Lighting.* London: Walter King, 1926.

Lee, Charles E. *The Evolution of Railways.* 2nd ed. London: Railway Gazette, 1943.

———. "Tyneside Tramroads of Northumberland: Some Notes on the Engineering Background of George Stephenson." *Transactions of the Newcomen Society* 26 (1953): 199–229.

Lemay, J. A. Leo. *Ebenezer Kinnersley: Franklin's Friend.* Philadelphia: University of Pennsylvania Press, 1964.

Lenher, Victor. "Selenium and Tellurium." *Journal of Industrial and Engineering Chemistry* 12, no. 6 (1920): 597–98.

Lester, Richard K. "A Roadmap for US Nuclear Energy Innovation." *Issues in Science and Technology* (Winter 2016): 4554.

Lester, Richard K., and David M. Harr. *Unlocking Energy Innovation: How America Can Build a Low-Cost, Low-Carbon Energy System.* Cambridge, MA: MIT Press, 2012.

Levere, Trevor H. "Dr. Thomas Beddoes: Chemistry, Medicine, and the Perils of Democracy." *Notes and Records of the Royal Society of London* 63 (2009): 215–29.

Lewis, Edward B. "Alfred Henry Sturtevant." *Biographical Memoirs of the National Academy of Sciences.* Vol. 73. Washington, DC: National Academies Press, 1998.

Lewis, M. J. T. *Early Wooden Railways.* London: Routledge & Kegan Paul, 1970.

Libby, Willard F. "Tritium in Nature." *Scientific American* 190, no. 4 (1954): 38–42.

Liebig, Justus. *Familiar Letters on Chemistry and Its Relation to Commerce, Physiology, and Agriculture.* Edited by John Gardner. London: Taylor and Walton, 1844.

———. *Organic Chemistry in Its Applications to Agriculture and Physiology.* Edited by Lyon Playfair. London: Taylor and Walton, 1840.

Lienhard, John H. *The Engines of Our Ingenuity: An Engineer Looks at Technology and Culture.* New York: Oxford University Press, 2000.

———. *How Invention Begins: Echoes of Old Voices in the Rise of New Machines.* New York: Oxford University Press, 2006.

Lindermuth, John R. *Digging Dusky Diamonds: A History of the Pennsylvania Coal Region.* Mechanicsburg, PA: Sunbury Press, 2013.

Lindsay, J. M. "The Iron Industry in the Highlands: Charcoal Blast Furnaces." *Scottish Historical Review* 56 (no. 161, pt. 1) (1977): 49–63.

Lloyd, William Foster. "W. F. Lloyd on the Checks to Population." *Population and Development Review* 6, no. 3 (1980): 473–96.

Loeb, Alan P. "Birth of the Kettering Doctrine: Fordism, Sloanism and the Discovery of Tetraethyl Lead." *Business and Economic History* 24, no. 1 (1995): 72–87.

Logsdon, Jeanne M. "Organizational Responses to Environmental Issues: Oil Refining Companies and Air Pollution." In *Research in Corporate Social Performance and Policy*. Vol. 7, edited by L. E. Preston. Greenwich, CT: JAI Press (1985), 47–71.

Lones, T. E. "A Précis of *Mettallum Martis* and an Analysis of Dud Dudley's Alleged Invention." *Transactions of the Newcomen Society* 20 (1941): 17–28.

Lord, Eleanor Louisa. *Industrial Experiments in the British Colonies of North America.* Baltimore: Johns Hopkins, 1898.

Loree, L. F. "The First Steam Engine of America." *Transactions of the Newcomen Society* 10 (1931): 15–27.

———. "The Four Locomotives Imported into America in 1829 by the Delaware & Hudson Company." *Transactions of the Newcomen Society* 4 (1925): 64–72.

Louchouarn, Patrick, Steven N. Chillrud, Stephane Houel, Beizhan Yan, Damon Chaky, Cornelia Rumpel, Claude Largeau, Gerard Bardoux, Dan Walsh, and Richard F. Bopp. "Elemental and Molecular Evidence of Soot- and Char-Derived Black Carbon Inputs to New York City's Atmosphere During the 20th Century." *Environmental Science & Technology* 41, no. 1 (2007): 82–87.

Lovins, Amory B. *Soft Energy Paths: Toward a Durable Peace.* New York: Harper & Row, 1977.

Lovins, Amory, L. Hunter Lovins, and Leonard Ross. "Nuclear Power and Nuclear Bombs." *Foreign Affairs* 58, no. 5 (1980): 1137–77.

Lowen, Rebecca S. "Entering the Atomic Power Race: Science, Industry, and Government." *Political Science Quarterly* 102, no. 3 (1987): 459–79.

Lucier, Paul. *Scientists & Swindlers: Consulting on Coal and Oil in America, 1820–1890.* Baltimore: Johns Hopkins University Press, 2008.

Luter, Paul. "Lord Dundonald." Oldcopper.org., 2005 (online).

Macaulay, Thomas Babington. *The History of England from the Accession of James II.* Facsimile of edition of 1849. Cambridge: Adamant Media, 2006.

Macfarlan, J. "George Dixon: Discoverer of Gas Light from Coal." *Transactions of the Newcomen Society* 5 (1924): 53–55 (plus plate).

MacLaren, Malcolm. *The Rise of the Electrical Industry During the Nineteenth Century.* Princeton, NJ: Princeton University Press, 1943.

Macy, Obed. *The History of Nantucket, Being a Compendious Account of the First Settlement of the Island by the English, Together with the Rise and Progress of the Whale Fishery.* Boston: Hilliard, Gray, 1835.

Maddox, John. *The Doomsday Syndrome.* New York: McGraw-Hill, 1972.

Madhaven, Guru. *Applied Minds: How Engineers Think.* New York: W. W. Norton, 2015.

Malm, Andreas. *Fossil Capital: The Rise of Steam Power and the Roots of Global Warming.* London: Verso, 2016.

Malthus, Thomas. *An Essay on the Principle of Population.* London: J. Johnson, 1798.

Mann, Charles C. *1493: Uncovering the New World Columbus Created.* New York: Knopf, 2011.

Marchetti, Cesare. "Energy Systems—The Broader Context." *Technological Forecasting and Social Change* 14 (1979): 191–203.

———. "Fifty-Year Pulsation in Human Affairs: Analysis of Some Physical Indicators." *Futures* 18, no. 3 (1986): 376–88.

———. "My CV as a Personal Story." Cesare Marchetti Web Archive, 2003 (online).

———. "On Decarbonization: Historically and Perspectively." *IIASA Interim Report,* 2005 (online).

———. "On the Long-Term History of Energy Markets and the Chances for Natural Gas." Working Paper 84-39, IIASA, 1984 (online).

———. *On Society and Nuclear Energy: A Historical Analysis of the Interaction Between Society and Nuclear Technology with Examples Taken from Other Innovations.* Final Report for contract no. PSS 0039/A between IIASA and the European Atomic Energy Commission. IIASA, 1988 (online).

———. "A Personal Memoir: From Terawatts to Witches: My Life with Logistics at IIASA." *Technological Forecasting and Social Change* 37 (1990): 409–14.

———. "Primary Energy Substitution Models: On the Interaction Between Energy and Society." *Technological Forecasting and Social Change* 10 (1977): 345–56.

———. "Renewable Energies in a Historical Context." Professional paper, International Institute for Applied Systems Analysis (IIASA), Laxenburg, Austria, December 1985 (online).

———. "Society as a Learning System: Discovery, Invention and Innovation Cycles Revisited." *IIASA Research Report* (repr.). IIASA, Laxenburg, Austria: RR-81-029. Reprinted from *Technological Forecasting and Social Change,* 18 (1980).

Marchetti, Cesare, and N. Nakicenovic. "The Dynamics of Energy Systems and the Logistic Substitution Model." Pt. 1, pt. 2. RR-79-13, IIASA, 1979 (online).

Markandya, Anil, and Paul Wilkinson. "Electricity Generation and Health." *Lancet* 370 (2007): 979–90.

Marsh, Arnold. *Smoke: The Problem of Coal and the Atmosphere.* London: Faber and Faber, 1947.

Martin, Richard. *Coal Wars: The Future of Energy and the Fate of the Planet.* New York: Palgrave Macmillan, 2015.

Martin, Thomas Commerford. "Electricity in the Modern City." *Journal of the Franklin Institute* 138 (September 1894): 198–211.

Martínez-Alier, J. "The Environment as a Luxury Good or 'Too Poor to Be Green'?" *Ecological Economics* 13 (1995): 1–10.

MIT Coal Energy Study. *The Future of Coal: Options for a Carbon-Constrained World—An Interdisciplinary MIT Study.* Cambridge, MA: MIT (2007) (online).

MIT Nuclear Energy Study. *The Future of Nuclear Power: An Interdisciplinary MIT Study.* Cambridge, MA: MIT (2003) (online).

Mason, W. W. "Trevithick's First Rail Locomotive." *Transactions of the Newcomen Society* 12 (1933): 85–103.

Massachusetts Historical Commission (MHC) (1984). MHC Reconnaissance Survey Town Report: Nantucket (online).

Mattausch, Daniel W. "David Melville and the First American Gas Light Patents." *Rushlight,* December 1998 (Rushlight Club online).

Mattingly, Garrett. *The Armada.* Boston: Houghton Mifflin, 1959.

Mawer, Granville Allen. *Ahab's Trade: The Saga of South Seas Whaling.* New York: St. Martin's Press, 1999.

Maxim, Hiram S. *My Life.* London: Methuen, 1915.

Maxim, Hiram Percy. *Horseless Carriage Days.* New York: Harper & Brothers, 1937.

Mayer, Ivan. "Human Consequences of Technological Change: Nuclear Power and Public Safety," International Atomic Energy Agency (IAEA) online, n.d.

McConnell, Curt. *Coast to Coast by Automobile: The Pioneering Trips, 1899–1908.* Stanford, CA: Stanford University Press, 2000.

McGill, Paul L., Frederick G. Sawyer, and Richard D. Cadle. "Smog: Fact and Fiction." *Proceedings, American Petroleum Institute Division of Refining Seventeenth Mid-Year Meeting.* San Francisco, CA, May 12–15, 1952.

McGlade, Christophe, and Paul Ekins. "The Geographical Distribution of Fossil Fuels Unused When Limiting Global Warming to 2°C." *Nature* 517 (2013): 187–90.

McGowan, Christopher. *The Rainhill Trials: The Greatest Contest of Industrial Britain and the Birth of Commercial Rail.* London: Little, Brown, 2004.

McJeon, Haewon, Jae Edmonds, Nico Bauer, Leon Clarke, Brian Fisher, Brian P. Flannery, Jérôme Hilaire et al. "Limited Impact on Decadal-Scale Climate Change from Increased Use of Natural Gas." *Nature* 514 (October 23, 2014): 482–85.

McKelvey, V. E. "Mineral Resource Estimates and Public Policy: Better Methods for Estimating the Magnitude of Potential Mineral Resources Are Needed to Provide the Knowledge That Should Guide the Design of Many Key Public Policies." *American Scientist* 60, no. 1 (1972): 32–40.

McLaughlin, Charles C. "The Stanley Steamer: A Study in Unsuccessful Innovation." *Explorations in Environmental History* 7, no. 1 (1954): 37–47.

McLaurin, John J. *Sketches in Crude-Oil: Some Accidents and Incidents of the Petroleum Development in All Parts of the Globe.* Franklin, PA: published by the author, 1902.

McMurray, Scott. *Energy to the World: The Story of Saudi Aramco.* Houston: Aramco Services, 2011.

McNeill, J. R. *Something New Under the Sun: An Environmental History of the Twentieth-Century World.* New York: W. W. Norton, 2000.

———. "Woods and Warfare in World History." *Environmental History* 9, no. 3 (2004): 388–410.

McPhee, John. *Encounters with the Archdruid.* New York: Farrar, Straus and Giroux, 1971.

McShane, Clay, and Joel A. Tarr. *The Horse in the City: Living Machines in the Nineteenth Century.* Baltimore: Johns Hopkins University Press, 2007.

Meadows, Donella H., Dennis L. Meadows, Jørgen Randers, and William W. Behrens III. *The Limits to Growth: A Report for the Club of Rome's Project on the Predicament of Mankind.* New York: Universe Books, 1972.

Meadows, Donella, Jørgen Randers, and Dennis Meadows. *Limits to Growth: The 30-Year Update.* White River Junction, VT: Chelsea Green Publishing, 2004.

Medhurst, Martin J. "Atoms for Peace and Nuclear Hegemony: The Rhetorical Structure of a Cold War Campaign." *Armed Forces & Society* 23, no. 4 (1997): 571–93.

Medvedev, Zhores. *The Legacy of Chernobyl.* New York: W. W. Norton, 1990.

Melville, Herman. *Moby-Dick.* San Bernardino, CA: Digireads (1851).

Mendenhall, T. C. *A Century of Electricity.* Boston: Houghton, Mifflin, 1890.

Mercer, Stanley. "Trevithick and the Merthyr Tramroad." *Transactions of the Newcomen Society* 26 (1953): 89–103.

Merrill, Karen R. *The Oil Crisis of 1972–1974: A Brief History with Documents.* Boston: Bedford/St. Martin's, 2007.

Meyer, William B. *Human Impact on the Earth.* Cambridge: Cambridge University Press, 1996.

Midgley, Thomas, Jr. "Tetraethyl Lead Poison Hazards." *Industrial and Engineering Chemistry* 17, no. 8 (1925): 827–28.

Miller, Albert H. "Technical Development of Gas Anesthesia." *Anesthesiology* 7, no. 2 (1941): 398–409.

Miller, David Philip, and Trevor H. Levere. "'Inhale It and See?' The Collaboration Between Thomas Beddoes and James Watt in Pneumatic Medicine." *Ambix* 55, no. 1 (March 2008): 5–28.

Miller, Donald L., and Richard E. Sharpless. *The Kingdom of Coal: Work, Enterprise, and Ethnic Communities in the Mine Fields.* Philadelphia: University of Pennsylvania Press, 1985.

Miller, R. L., and Gill, J. R. "Uranium from Coal." *Scientific American* 191, no. 4 (1954): 36–39.

Miller, Shawn William. *An Environmental History of Latin America.* Cambridge: Cambridge University Press, 2007.

Miner, Craig. *A Most Magnificent Machine: America Adopts the Railroad, 1825–1862.* Lawrence: University Press of Kansas, 2010.

Mitchell, Timothy. *Carbon Democracy: Political Power in the Age of Oil.* London: Verso, 2011.

Monier-Williams, G. W. *Power Alcohol: Its Production and Utilisation.* London: Henry Frowde and Hodder & Stoughton, 1922.

Montgomery, Scott L. *The Powers That Be: Global Energy for the Twenty-First Century and Beyond.* Chicago: University of Chicago Press, 2010.

Montgomery, Scott L., and Thomas Graham Jr. *Seeing the Light: The Case for Nuclear Power in the 21st Century.* Cambridge: Cambridge University Press, 2017.

Moran, Richard. *Executioner's Current: Thomas Edison, George Westinghouse, and the Invention of the Electric Chair.* New York: Knopf, 2002.

Morand, Paul. *1900 A. D.* Translated by Mrs. Romilly Fedden. New York: William Farquhar Payson, 1931.

Morris, Charles R. *The Dawn of Innovation: The First American Industrial Revolution.* New York: Public Affairs, 2012.

Morris, Eric. "From Horse Power to Horsepower." *Access* 30 (Spring 2007): 3–9.

Morison, Samuel Eliot. *History of United States Naval Operations in World War II.* Vol. 1, *The Battle of the Atlantic, September 1939–May 1943.* Urbana: University of Illinois Press, 1975.

Morone, Joseph G., and Edward J. Woodhouse. *The Demise of Nuclear Energy? Lessons for Democratic Control of Technology.* New Haven, CT: Yale University Press, 1989.

Mosley, Stephen. "Environmental History of Air Pollution and Protection." In *The Basic Environmental History*, edited by Mauro Agnoletti and Simone Neri Serneri. New York: Springer, 2014.

Mossman, David J., François Gauthier-Lafaye, Adriana Dutkiewicz, and Ralf Brüning. "Carbonaceous Substances in Oklo Reactors—Analogue for Permanent Deep Geologic Disposal of Anthropogenic Nuclear Waste." *Reviews in Engineering Geology* 19 (2008): 1–13.

Mott, R. A. "English Waggonways of the Eighteenth Century." *Tranactions of the Newcomen Society* 37 (1967): 1–33.

Mountford, C. E. *The History of John Bowes & Partners up to 1914.* Durham, UK: Durham University, 1967 (Durham E-Theses online).

Muirhead, James Patrick. *The Life of James Watt, with Selections from His Correspondence.* New York: D. Appleton, 1859.

Muller, Hermann J. "The Manner of Dependence of the 'Permissible Dose' of Radiation on the Amount of Genetic Damage." *Acta Radiologica* 41, no. 1 (1954): 5–20.

———. "The Production of Mutations." Nobel Lecture. Nobelprize.org, 1946 (online).

———. "Radiation Damage to the Genetic Material." *American Scientist* 38, no. 1 (1950): 32–59, 126.

———. "Radiation and Heredity." *American Journal of Public Health* 54, no. 1 (1964): 42–50.

Mulryne, J. R., and Margaret Shewring, eds. *Shakespeare's Globe Rebuilt.* Cambridge: Cambridge University Press, 1997.

Murdoch, William. "An Account of the Application of Gas from Coal to Economical Purposes." *Philosophical Transactions of the Royal Society of London* 98 (1808): 124–32.

Mushet, David. *Papers on Iron and Steel, Practical and Experimental.* London: John Weale, 1840.

Nakićenović, Nebojša. "Decarbonization: Doing More with Less." *Technological Forecasting and Social Change* 51 (1996): 1–17.

Nakićenović, Nebojša, and Arnulf Grübler, eds. *Diffusion of Technologies and Social Behavior.* Berlin: Springer-Verlag, 1991.

Nardizzi, Vin. "Shakespeare's Globe and England's Woods." *Shakespeare Studies* 39 (2011): 54–63.

Nash, Betty Joyce. "Economic History: Tar and Turpentine." *Federal Reserve Bank of Richmond Region Focus* (Fourth Quarter, 2011), 45–47.

Neal, Daniel. *The History of New England.* Vol. 2. London: A. Ward, 1747.

Neale, J. E. *Queen Elizabeth I.* Chicago: Academy Chicago, 1992.

Nef, John U. *The Rise of the British Coal Industry.* 2 vols. London: Routledge, 1966.

Nemery, Benoit, Peter H. M. Hoet, and Abderrahim Nemmar. "The Meuse Valley Fog of 1930: An Air Pollution Disaster." *Lancet* 357, no. 9257 (March 3, 2001): 704–8.

Nevell, Michael, and Terry Wyke, eds. "Bridgewater 250: The Archeology of the World's First Industrial Canal." *University of Salford Applied Archaeology Series 1.* Manchester, UK: Centre for Applied Archaeology, University of Salford, 2012.

"A New Oil-Field in Saudi Arabia." *Standard Oil Bulletin,* September 1–12, 1936 (online).

Newcomen Society. *The 1712 "Dudley Castle" Newcomen Engine.* Dudley, UK: Black Country Living Museum, 2012.

Newell, Richard G., and Kristian Rogers. "The U.S. Experience with the Phasedown of Lead in Gasoline." Discussion paper. Washington, DC: Resources for the Future, 2003.

Nicoll, Gayle. "Radiation Sources in Natural Gas Well Activities." *Occupational Health & Safety* online, last modified October 1, 2012.

Nicholls, H. G. *The Forest of Dean: An Historical and Descriptive Account, Derived from Personal Observation, and Other Sources, Public, Private, Legendary, and Local.* London: John Murray, 1858.

Nicholls, Robert. *Manchester's Narrow Gauge Railways: Chat Moss and Carrington Estates.* Huddersfield, UK: Narrow Gauge Railway Society, 1985.

Nichols, Elizabeth. "U.S. Nuclear Power and the Success of the American Anti-Nuclear Movement." *Berkeley Journal of Sociology* 32 (1987): 167–92.

Nickerson, Stanton P. "Tetraethyl Lead: A Product of American Research." *Journal of Chemical Education* 31, no. 11 (1954): 560–71.

Niering, William A. "Forces That Shaped the Forests of the Northeastern United States." *Northeastern Naturalist* 5, no. 2 (1998): 99–110.

[Nixon, George] *An Enquiry into the Reasons of the Advance of the Price of Coals, Within Seven Years Past.* London: E. Comyns, 1739.

Nordhaus, Ted, and Michael Shellenberger. *Breakthrough: From the Death of Environmentalism to the Politics of Possibility.* Boston: Houghton Mifflin, 2007.

Norman, Oscar Edward. *The Romance of the Gas Industry.* Chicago: A. C. McClurg, 1922.

Norris, Robert S., and Hans M. Kristensen. "Global Nuclear Weapons Inventories, 1945–2010." *Bulletin of the Atomic Scientists* 66, no. 4 (2010) online.

Novick, Sheldon. *The Electric War: The Fight over Nuclear Power.* San Francisco: Sierra Club Books, 1976.

Noxious Vapours, Great Britain, Royal Commission On. *Report of the Royal Commission on Noxious Vapours.* London: HMSO, 1878.

Nye, David E. *Consuming Power: A Social History of American Energies.* Cambridge, MA: MIT Press, 1998.

———. *Electrifying America: Social Meanings of a New Technology, 1880–1940.* Cambridge, MA: MIT Press, 1990.

Oberg, Barbara B., and J. Jefferson Looney, eds. *The Papers of Thomas Jefferson Digital Edition.* Charlottesville: University of Virginia Press, 2008–2016.

O'Connor, Peter A. "Energy Transitions." *Pardee Papers*, November 12, 2010 (online).

O'Dea, W. T. "Artificial Lighting Prior to 1800 and Its Social Effects." *Folklore* 62, no. 1 (1951): 312–24.

Odermatt, André A. *Welding: A Journey to Explore Its Past.* Troy, OH: Hobart Institute of Welding Technology, 2010.

Ogden, James. *A description of Manchester: giving an historical account of those limits in which the town was formerly included, some observations upon its public edifices,*

. . . *By a native of the town*. Manchester: Eighteenth Century Collections Online. (1783)

"The Oil Wells of Alsace." *New York Times* online, February 23, 1880.

Olien, Diana Davids, and Roger M. Olien. *Oil in Texas: The Gusher Age, 1895–1945*. Austin: University of Texas Press, 2002.

Oliver, Dave. *Against the Tide: Rickover's Leadership Principles and the Rise of the Nuclear Navy*. Annapolis: Naval Institute Press, 2014.

Olmsted, Frederick Law. *The Cotton Kingdom: A Traveller's Observations on Cotton and Slavery in the American Slave States*. New York: Da Capo Press, 1953.

Olwell, Russell B. *At Work in the Atomic City: A Labor and Social History of Oak Ridge, Tennessee*. Knoxville: University of Tennessee Press, 2004.

Orrell, John. "Building the Fortune." *Shakespeare Quarterly* 44 (1993): 127–44.

Orwell, George. *The Road to Wigan Pier*. New York: Harcourt, 1958. First published 1937.

Outerbridge, A. E., Jr. "The Smoke Nuisance and Its Regulation, with Special Reference to the Condition Prevailing in Philadelphia." *Journal of the Franklin Institute* 143, no. 66 (1897): 393–424.

Outland III, Robert B. "Suicidal Harvest: the Self-Destruction of North Carolina's Naval Stores Industry." *North Carolina Historical Review* 78, no. 3 (2001): 309–44.

Owen, David. *The Conundrum: How Scientific Innovation, Increased Efficiency, and Good Intentions Can Make Our Energy and Climate Problems Worse*. New York: Riverhead Books, 2011.

Page, Victor W. *The Model T Ford Car: Its Construction, Operation and Repair*. New York: Norman W. Henley, 1917.

Papin, Denis. "A New Method of Obtaining Very Great Moving Powers at Small Cost," 1690. Reprinted in translation in James Patrick Muirhead, *The Life of James Watt, with Selections from His Correspondence*. New York: D. Appleton, 1859.

Parker, Hershel. *Herman Melville: A Biography*. Vol. 1, *1819–1851*. Baltimore: Johns Hopkins University Press, 1996.

———. *Herman Melville: A Biography*. Vol. 2, *1851–1891*. Baltimore: Johns Hopkins University Press, 2002.

Paul, J. K., ed. *Ethyl Alcohol Production and Use as a Motor Fuel*. Park Ridge, NJ: Noyes Data, 1979.

Pease, Zeph. W., and George A. Hough. *New Bedford, Massachusetts: Its History, Industries, Institutions, and Attractions*. Edited by William L. Sayer. New Bedford, MA: New Bedford Board of Trade, 1889.

Pecci, Aurelio. *The Chasm Ahead*. New York: Macmillan, 1969.

———. *One Hundred Pages for the Future: Reflections of the President of the Club of Rome.* London: Futura, 1981.

Peebles, Malcolm W. H. *Evolution of the Gas Industry.* New York: New York University Press, 1980.

Pemberton, H. Earl. "The Curve of Culture Diffusion Rate." *American Sociological Review* 1, no. 4 (1936): 547–56.

Pendred, Loughnan St. L. "The Mystery of Trevithick's London Locomotives." *Transactions of the Newcomen Society* 1 (1922): 34–49.

Pera, Marcello. *The Ambiguous Frog: The Galvani-Volta Controversy on Animal Electricity.* Translated by Jonathan Mandelbaum. Princeton, NJ: Princeton University Press, 1992.

Perlin, John. *Forest Journey: The Role of Wood in the Development of Civilization.* New York: W. W. Norton, 1989.

———. *From Space to Earth: The Story of Solar Electricity.* Ann Arbor, MI: Aatec publications, 1999.

Perry, Percival. "The Naval-Stores Industry in the Old South, 1790–1860." *Journal of Southern History* 34, no. 4 (1968): 509–26.

Peters, Alan W., William H. Flank, and Burtron H. Davis. "The History of Petroleum Cracking in the 20th Century." Ch. 5 in *Innovations in Industrial and Engineering Chemistry,* edited by William H. Flank, Martin A. Abraham, and Michael A. Matthews. American Chemical Society Symposium Series. Washington, DC: American Chemical Society, 2009, 103–87.

Philbrick, Nathaniel. "The Nantucket Sequence in Crevecoeur's *Letters from an American Farmer.*" *New England Quarterly* 64, no. 3 (1991): 414–32.

Phillips, John Arthur. *A Treatise on Ore Deposits.* London: Macmillan, 1884.

Pittman, Walter E. "The One-Hundred Year War Against Air Pollution." *Quarterly Journal of Ideology* 26, nos. 1 and 2 (2003): 23 (online).

Pitts, James N., Jr., and Edgar R. Stephens. "Arie Jan Haagen-Smit, 1900–1977." *Journal of the Air Pollution Control Association* 28, no. 5 (1978): 516–17.

Podobnik, Bruce. *Global Energy Shifts: Fostering Sustainability in a Turbulent Age.* Philadelphia: Temple University Press, 2006.

Pontecorvo, Guido. "Hermann Joseph Muller, 1890–1967." *Biographical Memoirs of Fellows of the Royal Society* 14 (November 1968): 348–89.

Poore, Ben Perley. "Biographical Notice of John S. Skinner." *The Plough, the Loom, and the Anvil* 7, no. 1 (1854): 1–20.

Prentiss, Mara. *Energy Revolution: The Physics and the Promise of Efficient Technology.* Cambridge, MA: Harvard University Press, 2015.

Priestley, Joseph. *Historical Account of the Navigable Rivers, Canals, and Railways of Great Britain.* London: Longman, Rees, Orme, Brown & Green, 1831.

———. *The History and Present State of Electricity, with Original Experiments.* Vol. 1. 3rd ed. London: C. Bathurst et al., 1769.

President's Commission on the Accident at Three Mile Island. *The Need for Change: The Legacy of TMI.* Washington, DC, October 1979.

Price, Jerome. *The Antinuclear Movement.* Boston: Twain, 1982.

Prince, Morton B. "Early Work on Photovoltaic Devices at the Bell Telephone Laboratories." Ch. 33 in *Power for the World: The Emergence of Electricity from the Sun.* Edited by Wolfgang Palz. Singapore: Pan Stanford, 2011, 497–98.

Pritchard, R. E., ed. *Shakespeare's England: Life in Elizabethan & Jacobean Times.* Stroud, UK: History Press, 1999.

Prout, Henry G. *A Life of George Westinghouse.* New York: American Society of Mechanical Engineers, 1921.

Pumfrey, Stephen. "Who Did the Work? Experimental Philosophers and Public Demonstrators in Augustan England." *British Journal for the History of Science* 28, no. 2 (1995): 131–56.

Putnam, William Lowell. *The Kaiser's Merchant Ships in World War I.* Jefferson, NC: McFarland. Kindle edition, 2001.

de Quincey, Thomas. "The Nation of London," ch. 7, in *Autobiographic Sketches* (online).

Qvist, S. A., and B. W. Brook. "Potential for Worldwide Displacement of Fossil Fuel Electricity by Nuclear Energy in Three Decades Based on Extrapolation of Regional Deployment Data." *PLoS One* 10, no. 5 (2015): e0124074 (online).

Rackham, Oliver. *Trees and Woodland in the British Landscape.* London: J. M. Dent & Sons, 1976.

Raistrick, Arthur. *Dynasty of Iron Founders: The Darbys and Coalbrookdale.* Newton Abbot, UK: David & Charles, 1970.

Ramsay, William. *The Life and Letters of Joseph Black, M.D.* London: Constable, 1918.

Ransom, P. J. G. *The Victorian Railway and How It Evolved.* London: Heinemann, 1990.

Ratner, Michael. *21st Century US Energy Sources: A Primer.* Congressional Research Service Report R44854, 2017.

Reed, Brian. *The Rocket: Loco Profile* 7. Windsor, UK: Profile, 1970.

Reed, Terry. *Indy: The Race and the Ritual of the Indianapolis 500.* Washington, DC: Potomac Books, 2005.

Revelle, Roger. "Harrison Brown 1917–1986." *Biographical Memoirs of the National Academy of Sciences.* Washington, DC: National Academy of Sciences, 1994.

Rhodes, Richard. "A Demonstration at Shippingport: Coming On Line." *American Heritage* 32, no. 4 (1981) (online).

———. *John James Audubon: The Making of an American.* New York: Alfred A. Knopf, 2004.

———. *The Making of the Atomic Bomb.* New York: Simon & Schuster, 1986.

———. *Nuclear Renewal: Common Sense About Energy.* New York: Whittle/Viking, 1993.

———, ed. *Visions of Technology: A Century of Vital Debate About Machines, Systems and the Human World.* New York: Simon & Schuster, 1999.

Richards, Joseph W. "The Electro-Metallurgy of Aluminum. *Journal of the Franklin Institute* (May 1896): 357–81.

Richter, Burton. *Beyond Smoke and Mirrors: Climate Change and Energy in the 21st Century.* New York: Cambridge University Press, 2010.

Righter, Robert W. *Wind Energy in America: A History.* Norman: University of Oklahoma Press, 1996.

Riley, Joseph C. "The Pulsometer Steam Pump." *Technology Quarterly and Proceedings of the Society of Arts* 14 (1901): 243–54.

Ripy, Thomas B. *Federal Excise Taxes on Beverages: A Summary of Present Law and a Brief History.* Congressional Research Service Report RL30238. Washington, DC: Library of Congress, 1999.

Robert, Joseph C. *Ethyl: A History of the Corporation and the People Who Made It.* Charlottesville: University Press of Virginia, 1983.

Roberts, Peter. *The Anthracite Coal Industry; A Study of the Economic Conditions and Relations of the Cooperative Forces in the Development of the Anthracite Coal Industry of Pennsylvania.* New York: Macmillan, 1901.

Robertson, R. B. *Of Whales and Men.* New York: Knopf, 1954.

Robinson, Eric, and Douglas McKie. *Partners in Science: Letters of James Watt and Joseph Black.* London: Constable, 1970.

Robinson, Eric, and A. E. Musson. *James Watt and the Steam Revolution: A Documentary History.* New York: Augustus M. Kelley, 1969.

Robinson, H. W. "Denis Papin (1647–1712)." *Notes and Records of the Royal Society of London* 5, no. 1 (October 1947): 47–50.

Rockström, Johan, Will Steffen, Kevin Noone, Åsa Persson, F. Stuart Chapin III, Eric F. Lambin, Timothy M. Lenton et al. "A Safe Operating Space for Humanity." *Nature* 461, no. 7263 (September 24, 2009): 472–75.

Rockwell, Theodore. *Reflections on US Nuclear History.* London: World Nuclear Association, n.d. (online.)

———. *The Rickover Effect: How One Man Made a Difference.* Lincoln, NE: iUniverse, 2002.

Rogers, Naomi. "Germs with Legs: Flies, Disease, and the New Public Health." *Bulletin of the History of Medicine* (Winter 1989): 599–617.

Roland, Alex. "Bushnell's Submarine: American Original or European Import?" *Technology and Culture* 18, no. 2 (1977): 157–74.

Rolt, L. T. C. *George and Robert Stephenson: The Railway Revolution.* Stroud, UK: Amberley, 2016.

———. *Thomas Newcomen: The Prehistory of the Steam Engine.* Dawlish, UK: David and Charles, 1963.

Rose, Mark H. "Urban Environments and Technological Innovation: Energy Choices in Denver and Kansas City, 1900–1940." *Technology and Culture* 25, no. 3 (1984): 503–39.

Rosen, William. *The Most Powerful Idea in the World: A Story of Steam, Industry & Invention.* Chicago: University of Chicago Press, 2010.

Rosenbaum, Walter A. *The Politics of Environmental Concern.* 2nd ed. New York: Holt, Rinehart and Winston, 1977.

Rosenberg, Nathan. *Perspectives on Technology.* Cambridge: Cambridge University Press, 1976.

Ross, G. MacDonald. *Leibniz.* Oxford: Oxford University Press, 1984.

Rossin, A. David. "Marketing Fear: Nuclear Issues in Public Policy." Unpublished manuscript, n.d.

Rotch, William. "Memorandum Written by William Rotch in 1814 in the Eightieth Year of his Age." In Bullard, *The Rotches* (1947), 175–200, q.v.

Rotherham, Ian D., and David Egan. "The Economics of Fuel Wood, Charcoal and Coal: An Interpretation of Coppice Management of British Woodlands." In *History and Sustainability: Third International Conference of the European Society for Environmental History: Proceedings,* edited by Mauro Agnoletti, Marco Armiero, Stefania Barca, and Gabriella Corona. Florence: University of Florence, 2005, 100–104.

Roueché, Berton. "Annals of Medicine: The Fog." *New Yorker,* September 30, 1950: 33–51.

Ruebhausen, Oscar M., and Robert B. von Mehren. "The Atomic Energy Act and the Private Production of Atomic Power." *Harvard Law Review* 66, no. 8 (1953): 1450–96.

Rushmore, David B., and Eric A. Lof. *Hydro-Electric Power Stations.* 2nd ed. New York: John Wiley & Sons, 1923.

Russell, Ben. *James Watt: Making the World Anew.* London: Reaktion Books, 2014.

Russell, W. L. "Effect of Radiation Dose Rate on Mutation in Mice." Pt. 2. *Journal of Cellular and Comparative Physiology* 58, no. 3 (1961): 183–87.

———. "Genetic Hazards of Radiation." *Proceedings of the American Philosophical Society* 107, no. 1 (1963): 11–17.

———. "Reminiscences of a Mouse Specific-Locus Test Addict." *Environmental and Molecular Mutagenesis* 14, supp. 16 (1989): 16–22.

Rutkow, Eric. *American Canopy: Trees, Forests, and the Making of a Nation.* New York: Scribner, 2013.

Ryan, Harris J. "Developments in Electric Power Transmission." *Electrical Engineering* 53, no. 5 (1934): 712–14.

Sachs, Joseph. "Motor Road Vehicles." *Journal of the Franklin Institute* 144, no. 10 (1897): 286–305.

Sale, Kirkpatrick. *The Green Revolution: The American Environmental Movement, 1962–1992.* New York: Hill and Wang, 1993.

Salom, Pedro G. "Automobile Vehicles." *Journal of the Franklin Institute* 141, no. 4 (1896): 278–96.

Sanders, Alvin Howard. *A History of the Percheron Horse.* Chicago: Breeder's Gazette Print, 1917.

Sauder, Lee, and Skip Williams. "A Practical Treatise on the Smelting and Smithing of Bloomery Iron." *Historical Metallurgy* 36, no. 2 (2002): 122–31.

Saunders, Robert S. "Criticism and the Growth of Knowledge: An Examination of the Controversy over *The Limits to Growth.*" *Stanford Journal of International Studies* 9 (Spring 1974): 45–70.

Savery, Thomas. *The Miner's Friend; or, An Engine to Raise Water by Fire, Described. And of the Manner of Fixing It in Mines; With an Account of the Several Other Uses It is Applicable Unto; and an Answer to the Objections Made Against It.* London: S. Crouch, 1702.

Schallenberg, Richard H. *Bottled Energy: Electrical Engineering and the Evolution of Chemical Energy Storage.* Philadelphia: American Philosophical Society, 1982.

Schelling, Thomas C. *Costs and Benefits of Greenhouse Gas Reduction.* Washington: AEI Press, 1998.

———. *Strategies of Commitment and Other Essays.* Cambridge, MA: Harvard University Press, 2006.

Schiffer, Michael Brian. *Draw the Lightning Down: Benjamin Franklin and Electrical Technology in the Age of Enlightenment.* Berkeley: University of California Press, 2003.

———. *Power Struggles: Scientific Authority and the Creation of Practical Electricity Before Edison.* Cambridge, MA: MIT Press, 2008.

Schlesinger, Henry. *The Battery: How Portable Power Sparked a Technological Revolution.* New York: Harper, 2010.

Schmid, Sonja D. *Producing Power: The Pre-Chernobyl History of the Soviet Nuclear Industry.* Cambridge, MA: MIT Press, 2015.

Schrenk, H. H., Harry Heimann, George D. Clayton, W. M. Gafafer, and Harry Wexler. *Air Pollution in Donora, Pa.: Epidemiology of the Unusual Smog Episode of October 1948. Preliminary Report.* Public Health Bulletin no. 306. Washington, DC: Public Health Service, 1949.

Schumacher, E. F. *Small Is Beautiful: Economics as If People Mattered.* New York: Harper Perennial, 1973.

Schurr, Sam H., and Bruce C. Netschert. *Energy in the American Economy, 1850–1975: An Economic Study of Its History and Prospects.* Baltimore: Johns Hopkins Press, 1960.

Schurr, Sam H., Calvin C. Burwell, Warren S. Devine, and Sidney Sonenblum. *Electricity in the American Economy: Agent of Technological Progress.* New York: Greenwood Press, 1990.

Scott, Charlotte. "Dark Matter: Shakespeare's Foul Dens and Forests." *Shakespeare Survey* 1952 (2011) (online).

Scott-Warren, Jason. "When Theaters Were Bear-Gardens: Or, What's at Stake in the Comedy of Humors." *Shakespeare Quarterly* 54, no. 1 (2003): 63–82.

Scrivner, Lee. *Becoming Insomniac: How Sleeplessness Alarmed Modernity.* London: Palgrave Macmillan, 2014.

Sellers, Coleman. "The Utilization of the Power of Niagara Falls and Notes on Engineering Progress." *Journal of the Franklin Institute* (July 1891): 30–53.

Seltzer, Michael William. "The Technological Infrastructure of Science." PhD diss., Blacksburg, VA: Virginia Polytechnic Institute and State University, 2007.

Semmes, Raphael. *Memoirs of Service Afloat During the War Between the States.* Secaucus, NJ: Blue & Grey Press, 1987. First published 1868.

Severnini, Edson. "Impacts of Nuclear Plant Shutdown on Coal-Fired Power Generation and Infant Health in the Tennessee Valley in the 1980s." *Nature Energy* 21, no. 17051 (2017): 1–9.

Seyferth, Dietmar. "The Rise and Fall of Tetraethyllead," pt. 2, *Organometallics* 22, no. 25 (2003): 5154–78.

Shadley, Jeffery D., Veena Afzal, and Sheldon Wolff. "Characterization of the Adaptive Response to Ionizing Radiation Induced by Low Doses of X Rays to Human Lymphocytes." *Radiation Research* 111, no. 3 (1987): 511–17.

Shagena, Jack L. *Who Really Invented the Steamboat? Fulton's* Clermont *Coup.* Amherst, NY: Humanity Books, 2004.

Shapin, Steven. "The Invisible Technician." *American Scientist* 77, no. 6 (1989): 554–63.

Shapin, Steven, and Simon Schaffer. *Leviathan and the Air-Pump: Hobbes, Boyle, and the Experimental Life.* Princeton, NJ: Princeton University Press, 1985.

Shellenberger, Michael, and Ted Nordhaus. "Environmental Apocalypse Is a Myth." *San Francisco Chronicle* online, June 7, 2013.

Sheppard, Muriel Earley. *Cloud by Day: The Story of Coal and Coke and People.* Uniontown, PA: Heritage, 1947.

Sherman, William H. "Patents and Prisons: Simon Sturtevant and the Death of the Renaissance Inventor." *Huntington Library Quarterly* 72, no. 2 (2009): 239–56.

Shapiro, James. *A Year in the Life of William Shakespeare: 1599.* New York: Harper Perennial, 2006.

Shrader-Frechette, Kristin. "Conceptual Analysis and Special-Interest Science: Toxicology and the Case of Edward Calabrese." *Synthese* 177 (2010): 449–69.

Sieferle, Rolf Peter. *The Subterranean Forest: Energy Systems and the Industrial Revolution.* Cambridge: White Horse Press, 2001.

Silliman, Benjamin, Jr. *Report on the Rock Oil, or Petroleum, from Venango Co., Pennsylvania, with Special Reference to Its Use for Illumination and Other Purposes.* New Haven, CT: J. H. Benham, 1855.

Silliman Sr., Benjamin, ed. "Notice of a Fountain of Petroleum, Called the Oil Spring." *American Journal of Science and Arts* 23 (January 1833): 97–102.

Simon, Julian L. *The Ultimate Resource.* Princeton, NJ: Princeton University Press, 1981.

Simonson, R. D. *The History of Welding.* Morton Grove, IL: Monticello, 1969.

Skeat, W. O. *George Stephenson: The Engineer and His Letters.* London: Institution of Mechanical Engineers, 1973.

Sloan, Alfred P., Jr. *My Years with General Motors.* New York: Doubleday, 1963.

Smil, Vaclav. *Energy Transitions: History, Requirements, Prospects.* Santa Barbara, CA: Praeger, 2010.

Smiles, Samuel. *Industrial Biography: Iron Workers and Tool Makers.* Boston: Ticknor and Fields, 1864.

———. *The Life of George Stephenson, Railway Engineer.* 3rd ed., rev. London: John Murray, 1857.

———. *The Life of George Stephenson, and of His Son, Robert Stephenson; Containing Also a History of the Invention and Introduction of the Railway Locomotive.* New York: Harper & Brothers, 1864 (online).

Smith, Edgar C. "Pioneer Ships of the Atlantic Ferry." *Transactions of the Newcomen Society* 10 (1931): 46–54.

Smith, John. *A Description of New England, or, the Observations and Discoveries of Captain John Smith (Admiral of that Country), in the North of America, in the Year of Our Lord 1614.* London: Robert Clerke, 1616. (*American Colonial Tracts Monthly* Number One, May 1898. Rochester: George P. Humphrey.)

Smith, Robert Angus. *Air and Rain: The Beginnings of a Chemical Climatology.* London: Longmans, Green, 1872.

Smith, R. S. "England's First Rails: A Reconsideration." *Renaissance and Modern Studies* 4, no. 1 (1960): 119–34.

———. "Huntingdon Beaumont, Adventurer in Coal Mines." *Renaissance and Modern Studies* 1 (1957): 115–53.

Snow, Richard. *I Invented the Modern Age: The Rise of Henry Ford.* New York: Scribner, 2013.

Souder, William. *On a Farther Shore: The Life and Legacy of Rachel Carson.* New York: Crown, 2012.

Sousa, Luis de. "Marchetti's Curves." The Oil Drum: *Europe,* 2007 (online).

Spoerl, Edward. "The Lethal Effects of Radiation." *Scientific American* 185, no. 6 (1951): 22–25.

Sprague, Frank J. "The Solution of Municipal Rapid Transit." *American Institute of Electrical Engineers Transactions* 5 (1887): 352–99.

Spratt, H. Philip. "The Origin of Transatlantic Steam Navigation, 1819–1833." *Transactions of the Newcomen Society* 26 (1953): 131–43.

———. "The Prenatal History of the Steamboat." *Transactions of the Newcomen Society* 30 (1955): 13–23.

Stackpole, Edouard A. "Peter Folger Ewer: The Man Who Created the 'Camels.'" *Historic Nantucket* 33, no. 1 (July 1985) (accessed via Nantucket Historical Association online).

Standish, Arthur. *The Commons Complaint.* London: William Stansby, 1611.

———. *New Directions of Experience to the Commons Complaint, for the Planting of Timber and Firewood,* 1613. Early English Books Online.

Stanley, William. "Alternating-Current Development in America." *Journal of the Franklin Institute* 173, no. 6 (1912): 561–80.

Stansfield, Dorothy A., and Ronald G. Stansfield. "Dr. Thomas Beddoes and James Watt: Preparatory Work 1794–96 for the Bristol Pneumatic Institute." *Medical History* 30, no. 3 (July 1986): 276–302.

Starbuck, Alexander. *History of the American Whale Fishery from Its Earliest Inception to the Year 1876.* Waltham, MA: published by the author, 1878.

Stauffer, Robert C. "Speculation and Experiment in the Background of Oersted's Discovery of Electromagnetism." *Isis* 48, no. 1 (1957): 33–50.

Stegner, Wallace. *Discovery! The Search for Arabian Oil.* Vista, CA: Selwa Press, 2007. First published 1971.

Steinmueller, W. Edward. "The Preindustrial Energy Crisis and Resource Scarcity as a Source of Transition." *Research Policy* 42 (2013): 1739–48.

Stephens, Edgar R. "Smog Studies of the 1950s." *Eos* 68, no. 7 (1987): 89, 91–93.

Stephenson, Robert, and Joseph Locke. *Observations on the Comparative Merits of Locomotive and Fixed Engines, as Applied to Railways.* Philadelphia: Carey & Lea, 1831.

Stern, Jonathan. "UK Gas Security: Time to Get Serious." *Energy Policy* 32 (2004): 1967–79.

Stevenson, Robert Louis. *Strange Case of Dr. Jekyll and Mr. Hyde.* London: Longmans, Green, 1886.

Stommel, Henry, and Elizabeth Stommel. *Volcano Weather: The Story of 1816, the Year Without a Summer.* Newport: Seven Seas Press, 1983.

Stone, I. F. *The Haunted Fifties, 1953–1963.* Boston: Little, Brown, 1963.

Stone, Lawrence. "An Elizabethan Coalmine." *Economic History Review,* n.s. 3, no. 1 (1950): 97–106.

Storer, Jacob J. "Sanitary Care and Utilization of Refuse of Cities." *Journal of the Franklin Institute* 67 (1874): 48–55.

Stotz, L. *History of the Gas Industry.* New York: Stettiner Brothers, 1938.

Stow, John. *A Survey of London.* Rev. 1603. Henry Morley, ed. Thrupp, UK: Sutton, 1994. First printed 1598.

Stradling, David. *Smokestacks and Progressives: Environmentalists, Engineers, and Air Quality in America, 1881–1951.* Baltimore: Johns Hopkins University Press, 1999.

Strahler, Arthur N. *A Geologist's View of Cape Cod.* Garden City, NY: Natural History Press, 1966.

Strandh, Sigvard. *A History of the Machine.* New York: A&W, 1979.

Stratton, Ezra M. *The World on Wheels; or, Carriages, with Their Historical Associations from the Earliest to the Present Time.* New York: published by the author, 1878.

Strohl, Dan. "Ford, Edison and the Cheap EV That Almost Was." *Wired* 6 (2010) (online).

Stuart, Robert. *Historical and Descriptive Anecdotes of Steam-Engines, and of Their Inventors and Improvers.* 2 vols. London: Wightman and Cramp, 1829.

Sturtevant, A. H. "Social Implications of the Genetics of Man." *Science* 120, no. 3115 (September 10, 1954): 405–7.

Sugden, John. "Lord Cochrane, Naval Commander, Radical, Inventor (1775–1860): A Study of His Earlier Career, 1775–1818." PhD diss., Department of History, University of Sheffield, 1981.

Sullivan, Mark. *Our Times: The United States 1900–1925.* Vol. 1, *The Turn of the Century.* New York: Charles Scribner's Sons, 1927.

Summerside, Thomas. *Anecdotes, Reminiscences, and Conversations, of and With the Late George Stephenson, Father of Railways.* London: Bemrose and Sons, 1878.

Swank, James M. *History of the Manufacture of Iron in All Ages, and Particularly in the United States from Colonial Times to 1891. Also a Short History of Early Coal Mining in the United States and a Full Account of the Influences Which Long Delayed the Development of All American Manufacturing Industries.* Philadelphia: American Iron and Steel Association, 1892.

Swensrud, Sidney A. "Possibility of Converting the Large Diameter War Emergency Pipe Lines to Natural Gas Service After the War." Paper for presentation at the February 1944 meeting of the Petroleum Division of the American Institute of Mining and Metallurgical Engineers. New York City, 21 February 1944. *AIME Technical Publications & Contributions 1943–1944* (online).

Tabak, John. *Coal and Oil.* New York: Facts on File, 2009.

Tabuchi, Hiroko. "As Beijing Joins Climate Fight, Chinese Companies Build Coal Plants." *New York Times* online, July 1, 2017.

Tarbell, Ida M. *All in the Day's Work: An Autobiography.* Boston: G. K. Hall, 1985. First published 1939 by Macmillan.

———. *The History of the Standard Oil Company.* New York: McClure, Phillips, 1904.

Tassava, Christopher James. "Launching a Thousand Ships: Entrepreneurs, War Workers, and the State of American Shipbuilding, 1940–1945." PhD diss., Northwestern University, Evanston, IL, 2003.

Taylor, George Coffin. "Milton on Mining." *Modern Language Notes* 43, no. 1 (January 1930): 24–27.

Tenner, Edward. *Why Things Bite Back: Technology and the Revenge of Unintended Consequences.* New York: Alfred A. Knopf, 1996.

Testimonies Gathered by Ashley's Mines Commission. 1842. www.victorianweb.org (online).

Thomas, Brinley. "Was There an Energy Crisis in Great Britain in the 17th Century?" *Explorations in Economic History* 23 (1986): 124–52.

Thomas, Emyr. *Coalbrookdale and the Darby Family: The Story of the World's First Industrial Dynasty.* York, UK: Sessions Book Trust, 1999.

Thomas, Trevor. "Oil and Natural Gas: Discoveries and Exploration in the North Sea and Adjacent Areas." *Geography* 49, no. 1 (1964): 50–55.

Thomas, William. *Observations on Canals and Rail-Ways, Illustrative of the Agricultural and Commercial Advantages to Be Derived from an Iron Rail-way, Adapted to Common Carriages, Between Newcastle, Hexham, and Carlisle.* Newcastle upon Tyne, UK: G. Angus, 1825.

Thompson, Edward Stoops. The History of Illuminating Gas in Baltimore. Records of Phi Mu, Special Collections, University of Maryland Libraries online, 1928.

Thompson, Silvanus P. *Polyphase Electric Currents and Alternate-Current Motors.* New York: Spon & Chamberlain, 1895.

Thompson, William R. "Energy, K-Waves, Lead Economies, and Their Interpretation/Implications." *Social Studies Almanacs* online, 2012.

Thurston, Robert H. *A History of the Growth of the Steam-Engine.* 2nd rev. ed. New York: D. Appleton, 1884.

Tierie, Gerrit. "Cornelis Drebbel (1572–1633)." PhD diss., Leiden University online, 1982.

Titley, Arthur. "Richard Trevithick and the Winding Engine." *Transactions of the Newcomen Society* 10 (1931): 55–68.

Tomory, Leslie. *Progressive Enlightenment: The Origins of the Gaslight Industry, 1780–1820.* Cambridge, MA: MIT Press, 2012.

Tonkin, S. Morley. "Trevithick, Rastrick and the Hazledine Foundry, Bridgnorth." *Transactions of the Newcomen Society* 26 (1953): 171–83.

Tower, Walter S. *A History of the American Whale Fishery.* Philadelphia: University of Pennsylvania Press, 1907.

Toynbee, Arnold. *Lectures on the Industrial Revolution of the 18th Century in England.* San Bernardino, CA: Forgotten Books, 2012. First published 1887.

Trevithick, Francis. *Life of Richard Trevithick, with an Account of His Inventions.* 2 vols. London: E. & F. N. Spon, 1872.

Trinder, Barrie. *Coalbrookdale 1801: A Contemporary Description.* Museum Booklet no. 2003. Ironbridge Gorge, UK: Ironbridge Gorge Museum Trust, 1979.

Trollope, Anthony. *North America.* 3 vols. Leipzig: B. Tauchnitz, 1862.

Tubiana, Maurice. "Dose-Effect Relationship and Estimation of the Carcinogenic Effects of Low Doses of Ionizing Radiation: The Joint Report of the Academie Des Sciences (Paris) and of the Academie Nationale De Medecine." *International Journal of Radiation Oncology, Biology, Physics* 63, no. 2 (2005): 317–19.

Tussing, Arlon, and Bob Tippee. *The Natural Gas Industry: Evolution, Structure and Economics.* 2nd ed. Tulsa, OK: Pennwell Books, 1955.

Twitchell, K. S. *Saudi Arabia: With an Account of the Development of Its Natural Resources.* Princeton, NJ: Princeton University Press, 1947.

Twain, Mark. *Mark Twain's Letters.* Vol. 5, *1872–1873,* edited by Lin Salamo and Harriet Elinor Smith. Berkeley: University of California Press, 1997.

Tyler, Nick. "Trevithick's Circle." *Transactions of the Newcomen Society* 77 (2007): 101–113.

Uekoetter, Frank. *The Age of Smoke: Environmental Policy in Germany and the United States, 1888–1970.* Pittsburgh: University of Pittsburgh Press, 2009.

———. "The Environmentalists' Favorite Foe: Electric Power Plants in Germany." *Icon* 9 (2003): 44–61.

US Department of Transportation. *America's Highways, 1776–1976: A History of the*

Federal-Aid Program. Washington, DC: US Department of Transportation/Federal Highway Administration, 1976.

Volti, Rudi. *Cars and Culture: The Life Story of a Technology*. Baltimore: Johns Hopkins University Press, 2004.

Von Tschudi, John James. *Travels in Peru, on the Coast, in the Sierra, Across the Cordilleras and the Andes, into the Primeval Forests*. Translated by Thomasina Ross, New ed. New York: A. S. Barnes, 1854.

Von Tschudi, John James, and Mariano Edward Rivero. *Peruvian Antiquities*. Translated by Francis L. Hawks. New York: A. S. Barnes, 1855.

Uglow, Jenny. *The Lunar Men: Five Friends Whose Curiosity Changed the World*. New York: Farrar, Straus and Giroux, 2002.

United Nations. *New Sources of Energy. Proceedings of the United Nations Conference on New Sources of Energy: Solar Energy, Wind Power and Geothermal Energy. Rome, 21–31 August 1961*. New York: United Nations, 1964.

United States Federal Power Commission. *Natural Gas Investigation (Docket No. G-580). Report of Commissioner Nelson Lee Smith and Commissioner Harrington Wimberly*. Washington, DC: US Government Printing Office, 1948.

United States Bureau of Mines. *The Bureau of Mines Synthetic Liquid Fuels Program, 1944–55. Report of Investigations 5506*. Washington, DC: US Government Printing Office, 1959.

United States Department of Health, Education and Welfare Public Health Service Advisory Committee on Tetraethyl Lead to the Surgeon General. *Public Health Aspects of Increasing Tetraethyl Lead Content in Motor Fuel: A Report*. Washington, DC: Public Health Service Publication no. 712, 1959.

United States Department of the Interior. *Hearings Before the Secretary of the Interior on Leasing of Oil Lands and Natural-Gas Wells in Indian Territory and Territory of Oklahoma, May 8, 24, 25, and 29, and June 7 and 19, 1906*. Repr. New York: Arno Press, 1972.

United States House of Representatives (USHR). *Free Alcohol: Hearings Before the Committee on Ways and Means, 59th Congress, 1st Session*. February/March 1906. Washington: US Government Printing Office, 1910.

Uppenborn, Friedrich. *History of the Transformer*. London: E. & F. N. Spon, 1889.

Valenti, Phillip. "Leibniz, Papin and the Steam Engine: A Case Study of British Sabotage of Science." *American Almanac* online, 1996.

Varadi, Peter E. "Terrestrial Photovoltaic Industry—The Beginning." In *Power for the World*, edited by Wolfgang Palz. Singapore: Pan Stanford Publishing, 2011, 555–67.

Viana, Hélio Elael Bonini, and Paulo Alves Porto. "Thomas Midgley, Jr., and the

Development of New Substances: A Case Study for Chemical Educators." *Journal of Chemical Education* 90 (2013): 1632–38.

Vitalis, Robert. *America's Kingdom: Mythmaking on the Saudi Oil Frontier.* Stanford, CA: Stanford University Press, 2007.

Volti, Rudi. "A Century of Automobility." *Technology and Culture* 37, no. 4 (1996): 663–85.

Waerland, Are. "Marten Triewald and the First Steam Engine in Sweden." *Transactions of the Newcomen Society* 7, no. 1 (1926): 24–41.

Walker, J. Samuel. *Three Mile Island: A Nuclear Crisis in Historical Perspective.* Berkeley: University of California Press, 2004.

Walker, James. *Report to the Directors of the Liverpool and Manchester Railway, on the Comparative Merits of Locomotive and Fixed Engines, as a Moving Power.* (With Stephenson, Robert, and Joseph Locke, *Observations on the Comparative Merits of Locomotive and Fixed Engines, as Applied to Railways;* and Booth, Henry, *An Account of the Liverpool and Manchester Railway.*) Philadelphia: Carey & Lea, 1831.

Wallace, Anthony F. C. *St. Clair: A Nineteenth-Century Coal Town's Experience with a Disaster-Prone Industry.* Ithaca, NY: Cornell University Press, 1988.

Wallace, Charles William. "The First London Theatre, Materials for a History." *University Studies* 13, nos. 1, 2, 3 (1913): 1–35ff, passim.

Waples, David A. *The Natural Gas Industry in Appalachia: A History from the First Discovery to the Tapping of the Marcellus Shale.* 2nd ed. Jefferson, NC: McFarland, 2012.

Warde, Paul. *Energy Consumption in England & Wales 1560–2000.* Naples, Italy: Consiglio Nazionale delle Ricerche, Istituto di Studi sulle Società del Mediterraneo, 2007.

———. "Fear of Wood Shortage and the Reality of the Woodland in Europe, c. 1450–1850." *History Workshop Journal* 62 (2006): 28–57.

Warmington, Andrew. "Sir John Winter." *Oxford Dictionary of National Biography* (hereafter, *DNB*) online. Oxford: Oxford University Press, 2004–14.

Warren, Charles. "Parson Malthus Tolls the Bell." *Sierra Club Bulletin* 60, no. 3 (1975): 7–10, 24, 31.

Warren, J. G. H. *A Century of Locomotive Building by Robert Stephenson & Co., 1823–1923.* New York: Augustus M. Kelley, 1970.

Watkins, C. Malcolm. *Artificial Lighting in America 1830–1860.* Smithsonian Institution Publication 4080. Washington, DC: US Government Printing Office, 1952.

Watkins, Charles. *Trees, Woods and Forests: A Social and Cultural History.* London: Reaktion, 2014.

Weart, Spencer R. *The Discovery of Global Warming.* Cambridge, MA: Harvard University Press, 2008.

———. "The Idea of Anthropogenic Global Climate Change in the 20th Century." *Wiley Interdisciplinary Reviews: Climate Change* 1, no. 1 (2010): 67–81.

———. *The Rise of Nuclear Fear.* Cambridge, MA: Harvard University Press, 2012.

Webb, Sidney, and Beatrice Webb. *English Local Government: The Story of the King's Highway.* London: Longmans, Green, 1913.

Weinberg, Alvin M. *The First Nuclear Era: The Life and Times of a Technological Fixer.* New York: American Institute of Physics, 1994.

———. "Nuclear Energy and the Environment." *Bulletin of the Atomic Scientists* 26, no. 6 (1970): 69–74.

———. "Nuclear Energy: A Prelude to H. G. Wells's Dream." *Foreign Affairs* 49, no. 3 (1971): 407–18.

———. "Nuclear Energy: Salvaging the Atomic Age." *Wilson Quarterly* 3, no. 3 (1979): 88–112.

Wellock, Thomas Raymond. *Critical Masses: Opposition to Nuclear Power in California, 1958–1978.* Madison: University of Wisconsin Press, 1998.

Werner, M. M., D. K. Meyers, and D. P. Morrison. "Follow-up of CRNL Employees Involved in the NRX Reactor Clean-up." Paper presented at the Third Annual Meeting of the Canadian Radiation Protection Association, Vancouver, BC, 4 May 1982. Chalk River, Ont.: Chalk River Nuclear Laboratories.

West, Richard. *Daniel Defoe: The Life and Strange, Surprising Adventures.* New York: Carroll & Graf, 1998.

Westcott, Henry P. *Hand Book of Natural Gas.* Erie, PA: Metric Metal Works, 1915.

Westinghouse Educational Foundation. *The Future of Atomic Energy. The George Westinghouse Centennial Forum, May 16, 17, and 18, 1946.* New York: McGraw-Hill, 1946.

Wetherill, G. W., and M. G. Inghram. "Spontaneous Fission in Uranium and Thorium Ores." *Proceedings of the Conference on Nuclear Processes in Geologic Settings.* Williams Bay, WI.: National Research Council Committee on Nuclear Science, 1953.

Whilldin, J. K. "Memoranda on Electric Lights." *Journal of the Franklin Institute* (April 1862): 217–22.

White, Gerald T. *Scientists in Conflict: The Beginnings of the Oil Industry in California.* San Marino, CA: Huntington Library, 1968.

White, Gilbert. *The Natural History of Selborne.* New York: Harper & Brothers, 1841. First published 1789.

White, Kevin M., and Samuel H. Preston. "How Many Americans Are Alive Because of Twentieth-Century Improvements in Mortality?" *Population and Development Review* 22, no. 3 (1996): 415–29.

Whitehead, Alfred North. *The Concept of Nature.* Cambridge: Cambridge University Press, 1971.

———. *Modes of Thought.* New York: Free Press, 1938.

Whitehead, J. Rennie. *The Club of Rome and CACOR: Recollections,* n.d. (online).

Wilkinson, Paul. *The Historical Development of the Port of Faversham, Kent 1580–1780,* n.d., http://kafs.co.uk.

Willekens, Frans. "Demographic Transitions in Europe and the World." Max Planck Institute for Demographic Research Working Paper WP 2014-004, 2014 (online).

Williams, James H., Andrew DeBenedictis, Rebecca Ghanadan, Amber Mahone, Jack Moore, William R. Morrow III, Snuller Price, and Margaret S. Torn. "The Technology Path to Deep Greenhouse Gas Emissions Cuts by 2050: The Pivotal Role of Electricity." *Science* 335, no. 6064 (January 6, 2012): 53–59.

Williams, L. Pearce. *Michael Faraday: A Biography.* New York: Da Capo, 1965.

Williams, Michael. *Deforesting the Earth: From Prehistory to Global Crisis. An Abridgement.* Chicago: University of Chicago Press, 2006.

Williamson, George. *Memorials of the Lineage, Early Life, Education, and Development of the Genius of James Watt.* Greenock, Scot.: printed privately for the Watt Club, 1856.

Williamson, Harold F., and Arnold R. Daum. *The American Petroleum Industry: The Age of Illumination, 1859–1899.* Evanston, IL: Northwestern University Press, 1959.

Williamson, Harold F., Ralph L. Andreano, Arnold R. Daum, and Gilbert C. Klose. *The American Petroleum Industry: The Age of Energy 1899–1959.* Evanston, IL: Northwestern University Press, 1963.

Wilson, George. "On the Early History of the Air-Pump in England." *Edinburgh New Philosophical Journal* 46 (1849): 330–54.

Wilson, Leonard G., ed. *Benjamin Silliman and his Circle: Studies on the Influence of Benjamin Silliman on Science in America.* New York: Science History, 1979.

Wilson, Richard, Steven D. Colome, John D. Spengler, and David Gordon Wilson. *Health Effects of Fossil Fuel Burning: Assessment and Mitigation.* Cambridge, MA: Ballinger, 1980.

Winter, John. *A True Narrative concerning the Woods and Iron-Works of the Forrest of Deane, and how they have been disposed since the year 1635. As appears by records.* Hereford, UK: Herefordshire Archive and Records Centre, c. 1670 (online).

Wise, George. "William Stanley's Search for Immortality." *Invention and Technology* 4, no. 1 (1988) (online).

Witcover, Jules. *Sabotage at Black Tom: Imperial Germany's Secret War in America 1914–1917.* Chapel Hill, NC: Algonquin, 1989.

Wolfrom, Melville L. "Harold Hibbert 1877–1945." *Biographical Memoirs of the*

National Academy of Sciences. Washington, DC: National Academy of Sciences, 1958.

Wood, Nicholas. *A Practical Treatise on Rail-roads, and Interior Communication in General, Containing an Account of the Performance of the Different Locomotive Engines at and Subsequent to the Liverpool Contest; Upwards of Two Hundred and Sixty Experiments; With Tables of the Comparative Values of Canals and Rail-roads, and the Power of the Present Locomotive Engines.* First American, from the Second English Edition. Philadelphia: Carey & Lea, 1832.

Wordsworth, Dorothy. *Journals of Dorothy Wordsworth.* Edited by William Knight. Vol. 1. London: Macmillan, 1904.

Wrench, Kent, ed. *Tar Heels: North Carolina's Forgotten Economy: Pitch, Tar, Turpentine and Longleaf Pines.* Charleston, SC: CreateSpace, 2014.

Wright, Arthur W. "Biographical Memoir of Benjamin Silliman [Jr.], 1816–1885." *National Academy of Sciences Biographical Memoirs* 7. Washington, DC: National Academy of Sciences, 1911.

Wright, C. D. "Report to the President on Anthracite Coal Strike." *Bulletin of the Department of Labor* 43. Washington, DC: US Government Printing Office, November 1902.

Wrigley, E. A. *Energy and the English Industrial Revolution.* Cambridge: Cambridge University Press, 2010.

Yergin, Daniel. *The Prize: The Epic Quest for Oil, Money, and Power.* New York: Simon & Schuster, 1991.

Young, W. A. "Thomas Newcomen, Ironmonger: The Contemporary Background." *Transactions of the Newcomen Society* 20 (1941): 1–15.

Young, Matt, Michael Boland, and Don Hofstrand. "Current Issues in Ethanol Production." *Agricultural Marketing Resource Center Value-added Business Profile.* Ames: Iowa State University, 2007.

Zoellner, Tom. "The Uranium Rush." *Invention and Technology* 16, no. 1 (2000): 1–19 (online).

Zubrin, Robert. *Merchants of Despair: Radical Environmentalists, Criminal Pseudo-Scientists, and the Fatal Cult of Antihumanism.* New York: Encounter Books, 2013.

NOTES

ONE: NO WOOD, NO KINGDOM

1. Information about the events around 28 December 1598 comes from Melissa D. Aaron, "The Globe and Henry V as Business Document," *SEL: Studies in English Literature 1500–1900:* 40.2 (2000) 277–92; Vin Nardizzi, "Shakespeare's Globe and England's Woods," *Shakespeare Studies* 39 (2011): 54–63; James Shapiro, *A Year in the Life of William Shakespeare: 1599*. New York: Harper Perennial, 2006, 1–42; Charles William Wallace, "The First London Theatre: Materials for a History," *University Studies* 13, nos. 1, 2, 3 (1913): 1–35ff. passim.
2. J. R. Mulryne and Margaret Shewring, eds. *Shakespeare's Globe Rebuilt* (Cambridge: Cambridge University Press, 1997), 190.
3. Charles W. Eliot, ed. *The Harvard Classics,* vol. 35, *Chronicle and Romance: Froissart, Malory, Holinshed* (New York: P. F. Collier & Son, 1938), 293.
4. 1581 act of Parliament: Michael W. Flinn, "Timber and the advance of technology: A reconsideration," *Annals of Science,* 15:2, 110–11; John U. Nef, *The Rise of the British Coal Industry,* vol. 1 (London: Routledge, 1966), 158.
5. England's population, 1570–1600: Eric Rutkow, *American Canopy: Trees, Forests, and the Making of a Nation* (New York: Scribner, 2013), 12.
6. 2,500 loads of oak: Flinn (1959), 110, 4n.
7. English ship of the line: Robert Greenhalgh Albion, *The Timber Problem of the Royal Navy, 1652–1862* (Cambridge, MA: Harvard University Press, 1926), 4.
8. Sail dimensions: ibid., 28.
9. *The Diary of Samuel Pepys,* Sunday, 5 May 1667. www.pepysdiary.com.
10. Three hundred ironworks; three hundred thousand loads: Albion, *Timber Problem of Royal Navy,* 116–17.
11. Quoted in ibid., 118–19.
12. Daniel Neal, *The History of New England,* vol. 2 (London: A. Ward, 1747), 213 (spelling modernized).

13. Albion, *Timber Problem of Royal Navy*, 121.
14. Arthur Standish, *The Commons Complaint* (London: William Stansby, 1611), 1 (spelling modernized).
15. Eliot, *Chronicle and Romance*, 319–20.
16. Elizabeth sent brewer to prison: Nef, *Rise of British Coal*, vol. 1, 156.
17. Samuel Smiles, *Industrial Biography: Iron Workers and Tool Makers* (Boston: Ticknor and Fields, 1864), 59n.
18. Geoffrey Chaucer, "The Former Age," lines 27–30 (language modernized).
19. Georgius Agricola, *De re Metallica*, trans. Herbert Hoover and Lou Henry Hoover (1556; repr., New York: Dover, 1950), 7.
20. John Milton, *Paradise Lost*, bk. 1, lines 670–88.
21. Quoted in Galloway, *A History of Coal Mining in Great Britain* (London: Macmillan, 1882), 23.
22. Statistics; "the coal trade": Nef, *Rise of British Coal*, 25.
23. Sulfur content of Scottish versus English coals: Peter Brimblecombe, *The Big Smoke: A History of Air Pollution in London Since Medieval Times* (London: Methuen, 1987), 66, table 4.1.
24. "Historical Overview of London Population." http://www.londononline.co.uk/factfile/historical/.
25. Quoted in Benita Cullingford, *British Chimney Sweeps: Five Centuries of Chimney Sweeping* (Hove, UK: Book Guild, 2000), 9.
26. *Artificiall Fire, or, Coale for Rich and Poore:* Early English Books Online, Wing (2nd ed.) G623, British Library.
27. Coal shipments into London: Nef, *Rise of British Coal*, vol. 1, 21, table 2.
28. John Evelyn, *A Character of England* (London: Joseph Crooke, 1659), 29–30. Early English Books Online.
29. John Evelyn, *Fumifugium: or, the Inconvenience of the Aer, and Smoake of London Dissipated* (London: printed by W. Godbid, for Gabriel Bedel, and Thomas Collins, 1661; reprinted for B. White, 1672), 1–2.
30. Ibid., 3.
31. Ibid., 13.
32. Ibid., 17.
33. Ibid., 19–20.
34. Ibid., 24–25.
35. Ibid., 36.
36. Ibid., 37.
37. Ibid., 47.
38. Ibid., 48–49.

TWO: RAISING WATER BY FIRE

1. John Holland, *The History and Description of Fossil Fuel, the Collieries, and Coal Trade of Great Britain,* 2nd ed. (London: Whittaker, 1841), 178.
2. Sinking and boring: John Hatcher, *The History of the British Coal Industry,* vol. 1, *Before 1700: Towards the Age of Coal* (Oxford: Clarendon Press, 1993), 196–97.
3. Of coal borings, communicated by Dr. Martin Lister, Fell. Coll. Phys. & R. S. which role or record he had from Mr. Maleverer, of Arncliffe in Yorkshire. *Philosophical Transactions of the Royal Society of London* 21(1699): 73–78.
4. Robert Galloway, *Annals of Coal Mining and the Coal Trade: The Invention of the Steam Engine and the Origin of the Railway* (London: Colliery Guardian, 1898), 56.
5. *Testimonies Gathered by Ashley's Mines Commission.* www.victorianweb.org.
6. Damps: Galloway, *Annals of Coal Mining,* 160.
7. Ibid., 214.
8. Ibid., 220.
9. Ibid., 221.
10. Ibid., 222.
11. Ibid., 157.
12. Boyle read of experiment: Boyle (1660), 5.
13. Ibid., 22.
14. Quoted in Gerrit Tierie, "Cornelis Drebbel (1572–1633)" (PhD diss., Leiden University, 1982), 17 (online).
15. Drebbel's retort pump: Tierie, "Cornelis Drebbel," 32–33.
16. Quoted in Tierie, "Cornelis Drebbel," 28.
17. Drebbel generated oxygen: see discussion in ibid., 70.
18. Royal navy at La Rochelle: ibid., 72.
19. Quoted in C. D. Andriesse, *Huygens: The Man Behind the Principle,* trans. Sally Miedema (Cambridge: Cambridge University Press, 2005), 229.
20. Quoted in Phillip Valenti, "Leibniz, Papin, and the Steam Engine: A Case Study of British Sabotage of Science," *American Almanac* online, 1996, n.p.
21. Leibniz report on Von Guericke: Antognazza (2009), 141.
22. Quoted in Andriesse, *Huygens: Man Behind the Principle,* 278.
23. Quoted in Valenti, "Leibniz, Papin, and the Steam Engine," 3.
24. Quoted and illustrated in ibid., 6.
25. Demonstrators not accorded standing: see Steven Shapin, "The Invisible Technician," *American Scientist* 77, no. 6 (1989): 554–63.
26. Denis Papin, "A New Method of Obtaining Very Great Moving Powers at Small Cost," reprinted in translation in James Patrick Muirhead, *The Life of James Watt,*

with Selections from His Correspondence* (New York: D. Appleton, 1859), 136–42, q.v. (1690), 105–6.

27. Ibid., 108–9 (tran. ed.).
28. Landgrave fountain project: Sigvard Strandh, *A History of the Machine* (New York: A&W, 1979), 115.
29. Quoted in Valenti, "Leibniz, Papin, and the Steam Engine," 10.
30. Ibid.
31. Papin, "A New Method of Obtaining," in Muirhead, *The Life of James Watt,* 106.
32. Quoted in Valenti, "Leibniz, Papin, and the Steam Engine," 10.
33. Papin's book: *Recueil de diverses Pieces touchant quelques nouvelles Machines, &c.* Par Mr. D. Papin, Dr. en., Med. &c. A Cassel, 1695.
34. Papin's book reviewed: *Philosophical Transactions of the Royal Society of London* 19 (1695–1697): 481.
35. Papin superheating steam: Valenti, "Leibniz, Papin, and the Steam Engine," 11.
36. 14 June 1699: Thomas Savery, *The Miner's Friend; or, An Engine to Raise Water by Fire, Described. And of the Manner of Fixing It in Mines; With an Account of the Several Other Uses It is Applicable Unto; and an Answer to the Objections Made Against It* (London: S. Crouch, 1702), A4.

THREE: A GIANT WITH ONE IDEA

1. Bernard Forest de Belidor, *Architecture hydraulique, ou L'art de Conduire, d'Élever et de Ménager les Eaux,* vol. 2 (Paris: Chez L. Cellot, 1782), 309.
2. Papin realized: as he wrote to Leibniz on 23 July 1705: See translation, Valenti, "Leibniz, Papin, and the Steam Engine," 14.
3. Efficiency of Savery's engine: Landes (1969), 101.
4. Valenti, "Leibniz, Papin, and the Steam Engine," 19 (my italics).
5. Papin steamboat proposal: H. W. Robinson, "Denis Papin (1647–1712)," *Notes and Records of the Royal Society of London* 5, no. 1 (October 1947): 49.
6. Rhys Jenkins, "The Heat Engine Idea in the Seventeenth Century: A Contribution to the History of the Steam Engine," *Transactions of the Newcomen Society* 17 (1937): 9.
7. Savery's various inventions: Robinson, "Denis Papin," 49.
8. Richard L. Hills, *Power from Steam: A History of the Stationary Steam Engine* (Cambridge: Cambridge University Press, 1989), 16.
9. Ibid.; Savery, *Miner's Friend.*
10. Quoted in Rhys Jenkins, "Savery, Newcomen and the Early History of the Steam Engine," pt. 2, *Transactions of the Newcomen Society* 4, no. 1 (1923): 116.
11. York Buildings, "Queen Anne's Palace": Hills, *Power from Steam,* 16.

12. Galloway, *Annals of Coal Mining*, 198.
13. Savery abandoned mine drainage after 1705: L. T. C. Rolt, *Thomas Newcomen: The Prehistory of the Steam Engine* (Dawlish, UK: David and Charles, 1963), 39.
14. Quoted in Galloway, *Annals of Coal Mining*, 175.
15. Newcomen's background: Rolt, *Thomas Newcomen*, 42–48.
16. See Are Waerland, "Marten Triewald and the First Steam Engine in Sweden," *Transactions of the Newcomen Society* 7, no. 1 (1926): 24ff.
17. Quoted in Rolt, *Thomas Newcomen*, 51.
18. Thurston's review of advantages: Robert H. Thurston, *A History of the Growth of the Steam-Engine*, 2nd rev. ed. (New York: D. Appleton, 1884), 60.
19. Quoted in Hills, *Power from Steam*, 25.
20. Dorothy Wordsworth, *Journals of Dorothy Wordsworth*, ed. William Knight, vol. 1 (London: Macmillan, 1904), 177.
21. Newcomen continued with Proprietors: Jenkins, "Savery, Newcomen," pt. 2., 119.
22. First full-scale commercial Newcomen engine: Galloway, *Annals of Coal Mining*, 238–39; Jenkins, "Savery, Newcomen," pt. 2., 119; John S. Allen, "Thomas Newcomen," *DNB* online, 4.
23. 1720 Cornwall engine: Anthony Burton, *Richard Trevithick: Giant of Steam* (London: Aurum Press, 2000), 9.
24. Galloway on horse power versus Newcomen engine: Galloway, *Annals of Coal Mining*, 241. Galloway instances Griff Colliery in Warwickshire.
25. Quoted in ibid., 241.
26. Newcomen engine revitalized mining industry: ibid.
27. 104 Newcomen engines: John S. Allen, "The Introduction of the Newcomen Engine, 1710–1733: Second Addendum," *Transactions of the Newcomen Society* 45 (1972): 223.
28. Harry Kitsikopoulos, "The Diffusion of Newcomen Engines, 1706–73: A Reassessment," Economic History Association online, 2013, 9.
29. Galloway, *Annals of Coal Mining*, 151.
30. Three ancient obligations—the *trinoda necessitas* of Saxon England: Sidney Webb and Beatrice Webb, *English Local Government: The Story of the King's Highway* (London: Longmans, Green, 1913), 5.
31. *Fyne Morrison's Itinerary, or Ten Years' Travel throughout Great Britain and other Parts of Europe* (1617), quoted in Ezra M. Stratton, *The World on Wheels; or, Carriages, with Their Historical Associations from the Earliest to the Present Time* (New York: published by the author, 1878), 272.
32. Quoted in Webb and Webb, *English Local Government*, 68.

33. Quoted in Galloway, *Annals of Coal Mining*, 156.
34. Quoted in Charles E. Lee, *The Evolution of Railways*, 2nd ed. (London: Railway Gazette, 1943), 29.
35. Twenty thousand carts and cart horses: Galloway, *Annals of Coal Mining*, 169.
36. Durham, Northumberland, total coal production: M. J. T. Lewis, *Early Wooden Railways* (London: Routledge & Kegan Paul, 1970), 86.
37. John Taylor (1623), *The World runnes on Wheeles; or, Oddes betwixt Carts and Coaches* (London: Henry Gosson), quoted in Stratton, *The World on Wheels*, 278.
38. Moving cart on rails one-sixth effort: Nef, *Rise of British Coal*, 28.
39. Quoted in Joe Earp, "Huntingdon Beaumont and Britain's First Railway," Nottingham Hidden History Team (online), last modified September 3, 2012.
40. Quoted in Galloway, *Annals of Coal Mining*, 224.
41. Quoted in Lee, *Evolution of Railways*, 65.
42. Ibid., 28.
43. Galloway, *Annals of Coal Mining*, 250.
44. Quoted in ibid., 249.
45. Wood scarcity and iron demand: E. A. Wrigley, *Energy and the English Industrial Revolution* (Cambridge: Cambridge University Press, 2010), 16.
46. Abraham Darby background, patent: Emyr Thomas, *Coalbrookdale and the Darby Family: The Story of the World's First Industrial Dynasty* (York, UK: Sessions Book Trust, 1999), 4–5.
47. Abiah Darby comparison: quoted in ibid., 13.
48. Quoted in ibid., 44.
49. Iron replaced wood: Galloway, *Annals of Coal Mining* , 259.
50. First iron railway wheels: Thomas, *Coalbrookdale and the Darby Family*, 28.
51. Cast-iron plates: Lee, *Evolution of Railways*, 57.
52. Major Coalbrookdale product: Thomas, *Coalbrookdale and the Darby Family*, 28.
53. Galloway, *Annals of Coal Mining*, 248.
54. Thomas, *Coalbrookdale and the Darby Family*, 35.
55. Galloway, *Annals of Coal Mining*, 259.
56. Ibid., 245.

FOUR: TO MAKE FOR ALL THE WORLD

1. Optician and instrument maker: David Clarke, *Reflections on the Astronomy of Glasgow: A Story of Some Five Hundred Years* (Edinburgh: Edinburgh University Press, 2013), 57.
2. Second cousin: As far as I can make out, George Muirheid (1715–1773) was a kinsman of Watt's mother, Agnes, neé Muirheid.

3. Quoted in Jennifer Tann, "James Watt," *DNB* online, 2.
4. Quoted in Hills, *Power from Steam*, 57.
5. *Philosophical Transactions of the Royal Society of London* 38 (1733): 303.
6. University paid him well: Hills, *Power from Steam*, 64.
7. Clarke, *Reflections on the Astronomy of Glasgow*, 58.
8. Eric Robinson and A. E. Musson, *James Watt and the Steam Revolution: A Documentary History* (New York: Augustus M. Kelley, 1969), 24.
9. Muirhead, *Life of James Watt*, 48.
10. Ibid., 46.
11. Anderson eight years older: George Williamson, *Memorials of the Lineage, Early Life, Education, and Development of the Genius of James Watt* (Greenock, Scot.: printed privately for the Watt Club, 1856), 162.
12. William Rosen, *The Most Powerful Idea in the World: A Story of Steam, Industry & Invention* (Chicago: University of Chicago Press, 2010), 97n.
13. £2, Jonathan Sisson: Clarke, *Reflections on the Astronomy of Glasgow*, 58.
14. Quoted in Eric Robinson and Douglas McKie, *Partners in Science: Letters of James Watt and Joseph Black* (London: Constable, 1970), 411.
15. Quoted in Muirhead, *Life of James Watt*, 59.
16. Experiment with Papin's digester: ibid.
17. Quoted in ibid., 59–60.
18. Ibid.
19. The demands of his business: Muirhead, *Life of James Watt*, 60.
20. Robinson and McKie, *Partners in Science*, 431.
21. Robison returned in 1761: Jones (2014), 197.
22. Robinson and McKie, *Partners in Science*, 434.
23. Watt's new Glasgow residence: Muirhead, *Life of James Watt*, 37.
24. Sixteen employees in 1764; £600 annual gross: ibid., 33.
25. Margaret "Peggy" Millar: Hills (2002), vol. 1, 13.
26. Quoted in Muirhead, *Life of James Watt*, 52.
27. Ibid., 63–64.
28. Boiling off a quantity of water: Joseph Black, *Lectures on the Elements of Chemistry: Delivered at the University of Edinburgh*, ed., John Robison, vol. 1 (Philadelphia: Mathew Carey, 1807), 150.
29. Black, *Lectures*, vol. 1, 151.
30. Quoted in Muirhead, *Life of James Watt*, 64.
31. Jones (2014), 203.
32. April 1765: Hills (2002), 340. Watt wrote a letter to a friend describing his idea of a separate condenser on 29 April 1765.

33. Watt's Glasgow Green insight: Muirhead, *Life of James Watt*, 67.
34. Ibid., 74. Original italics.
35. Ibid., 52–53. Muirhead has "and *hot* water injected if I please," which can't be right.
36. Ibid., 53.
37. Ibid., 54.
38. Quoted in ibid., 67.
39. Boulton to Watt, 7 February 1769, in Robinson and Musson, *James Watt and the Steam Revolution*, 62.
40. Britain's 1700 and 1800 energy consumption: Wrigley, *Energy and the English Industrial Revolution*, 38.
41. *Dublin Weekly Journal*, 9 August 1729. Reprinted in *The Works of Jonathan Swift*, ed. Walter Scott. Boston: Houghton, Mifflin, vol. VII, p. 217.
42. John Farey, *A Treatise on the Steam Engine, Historical, Practical, and Descriptive* (London: Longman, Rees, Orme, Brown, and Green, 1827), 444n.
43. Galloway, *Annals of Coal Mining*, 301.
44. Frank Dawson, *John Wilkinson: King of the Ironmasters,* ed. David Lake (Stroud, UK: History Press, 2012), 67.
45. Quoted in Smiles, *Industrial Biography*, 128.
46. First US steam engine, 1755: L. F. Loree, "The First Steam Engine of America." *Transactions of the Newcomen Society* 10 (1931): 21; John Fitch: Shagena, Jack L. *Who Really Invented the Steamboat? Fulton's* Clermont *Coup* (Amherst, NY: Humanity Books, 2004), 171ff.

FIVE: CATCH ME WHO CAN

1. The retreat of a heartbroken duke: Francis Espinasse, *Lancashire Worthies* (London: Simpkin, Marshall, 1874), 272.
2. Risking Gibraltar: Rolt (1963), 29.
3. March 1759: Glen Atkinson, "The Bridgewater's Beginning" in "Bridgewater 250: The Archeology of the World's First Industrial Canal," University of Salford Applied Archaeology Series 1, ed. Michael Nevell and Terry Wyke (Manchester, UK: Centre for Applied Archaeology, University of Salford, 2012).
4. Model made of cheddar: Glen Atkinson, "Barton's Aqueducts," in ibid.
5. Quoted in Atkinson, ibid., 78.
6. Quoted in Cyril T. G. Boucher, *James Brindley, Engineer, 1716–1772* (Norwich, UK: Goose and Son, 1968), 51–52.
7. Fifty-two-mile underground canal system: Nevell and Wyke, "Bridgewater 250," 7. Boucher, *James Brindley*, 63, has forty-six miles.

8. Quoted in Nevell and Wyke, "Bridgewater 250," 4.
9. Increasing cost of charcoal: Charles K. Hyde, *Technological Change and the British Iron Industry, 1700–1870* (Princeton, NJ: Princeton University Press, 1977), 62.
10. Fuel concentration from point sources: Wrigley, *Energy and the English Industrial Revolution*, 102–3.
11. Quoted in Arthur Raistrick, *Dynasty of Iron Founders: The Darbys and Coalbrookdale* (Newton Abbot, UK: David & Charles, 1970), 176.
12. First feeder railway: Lee, *Evolution of Railways*, 61.
13. Janet M. Hartley, Paul Keenan, and Dominic Lieven, eds., *Russia and the Napoleonic Wars (War, Culture and Society, 1750–1850)* (London: Palgrave Macmillan, 2015), 53.
14. British cavalry horse-buying practice: Anthony Dawson, personal communication, 25 February 2016.
15. 1797 election, mine list, marriage: Francis Trevithick, *Life of Richard Trevithick, with an Account of His Inventions*, vol. 1 (London: E. & F. N. Spon, 1872), 63–64.
16. Plunger pump: ibid., 67ff.
17. Water-pressure engine: Ibid., 73.
18. 1792 Boulton & Watt engine: Burton, *Richard Trevithick*, 59.
19. Trevithick to London in 1796; first time out of Cornwall: Trevithick, *Life of Richard Trevithick*, vol. 1, 93.
20. Quoted in ibid., 63.
21. Quoted in Roland Thorne, "Francis Basset," *DNB* online.
22. Trevithick, *Life of Richard Trevithick*, vol. 1, 103.
23. H. W. Dickinson and Arthur Titley, *Richard Trevithick: The Engineer and the Man* (Cambridge: Cambridge University Press, 1934), 46–47.
24. Grades on British railroads: "The failing to see or to understand what Trevithick's working experiments clearly pointed out, has caused millions of money to be wasted in avoiding inclines and curves on railways that could easily have been passed by the locomotive engine and its train of carriages." Trevithick, *Life of Richard Trevithick*, vol. 1, 139.
25. Quoted in ibid., 107.
26. Quoted in Dickinson and Titley, *Richard Trevithick*, 48.
27. Trevithick and Vivian 1802 patent reprinted in full in ibid., appendix 1, 269ff.
28. Quoted in ibid., 54.
29. Falmouth to London in August: ibid., 57, quoting Vivian's account records.
30. Autumn 1803: the date is uncertain, but it can't have been "May or June," as Joan M. Eyles has it in "William Smith, Richard Trevithick, and Samuel Homfray: Their Correspondence on Steam Engines, 1804–1806," *Transactions of the*

Newcomen Society 43, no. 1 (1970): 138, nor "January 1803," as Francis Trevithick says in *Life of Richard Trevithick*, vol. 1, 144, since the engine and boiler were shipped to London in August 1803.

31. £207: Dickinson and Titley, *Richard Trevithick*, 57.
32. Six months of London trials: Trevithick, *Life of Richard Trevithick*, vol. 1, 144.
33. "Trevithick's Dragon": Dickinson and Titley, *Richard Trevithick*, 51.
34. Quoted in ibid., 57.
35. Jarring ride: Burton, *Richard Trevithick*, 81.
36. Mrs. Humblestone's narrative: Trevithick, *Life of Richard Trevithick*, vol. 1, 143–44.
37. Quoted in Loughnan St. L. Pendred, "The Mystery of Trevithick's London Locomotives," *Transactions of the Newcomen Society* 1 (1922): 35.
38. Trevithick, *Life of Richard Trevithick*, vol. 1, 143.
39. Dickinson and Titley, *Richard Trevithick*, 60.
40. Ibid.
41. Samuel Homfray, quarter of patent, £10,000: ibid., 61.
42. Burton, *Richard Trevithick*, 87.
43. Tramway opened 1799: Stanley Mercer, "Trevithick and the Merthyr Tramroad," *Transactions of the Newcomen Society* 26 (1953): 102.
44. Crawshay-Homfray bet: Burton, *Richard Trevithick*, 89. Dickinson and Titley, *Richard Trevithick*, 63, ascribe the bet to Homfray and Anthony Hill, another ironmaster, but according to Burton, Hill was, in fact, the purse holder and referee.
45. Burton, Richard Trevithick, 89–90.
46. Quoted in ibid., 90–91.
47. Quoted in Dickinson and Titley, *Richard Trevithick*, 64–65.
48. Quoted in Burton, *Richard Trevithick*, 94.
49. Quoted in Dickinson and Titley, *Richard Trevithick*, 67.
50. Twelve winding engines by 1803, others up to 1808: Arthur Titley, "Richard Trevithick and the Winding Engine," *Transactions of the Newcomen Society* 10 (1929): 55–68, p. 61.
51. *London Times* advertisement: Quoted in Dickinson and Titley, *Richard Trevithick*, 107.
52. Quoted in ibid., 109.
53. Ibid.

SIX: UNCONQUERED STEAM!

1. Wrigley, *Energy and the English Industrial Revolution*, 100.
2. Ibid., 111.

3. Ibid., 71.
4. From William Wordsworth, "Composed upon Westminster Bridge, September 3, 1802."
5. *Dorothy Wordsworth's Journals,* vol. 6, 144.
6. Quoted in Webb and Webb, *English Local Government,* 144. Italics in original.
7. Thomas de Quincey, "The Nation of London," *Autobiographic Sketches.* Ch. 7.
8. Eight bales in 1784: Stephenson and Locke in James Walker, *Report to the Directors of the Liverpool and Manchester Railway, on the Comparative Merits of Locomotive and Fixed Engines, as a Moving Power* (with Stephenson, Robert, and Joseph Locke, *Observations on the Comparative Merits of Locomotive and Fixed Engines, as Applied to Railways;* and Booth, Henry, *An Account of the Liverpool and Manchester Railway*) (Philadelphia: Carey & Lea, 1831), 126. Fifty-three percent in 1806: Stanley Chapman, "British exports to the USA, 1776–1914: Organization and Strategy (3) Cottons and Printed Textiles," table 2, 34.
9. 68 percent of value added in 1801: Wrigley, *Energy and the English Industrial Revolution,* 36.
10. Quantity of iron doubled: Galloway, *Annals of Coal Mining,* 304.
11. Smiles, *Industrial Biography,* 108.
12. Erasmus Darwin, *The Botanic Garden,* pt. 1, lines 289–92.
13. To George Washington from James Rumsey, March 10, 1785, Founders Online, National Archives, source: *The Papers of George Washington,* Confederation Series, vol. 2, *18 July 1784–18 May 1785,* ed. W. W. Abbot. Charlottesville: University Press of Virginia, 1992, 425–29.
14. Wrigley, *Energy and the English Industrial Revolution,* 3.
15. 11 February 1805: Tomlinson (1858), has 1800, as do later sources; but William Thomas's earlier publication, *Observations on Canals and Rail-Ways, Illustrative of the Agricultural and Commercial Advantages to be Derived from an Iron Rail-way, Adapted to Common Carriages, Between Newcastle, Hexham, and Carlisle* (Newcastle upon Tyne, UK: G. Angus, 1825), which Tomlinson references, has 1805: "In the year 1805, the late Mr. Thomas, then of Denton Hall, Northumberland, conceiving by the construction of an iron rail-way . . . that many of the obstacles to the establishment of a navigable canal would be avoided, whilst the mode of conveyance . . . would be greatly improved; Mr. Thomas drew up a report and estimate thereon, which was read to the Literary and Philosophical Society of Newcastle upon Tyne" (editor's introduction, *iv*).
16. Thomas, *Observations on Canals and Rail-Ways,* 10.
17. Ibid., 12.
18. Ibid., 18.

19. Ibid., 26.
20. Ibid., 28.
21. Quoted in William H. Brown, *The History of the First Locomotives in America from Original Documents and the Testimony of Living Witnesses* (New York: D. Appleton, 1871), 37.
22. 25 March 1807, Oystermouth tramroad: Lee, *Evolution of Railways*, 75.
23. May 1809, Bewick Main Colliery, and so on: Galloway, *Annals of Coal Mining*, 369–70.
24. Quoted in ibid., 382.
25. Blenkinsop and other engines: ibid., 379.
26. Blackett and handcar experiments: Brown. *History of the First Locomotives*, 45.
27. 9 June 1781: Thomas Summerside, *Anecdotes, Reminiscences, and Conversations, of and with the Late George Stephenson, Father of Railways* (London: Bemrose and Sons, 1878), 2, has July, but both the *DNB* and W. O. Skeat, *George Stephenson: The Engineer and His Letters* (London: Institution of Mechanical Engineers, 1973), 13, have June.
28. Summerside, *Anecdotes, Reminiscences, and Conversations, of and with the Late George Stephenson*, 2.
29. Ibid., 5.
30. Stephenson studied the engine: L. T. C. Rolt, *George and Robert Stephenson: The Railway Revolution* (Stroud, UK: Amberley, 2016), 23–24.
31. Ibid., 24–25.
32. *Blucher*, flanging: Lee, *Evolution of Railways*, 89; 25 July 1814: Summerside, *Anecdotes, Reminiscences, and Conversations, of and with the Late George Stephenson*, 8.
33. Galloway, *Annals of Coal Mining*, 384.
34. Quoted in Samuel Smiles, *The Life of George Stephenson, Railway Engineer.* 3rd ed., rev. (London: John Murray, 1857), 86–87.
35. 43 percent of haulage in 1828: Christopher McGowan, *The Rainhill Trials: The Greatest Contest of Industrial Britain and the Birth of Commercial Rail* (London: Little, Brown, 2004), 320, n. 29.
36. Galloway, *Annals of Coal Mining*, 367.
37. Thirteen million tons of coal in 1815: ibid., 444.
38. Robert Stevenson, "Report of a Proposed Railway from the Coal-field of Mid-Lothian to the City of Edinburgh" is quoted in full in Chapter 7 of David Stevenson, *Life of Robert Stevenson* (Edinburgh: Adam and Charles Black, 1878).
39. Quoted in Lee, *Evolution of Railways*, 89.
40. Stockton & Darlington half wrought iron, half cast iron: Galloway, *Annals of Coal Mining*, 452–53.

41. Duke of Cleveland's fox cover: *Notes and Extracts on the History of the London & Birmingham Railway,* ch. 4 (http://gerald-massey.org.uk/Railway/c04_route.htm).
42. Quoted in McGowan, *Rainhill Trials,* 144.
43. Quoted in Rolt, *George and Robert Stephenson,* 110.
44. Smiles, *Life of George Stephenson,* 222. Original italics.
45. 25 April 1825, "eight or ten barristers," ibid., 226.
46. Great Britain, *Proceedings of the Committee of the House of Commons on the Liverpool and Manchester Railroad Bill: Sessions, 1825,* 205.
47. Ibid., 242.
48. Chat Moss: this description adapted from "Chat Moss Crossing, L&M Railway," Engineering Timelines online.
49. Walker, *Report to the Directors of the Liverpool and Manchester Railway,* 170; Smiles, *Life of George Stephenson,* 235.
50. Smiles, *Life of George Stephenson,* 250–51.
51. Ibid., 251.
52. Ibid., 253–54.
53. Smiles, *Industrial Biography,* 285ff.
54. Walker, *Report to the Directors of the Liverpool and Manchester Railway,* 137.
55. Ibid., Appendix: General Abstract of Expenditure to 31st May 1830, 199.
56. Ibid., 205.
57. Skeat, *George Stephenson: Letters,* 116.
58. Walker, *Report to the Directors of the Liverpool and Manchester Railway,* 51.
59. Stephenson's tour, engineers' tour: ibid., 51–52.
60. Quoted in Skeat, *George Stephenson: Letters,* 118.
61. Quoted in McGowan, *Rainhill Trials,* 321.
62. Quoted in ibid.
63. Quoted in ibid., 321–22.
64. Ibid., 322.
65. *Cycloped,* two horses: see Dendy Marshall, "The Rainhill Locomotive Trials of 1829," *Transactions of the Newcomen Society* 9, no. 1 (1928): 85. Illustrations of the event typically show only one.
66. Four other Rainhill entrants: Booth, *An Account of the Liverpool and Manchester Railway,* 182, in Walker, *Report to the Directors of the Liverpool and Manchester Railway,* 182.
67. Multitube boiler mechanism: Richard Gibbon, *Stephenson's Rocket and the Rainhill Trials* (Oxford: Shire, 2010), 20–21.
68. *Rocket* first trial: Marshall, "Rainhill Locomotive Trials of 1829," 87.

69. Quoted in McGowan, *Rainhill Trials,* 197–98.
70. Quoted in Marshall, "Rainhill Locomotive Trials of 1829," 88.
71. Walker, *Report to the Directors of the Liverpool and Manchester Railway,* 194–97.

SEVEN: RUSHLIGHT TO GASLIGHT

1. Gilbert White, *The Natural History of Selborne* (New York: Harper & Brothers, 1841; first published 1789), 231–32.
2. Ibid., 232.
3. Ibid.
4. Ibid., 232–33.
5. W. T. O'Dea, "Artificial Lighting Prior to 1800 and Its Social Effects," *Folklore* 62, no. 1 (1951): 315.
6. The Lunar Society: see Jenny Uglow, *The Lunar Men: Five Friends Whose Curiosity Changed the World* (New York: Farrar, Straus and Giroux, 2002).
7. Quoted in Jane Brox, *Brilliant: The Evolution of Artificial Light* (Boston: Houghton Mifflin Harcourt, 2010), 21.
8. John Stow, *A Survey of London,* rev. 1603 (Dover, UK: Alan Sutton, 1994), 125.
9. Darwin using phosphorescent fish heads: Desmond King-Hele, *Erasmus Darwin: A Life of Unequalled Achievement* (London: DLM, 1999), 18.
10. Quoted in Brox, *Brilliant,* 29.
11. Alec Campbell, "Archibald Cochrane," *DNB* online, p. 1; nine patents: Luter (2005), p. 3.
12. Campbell, 1–3.
13. Ibid., 5.
14. Quoted in John Sugden, "Lord Cochrane, Naval Commander, Radical, Inventor (1775–1860): A Study of His Earlier Career, 1775–1818" (PhD diss., Department of History, University of Sheffield, 1981), 14.
15. Quoted in Charles Hunt, *A History of the Introduction of Gas Lighting* (London: Walter King, 1907), 18.
16. Estate sale 1798: Sugden, "Lord Cochrane," 28.
17. Campbell, 4.
18. Boulton dined with Lady Dundonald: Archibald Clow and Nan Clow, "Lord Dundonald," *Economic History Review* 12: nos. 1 and 2 (1942): 49.
19. Jan-Pieter Minckelers: Leslie Tomory, *Progressive Enlightenment: The Origins of the Gaslight Industry, 1780–1820* (Cambridge, MA: MIT Press, 2012), 29.
20. Quoted in ibid., 32.
21. M. Amboise, Benjamin Healy: *Gas Age-Record,* 13 December 1924, 828.
22. Boswell visited Soho in 1776: John C. Griffiths, "William Murdoch," *DNB* online.

23. Quoted in John Griffiths, *The Third Man: The Life and Times of William Murdoch, 1754–1839, the Inventor of Gas Lighting* (London: Andre Deutsch, 1992), 102.
24. Quoted in Griffiths, "William Murdoch," *DNB* online.
25. Quoted in Tomory, *Progressive Enlightenment*, 74.
26. Moved on to coal in 1791, Charles Hunt's reasonable estimate: Hunt, *Introduction of Gas Lighting*, 40.
27. William Murdoch, "An Account of the Application of Gas from Coal to Economical Purposes," *Philosophical Transactions of the Royal Society of London* 98 (1808): 130.
28. The best evidence: Griffiths, *Third Man*, 242–52.
29. Quoted in ibid., 245.
30. Quoted in Hunt, *Introduction of Gas Lighting*, 42. Hunt (41) thinks the memoirist was Henry Creighton.
31. Gun barrel light: Griffiths, *Third Man*, 244; portable light: ibid., 248.
32. Quoted in Hunt, *Introduction of Gas Lighting*, 49.
33. Quoted in Trevor H. Levere, "Dr. Thomas Beddoes: Chemistry, Medicine, and the Perils of Democracy," *Notes and Records of the Royal Society of London* 63 (2009): 216.
34. Dorothy A. Stansfield and Ronald G. Stansfield, "Dr. Thomas Beddoes and James Watt: Preparatory Work 1794–96 for the Bristol Pneumatic Institute," *Medical History* 30, no. 3 (July 1986): 283.
35. Quoted in ibid.
36. Ibid.
37. Quoted in ibid., 284.
38. Thomas Beddoes and James Watt, *Considerations on the Medicinal Use of Factitious Airs, and on the Manner of Obtaining Them in Large Quantities* (Bristol, UK: J. Johnson, 1794), 2.
39. Ibid.
40. Watt's apparatus: ibid., 3.
41. Quoted in Mike Jay, *The Atmosphere of Heaven: The Unnatural Experiments of Dr. Beddoes and His Sons of Genius* (New Haven, CT: Yale University Press, 2009), 172.
42. Ibid., 183.
43. Ibid., 176.
44. Ibid., 192.
45. Ibid., 212.
46. Quoted in ibid.
47. Ibid., 214.

48. Quoted in Tomory, *Progressive Enlightenment,* 75.
49. Ibid., 7–8.
50. 1792, National Reward: Hunt, *Introduction of Gas Lighting,* 50.
51. Quoted in Griffiths, *Third Man,* 241.
52. Ibid.
53. Quoted in Tomory, *Progressive Enlightenment,* 76.
54. Celebration already planning: ibid., 77.
55. Quoted in Hunt, *Introduction of Gas Lighting,* 63–64.
56. George Lee to Soho 1800: Griffiths, *Third Man,* 249.
57. George Lee, new mill: J. J. Mason, "George Augustus Lee," *DNB* online.
58. "Significant Scots: William Murdoch," *Electric Scotland* online.
59. Quoted in Griffiths, *Third Man,* 255.
60. Lebon stabbed to death: Hunt, *Introduction of Gas Lighting,* 61.
61. Quoted in ibid., 129.
62. Quoted in Edward Stoops Thompson, *The History of Illuminating Gas in Baltimore* (Records of Phi Mu, Special Collections, University of Maryland Libraries online, 1928), 5.
63. David Melville history: Christopher J. Castaneda, *Invisible Fuel: Manufactured and Natural Gas in America, 1800–2000* (New York: Twayne, 1999), 14–15.
64. Melville's design, fumes complaints, pine tar: ibid., 16–17.
65. Baltimore 13 June 1816 opening night: Thompson, *History of Illuminating Gas in Baltimore,* 9.
66. Baltimore's first gas company: ibid.
67. Christopher Castaneda, "Manufactured and Natural Gas Industry," EH.Net online.

EIGHT: PURSUING LEVIATHAN

1. Thomas Jefferson, "Observations on the Whale Fishery," 1791, 12, Jefferson Papers, Avalon Project, Yale Law School Lillian Goldman Law Library online.
2. Quoted in David Tedone, ed., *A History of Connecticut's Coast,* US Department of Commerce pamphlet, 14 (online).
3. Mayhew sold Nantucket: for these details, see copy of original deed in Lydia S. Hinchman, *Early Settlers of Nantucket: Their Associates and Descendants,* 2nd ed. and enl. ed. (Philadelphia: Ferris & Leach, 1901), 3–4.
4. Mayhew bought from two Wampanoag sachems: ibid., 7–8.
5. Clear the remaining rights: ibid., 9–11.
6. Wampanoag oral tradition: according to Nanepashemet, director of the Wampanoag Indian Program at Plimoth Plantation. See Nantucket Historical Association

(NHA) online, "A Report on the NHA Symposium: Nantucket and the Native American Legacy of New England," *Historic Nantucket* 44, no. 3 (Winter 1996): 98–100.

7. 358 Wampanoags, 222 deaths, relapsing fever: Timothy J. Lapore in ibid.
8. Alexander Starbuck, *History of the American Whale Fishery from Its Earliest Inception to the Year 1876* (Waltham, MA: published by the author, 1878), 23.
9. Whaling technology advances: Lance E. Davis, Robert E. Gallman, and Karin Gleiter, *In Pursuit of Leviathan: Technology, Institutions, Productivity, and Profits in American Whaling, 1816–1906* (Chicago: University of Chicago Press, 1997), 36–37.
10. 1774 statistics: Starbuck, *History of the American Whale Fishery*, 57.
11. Gerald S. Graham, "The Migrations of the Nantucket Whale Fishery: An Episode in British Colonial Policy," *New England Quarterly* 8, no. 2 (1935): 182.
12. Starbuck, *History of the American Whale Fishery*, 77.
13. Loss of nearly £8 per ton: William Rotch, memorandum written by William Rotch in 1814 in the eightieth year of his age, in John M. Bullard, *The Rotches* (printed privately, 1947), 187; four thousand tons sperm oil, £300,000: Graham, "Migrations of the Nantucket Whale Fishery," 184.
14. Bullard, *The Rotches*, 188.
15. Rotch advised to delay: Eric Jay Dolin, *Leviathan: The History of Whaling in America* (New York: W. W. Norton, 2007), 173.
16. Bullard, *The Rotches*, 188.
17. Chancellor of the exchequer: Bullard, *The Rotches*, 189.
18. John Adams, *The Works of John Adams*, vol. 8., ed. Charles Francis Adams (Boston, Little, Brown, 1853), 308.
19. Ibid.
20. Ibid., 313.
21. Bullard, *The Rotches*, 189.
22. Ibid.
23. Ibid., 190. From this text to "'Yes,' I replied, 'but with regret'": ibid., 190–91.
24. Quoted in Starbuck, *History of the American Whale Fishery*, 80n.
25. Thomas Jefferson (1791). "Report on the American Fisheries by the Secretary of State," 3. Hereafter "Jefferson Report (1791)."
26. Terms not liberal enough; according to Jefferson: ibid., 10.
27. Thomas Jefferson (1789), "Observations on the Whale Fishery," Avalon Project, Yale Law School, n.p. (online).
28. Ibid.
29. Recovery of market, extension of hunt: Davis, Gallman, and Gleiter, *In Pursuit of Leviathan*, 37.

30. The *Beaver* rounds Cape Horn: according to Dolin, *Leviathan,* 418, n. 38, citing nineteenth-century sources. This was not the Rotches' Boston Tea Party *Beaver.* The *Rebecca,* of New Bedford, also made the passage in 1791.
31. Number of Nantucket and New Bedford whaling ships, 1805–1807: Michael Dyer, senior maritime historian, New Bedford Whaling Museum, personal communication, 21 June 2016.
32. A thousand deserters out of ten thousand impressments: "Embargo of 1807," Monticello online, www.monticello.org/site/research-and-collections/embargo-1807, citing *Oxford Companion to American History,* 404.
33. Nantucket lost half her shipping: Graham, "Migrations of the Nantucket Whale Fishery," 199; Starbuck, *History of the American Whale Fishery,* 95, says Nantucket's fleet had been reduced by the war "from forty-six to twenty-three."
34. Starbuck, *History of the American Whale Fishery,* 95.
35. More than fifty ships hunting offshore ground: ibid., 56.
36. Alice Ford, *The 1826 Journal of John James Audubon* (Norman: University of Oklahoma Press, 1967), 69.
37. Peter Ewer and the Nantucket Bar: details from Edouard A. Stackpole, "Peter Folger Ewer: The Man Who Created the 'Camels,'" *Historic Nantucket* 33, no. 1 (July 1985): n.p.

NINE: BURNING FLUIDS

1. Camphene on Minnesota frontier, "a possible source of supply": Evadene A. Burris, "Keeping House on the Minnesota Frontier," *Minnesota History* 14, no. 3 (1933): 266–67.
2. Whaling fleet in 1846: Teresa Dunn Hutchins, "The American Whale Fishery, 1815–1900: An Economic Analysis" (PhD diss., Department of Economics, University of North Carolina at Chapel Hill, 1988), 46.
3. Composition of burning fluid: B. Ghobadian and H. Rahimi, "Biofuels—Past, Present and Future Perspective," *Proceedings of the Fourth International Iran & Russia Conference in Agriculture and Natural Resources* (Shahrekord, Iran: University of Shahrekord, 2004), 782.
4. 400,000 acres: W. W. Ashe, *The Forests, Forest Lands, and Forest Products of Eastern North Carolina. North Carolina Geological Survey Bulletin No.* 5 (Raleigh, NC: Josephus Daniels, State Printer and Binder, 1894), 59.
5. Boxing pine trees: ibid., 73–74.
6. Quoted in Kent Wrench, ed., *Tar Heels: North Carolina's Forgotten Economy: Pitch, Tar, Turpentine and Longleaf Pines* (Charleston, SC: CreateSpace, 2014), 19.

7. $7.5 million: Ashe, *Forests, Forest Lands, and Forest Products,* 76; North Carolina more than $5 million: Wrench, *Tar Heels,* 26.
8. Pine forests harvested: Ashe, *Forests, Forest Lands, and Forest Products,* 76.
9. Gesner's horse-trading: Gray, (2008), 11.
10. Gesner's education: Kendall Beaton, "Dr. Gesner's Kerosene: The Start of American Oil Refining." *Business History Review* 29, no. 1 (1955): 31–32.
11. Coal oil from bitumen: ibid., 34.
12. Gesner 1846 public lectures: Abraham Gesner, *A Practical Treatise on Coal, Petroleum, and Other Distilled Oils,* 2nd ed., rev. and enl. by George Weltden Gesner (New York: Bailliere Brothers, 1865), 9; Paul Lucier, *Scientists & Swindlers: Consulting on Coal and Oil in America, 1820–1890* (Baltimore: Johns Hopkins University Press, 2008), 43, has "in the summer of 1847."
13. Dundonald land around Pitch Lake, 1851: Beaton, "Dr. Gesner's Kerosene," 37.
14. Patent application, kerosene, kerosene gas: Lucier, *Scientists & Swindlers,* 43.
15. Quoted in Beaton, "Dr. Gesner's Kerosene," 38–39.
16. Quoted in Lucier, *Scientists & Swindlers,* 147.
17. Beaton, "Dr. Gesner's Kerosene," 43.
18. Quoted in ibid., 43–44.
19. Fighting its way into the market: Gesner, *Practical Treatise on Coal, Petroleum, and Other Distilled Oils,* 10.
20. Details of kerosene production in 1859: Beaton, "Dr. Gesner's Kerosene," 50.
21. Sixty to seventy-five coal-oil plants by 1860; two hundred coal gas companies: Lucier, *Scientists & Swindlers,* 155.
22. Early 1860 coal-oil production: ibid., 156.
23. 10.3 million gallons whale oil: A. Howard Clark, "The American Whale-Fishery 1877–1886," *Science* ns-9, no. 217S (1887): 321.
24. Quoted in William R. Brice, *Myth Legend Reality: Edwin Laurentine Drake and the Early Oil Industry* (Oil City, PA: Oil City Alliance, 2009), 56.
25. Benjamin Silliman Sr., ed., "Notice of a Fountain of Petroleum, Called the Oil Spring," *American Journal of Science and Arts* 23 (January 1833): 100.
26. Ibid., 101–2.
27. Bissell biography: J. T. Henry, *The Early and Later History of Petroleum, with Authentic Facts in Regard to Its Development in Western Pennsylvania* (Philadelphia: Jas. B. Rodgers, 1873), 346–47.
28. Paul H. Giddens, ed., *Pennsylvania Petroleum 1750–1872: A Documentary History* (Titusville: Pennsylvania Historical and Museum Commission, 1947), 46.
29. Hubbard thought petroleum scarce: ibid., 45–46.
30. Ibid., 46.

31. Albert Crosby: Charles Henry Bell, *The Bench and Bar of New Hampshire: Including Biographical Notices of Deceased Judges of the Highest Court, and Lawyers of the Province and State, and a List of Names of Those Now Living* (Boston: Houghton Mifflin, 1894), 292–93.
32. Giddens, *Pennsylvania Petroleum 1750–1872*, 46–47.
33. New Haven group required Silliman report: ibid., 54.
34. Henry, *Early and Later History of Petroleum*, 67.
35. Silliman rents Yale president's house: Brooks Mather Kelley, *Yale: A History* (New Haven, CT: Yale University Press, 1974), 183.
36. Noted in Leonard G. Wilson, ed., *Benjamin Silliman and his Circle: Studies on the Influence of Benjamin Silliman on Science in America* (New York: Science History, 1979), 175.
37. Silliman 1854 salary: ibid., 176.
38. Quoted in Lucier, *Scientists & Swindlers*, 197.
39. Fractional distillation: Benjamin Silliman Jr., *Report on the Rock Oil, or Petroleum, from Venango Co., Pennsylvania, with Special Reference to Its Use for Illumination and Other Purposes* (New Haven, CT: J. H. Benham, 1855), 6–7.
40. Ibid., 7–8.
41. Ibid., 9.
42. Quoted in Lucier, *Scientists & Swindlers*, 198 (Silliman's italics).
43. Ibid., 200.
44. Seneca Oil Company, 1858: Brice, *Myth Legend Reality*, 264.
45. Edwin L. Drake: ibid., 91–96, 116–22, passim.
46. Trench enlargement, six gallons: ibid., 226.
47. Drake appointed "superintendent," $1,000 salary (later raised to $1,200); according to Townsend: Giddens, *Pennsylvania Petroleum*, 56.
48. Drake stopped in Syracuse: Brice, *Myth Legend Reality*, 258.
49. Drake's wife, Laura, on his rank: ibid., 223.
50. One recruit died: ibid., 302.
51. Quoted in ibid., 285.
52. Ibid., 302.
53. Ibid., 306.
54. Ibid.
55. Ibid., 309.
56. Uncle Billy Smith arrangements, family life: ibid., 306–7, 309–11.
57. Location of first Drake well: see contemporary map, 154.
58. Drake decided to bore: Brice, *Myth Legend Reality*, 501.
59. Giddens, *Pennsylvania Petroleum*, 75.

60. Quoted in ibid., 68.
61. 2 Aug., 9 Aug.: Brice, *Myth Legend Reality*, 312–13; Giddens, *Pennsylvania Petroleum 1750–1872*, 75.
62. Quoted in Brian Black, *Petrolia: The Landscape of America's First Oil Boom* (Baltimore: Johns Hopkins University Press, 2000), 32.
63. Sixty-nine and a half feet: Brice, *Myth Legend Reality*, 321.

TEN: WILD ANIMALS

1. Quoted in Paul H. Giddens, *The Birth of the Oil Industry* (New York: Macmillan, 1938), 90.
2. Tarbell, *History of Standard Oil*, 15.
3. A thousand barrels a day: Harold F. Williamson and Arnold R. Daum, *The American Petroleum Industry: The Age of Illumination, 1859–1899* (Evanston, IL: Northwestern University Press, 1959), 107–8. Only a small part shipped; leakage: ibid., 108.
4. Giddens, *Birth of the Oil Industry*, 105–6.
5. Leakage: ibid., 108.
6. 262,500 barrels in 1860: Williamson and Daum, *American Petroleum Industry*, 108.
7. Details of 17 April gusher and fire: Giddens, *Birth of the Oil Industry*, 76–78.
8. Railroad connection and haulage: ibid., 111–12.
9. Pipelines, 1,500 teamsters: ibid., 144–45.
10. Federal taxes on refined petroleum, 1862–1865: ibid., 95.
11. Union blockade of turpentine: Percival Perry, "The Naval-Stores Industry in the Old South, 1790–1860," *Journal of Southern History* 34, no. 4 (1968): 525.
12. Thirteen million gallons grain alcohol: Rufus Frost Herrick, *Denatured or Industrial Alcohol: A Treatise on the History, Manufacture, Composition, Uses, and Possibilities of Industrial Alcohol in the Various Countries Permitting its Use, and the Laws and Regulations Governing the Same, Including the United States* (New York: John Wiley & Sons, 1907), 16.
13. *37th Congress, Sess. II, Ch. 119, July 1, 1862*, 432.
14. Ibid., 447; increased to $2: Thomas B. Ripy, *Federal Excise Taxes on Beverages: A Summary of Present Law and a Brief History*, Congressional Research Service Report RL30238 (Washington, DC: Library of Congress, 1999), 4.
15. Camphene price increase: Herrick, *Denatured or Industrial Alcohol*, 16.
16. Kerosene sales 200 million gallons by 1870: William Kovarik, "Henry Ford, Charles F. Kettering, and the Fuel of the Future," *Automotive History Review* 32 (Spring 1998): n. 22.

17. Civil War decline in whaling: Davis, Gallman, and Gleiter, *In Pursuit of Leviathan,* 38.
18. From a poem, "Captain Semmes, C. A. S. N.," by George H. Boker, published originally in the *Philadelphia Press,* reproduced in *Mr. Merryman's Monthly,* a New York humor magazine, October 1864, 265.
19. Semmes captured ten merchant ships: these details and those following concerning Semmes from Stephen Fox, *Wolf of the Deep: Raphael Semmes and the Notorious Confederate Raider* CSS Alabama (New York: Vintage, 2007), unless otherwise specified.
20. Ibid., 45.
21. Ibid., 46. Fox quotes a *Times* of London correspondent who observed the *Sumter* in Gibraltar.
22. Raphael Semmes, *Memoirs of Service Afloat During the War Between the States* (Secaucus, NJ: Blue & Grey Press, 1987; first published 1868), 345.
23. Ibid., 423; Fox, *Wolf of the Deep,* 15, says the *Alabama* displayed a British flag.
24. Semmes, *Memoirs of Service,* 431.
25. Confederate raider *Shenandoah:* Baldwin and Powers (2007) tell this story: 247–51.
26. Bering Strait losses: Zeph W. Pease and George A. Hough, *New Bedford, Massachusetts: Its History, Industries, Institutions, and Attractions,* ed. William L. Sayer (New Bedford, MA: New Bedford Board of Trade, 1889), 31.
27. 722 ships in 1846, 124 by 1886: Clark, "American Whale-Fishery 1877–1886," 321.
28. Legal status of underground resources: See Black, *Petrolia,* ch. 2, for a full discussion of right of capture and Hardin's "The Tragedy of the Commons."
29. Fixing water's location: Joseph W. Dellapenna, "The Rise and Demise of the Absolute Dominion Doctrine for Groundwater," *University of Arkansas at Little Rock Law Review* 35, no. 2 (2013): 273.
30. Case brief, *Pierson v. Post,* 3 Cai. R. 175, 2 Am. Dec. 264 (N.Y. 1805), www.lawnix.com.
31. Quoted in Joseph W. Dellapenna, "A Primer on Groundwater Law," *Idaho Law Review* 49 (2012): 272, from *Westmoreland Cambria Nat. Gas Co. v. Dewitt,* 18 A. 724, 725 (Pa. 1889).
32. *Acton v. Blundell,* (1843) 152 Eng. Re1223 (Exch. Chamber). *The English Reports: Exchequer,* ed. W. Green, 1915, 1233.
33. Garrett Hardin, "The Tragedy of the Commons," *Science* 162, no. 3859 (December 13, 1968): 1243–48.
34. Ibid., 1244.

35. Ida M. Tarbell, *All in the Day's Work: An Autobiography* (Boston: G. K. Hall, 1985; first published 1939 by Macmillan), 9.
36. Oil production in 1870: Giddens, *Birth of the Oil Industry*, 192–93.

ELEVEN: GREAT FORCES OF NATURE

1. Franklin to Collinson, 28 March 1747, in Online Library of Liberty: *The Works of Benjamin Franklin*, vol. 2, *Letters and Misc. Writings 1735–1753*, 125.
2. Benjamin Franklin, *Experiments and Observations on Electricity Made at Philadelphia in America* (London: E. Cave, 1751), 34.
3. Ibid., 34–35.
4. Benjamin Franklin, *The Papers of Benjamin Franklin*, vol. 5, *July 1, 1953, Through March 31, 1755*, ed. Leonard W. Labaree (New Haven, CT: Yale University Press, 1962), 126.
5. Joseph Priestley, *The History and Present State of Electricity, with Original Experiments*, vol. 1., 3rd ed. (London: C. Bathurst et al., 1769), 204.
6. Luigi Galvani, *Commentary on the Effect of Electricity on Muscular Motion (De Viribus Electricitatis in Motu Musculari Commentarius)*, trans. Robert Montraville Green (Cambridge, MA: Elizabeth Licht, 1953; first published 1791). I have edited this translation for greater clarity.
7. Volta responded to Galvani's report: this discussion follows Marcello Pera, *The Ambiguous Frog: The Galvani-Volta Controversy on Animal Electricity*, trans. Jonathan Mandelbaum (Princeton, NJ: Princeton University Press, 1992), particularly ch. 4.
8. Quoted in ibid., 101.
9. Ibid., 105–6.
10. Ibid., 111.
11. Ibid.
12. Anon., "Royal Institution." *The Philosophical Magazine* 35 (1810) 146: 463.
13. Oersted 1813 Paris paper: *Recherches sur l'identité des forces chimiques et electriques.*
14. Oersted anticipated difficulty: as he testified later, writing of himself in the third person, "For a long time he imagined that it would be more difficult to confirm this idea by experiments than it later turned out to be." Quoted in Robert C. Stauffer, "Speculation and Experiment in the Background of Oersted's Discovery of Electromagnetism," *Isis* 48, no. 1 (1957): 44.
15. Quoted in ibid., 45.
16. Ibid., 44.
17. Quoted in ibid., 45.

18. "feeble": Dibner (1962), 71; "unmistakable . . . confused": Stauffer, "Oersted's Discovery of Electromagnetism," 45.
19. Davy hiring Faraday: L. Pearce Williams, *Michael Faraday: A Biography* (New York: Da Capo, 1965), 28–30.
20. Quoted in Michael Brian Schiffer, *Power Struggles: Scientific Authority and the Creation of Practical Electricity Before Edison* (Cambridge, MA: MIT Press, 2008), 50.
21. Ibid., 51.

TWELVE: A CADENCE OF WATER

1. Mohawk *ohnyá·kara î:* "Niagara," *Oxford English Dictionary.*
2. 1679, "a vast and prodigious": cited and quoted in Edward Dean Adams, *Henry Adams of Somersetshire, England, and Braintree, Mass.: His English Ancestry and Some of His Descendants* (New York: printed privately, 1927), 6.
3. Almost eighty-eight thousand square miles: Adams, *Henry Adams,* 16; two feet variation: ibid., 19.
4. Niagara River development, John Steadman, hotels: ibid., 41–48.
5. W. Grylls Adams, "The Scientific Principles Involved in Electric Lighting." *Journal of the Franklin Institute* (November 1881): 364.
6. Ibid., 279.
7. US energy consumption: Peter A. O'Conner, "Energy Transitions," *Pardee Papers* online, November 12, 2010, 3.
8. Adams, "Scientific Principles" (November 1881), 364.
9. Quoted in Laurence A. Hawkins, *William Stanley (1858–1916)—His Life and Work* (New York: Newcomen Society of North America, 1951), 22.
10. William Stanley, "Alternating-Current Development in America," *Journal of the Franklin Institute* 173, no. 6 (1912): 561–62.
11. Henry G. Prout, *A Life of George Westinghouse* (New York: American Society of Mechanical Engineers, 1921), 91.
12. Edison's model gas light: Robert Friedel and Paul Israel, *Edison's Electric Light: Biography of an Invention* (New Brunswick, NJ: Rutgers University Press, 1986), 177.
13. Elihu Thomson and Edwin J. Houston, "On the Transmission of Power by Means of Electricity," *Journal of the Franklin Institute* 77 (January 1879): 36.
14. Edison 3-wire system: Stanley, "Alternating-Current Development," 561.
15. Great Barrington, Stanley's childhood: Hawkins, *William Stanley,* 10.
16. Quoted in ibid.
17. Stanley, "Alternating-Current Development," 566–67.

18. Quoted in George Wise, "William Stanley's Search for Immortality," *Invention and Technology* 4, no. 1 (1988): (online), n.p.
19. Edison's fundamental electric light patent: US patent 223,898, filed 4 November 1879. Cited in Friedel and Israel, *Edison's Electric Light*, 105.
20. Maxim ELC chief engineer: Hiram S. Maxim, *My Life* (London: Methuen, 1915), 120.
21. Quoted in Hawkins, *William Stanley*, 12.
22. Casswell-Massey lighting system: ibid., 12–13.
23. Maxim to France in 1881: Maxim, *My Life*, 152.
24. Herman Westinghouse met Stanley on a train: Moran (2002), 48; Prout, *George Westinghouse*, 92 (without the train).
25. Westinghouse investigating AC: Prout, *George Westinghouse*, 92.
26. Leucine lamp filament: US patent no. 316,302, filed 4 September 1884, issued 21 April 1885; Hawkins, *William Stanley*, 15.
27. Stanley, "Alternating-Current Development," 567.
28. Ibid.
29. Ibid.
30. Ibid., 568.
31. Ibid.
32. Ibid.
33. Ibid., 569.
34. Ibid., 570.
35. Ibid.
36. Twenty-six transformers: ibid., 571; "fastened to insulators": ibid., 570; four thousand feet of line: ibid., 572; 500 volts to 3,000 volts to 500 volts: ibid., 572, says "the generating electromotive force [was] transformed from 500 to 3,000 volts and from 3,000 back to 500 volts, *and then sent over the line downtown*." (My emphasis.) Later histories have followed this sequence, but it makes no sense: Why step up the power only to step it back down *before* transmitting it? The fact that Stanley had transformers *for stepping down the current* installed in the basements of the downtown buildings refutes the sequence he reports. Assuming he simply misspoke, I have corrected it here. See also Cummings C. Chesney, "Some Contributions to the Electrical Industry," *Electrical Engineering* 52, no. 12 (1933): 727: "In 1886, William Stanley, in the first alternating current plant in America, which was engineered and built by him at Great Barrington, Mass., demonstrated how electric power could be generated at a low voltage, transformed into a higher voltage, *transmitted at the higher voltage*, retransformed to a lower voltage, and used at this voltage as might be required." (My emphasis.)

37. Stanley, "Alternating-Current Development," 572.
38. Quoted in Hawkins, *William Stanley*, 20.
39. Ibid.
40. Ibid.
41. Stanley, "Alternating-Current Development," 574.
42. Cerchi, Italy, installation: David B. Rushmore and Eric A. Lof. *Hydro-Electric Power Stations*, 2nd ed. (New York: John Wiley & Sons, 1923), 5.
43. Chesney, "Some Contributions to the Electrical Industry," 728.
44. Willamette Falls Electric: Rushmore and Lof, *Hydro-Electric Power Stations*, 7.
45. San Bernardino, Telluride, Pomona, Redlands, Hartford: ibid., 7–8.
46. Ibid., 9.
47. Hydraulic canal: Adams, *Henry Adams*, vol. 1, 69ff; Buffalo sewerage scheme: ibid., 97.
48. Ibid., 101.
49. Niagara State Reservation background: ibid., 101–5.
50. Sixty-six thousand visitors in 1887: ibid., 106, n. 1.
51. "Power from Niagara" prospectus: Quoted in ibid., 122–23.
52. Ibid., 144.
53. "Golden Jubilee," *Time* online, 27 May 1929, n.p.
54. Adams biography: Irene D. Neu, "Adams, Edward Dean," *American National Biography Online*; father a grocer: *US Census*, 1880.
55. "Golden Jubilee"; six-month option: ibid.; Cataract Construction Company, August, half-interest: Adams, *Henry Adams*, vol. 1, 141.
56. Quoted in Adams, *Henry Adams*, vol. 1, 149.
57. Quoted in "Golden Jubilee," *Time* online, 27 May 1929, n.p.
58. Edison's plan: Adams, *Henry Adams*, vol. 1, 146–47.
59. Tunnel divert ~3% of water: ibid., 403.
60. Niagara Falls Power, $2,630,000 for tunnel construction: ibid., 156–60.
61. Adams in Europe February–May: ibid., 171–72.
62. Ibid., 172–73.
63. Coleman Sellers, "The Utilization of the Power of Niagara Falls and Notes on Engineering Progress," *Journal of the Franklin Institute* (July 1891): 34.
64. Quoted in Adams, *Henry Adams*, vol. 1, 365.
65. Niagara power to New York in 1896; 10 percent of US total by 1904: Resources: "Adams Plant," Tesla at Niagara Museum online, www.teslaniagara.org.
66. Second powerhouse: Adams, *Henry Adams*, vol. 2, 75; 100,000 hp total US power elsewhere: ibid., 181.

THIRTEEN: AN ENORMOUS YELLOW CHEESE

1. 130,000 horses in Manhattan in 1900: Clay McShane and Joel A. Tarr, *The Horse in the City: Living Machines in the Nineteenth Century* (Baltimore: Johns Hopkins University Press, 2007), 16.
2. Ann Norton Greene, *Horses at Work: Harnessing Power in Industrial America* (Cambridge, MA: Harvard University Press, 2008), 184.
3. Thirty companies, seven hundred buses, limited clientele: McShane and Tarr, *Horse in the City*, 60.
4. Flush rails: Greene, *Horses at Work*, 179.
5. Steam locomotives barred; number of street railway passengers in 1840, 1859: McShane and Tarr, *Horse in the City* (2007), 63–64.
6. No traffic controls until twentieth century: Greene, *Horses at Work* (2008), 178.
7. McShane and Tarr, *Horse in the City*, title page.
8. No evidence for Percheron Arabian blood: Alvin Howard Sanders, *A History of the Percheron Horse* (Chicago: Breeder's Gazette Print, 1917), 34–54.
9. Feeding horses: calories per day, tons per year, acres of farmland: McShane and Tarr, *Horse in the City*, 128–29.
10. Hay transport limited to twenty to thirty miles: ibid., 133–34.
11. 1879, 1909 statistics: ibid., 135–36.
12. "Guano" derivation: Gregory T. Cushman, *Guano and the Opening of the Pacific World: A Global Ecological History* (Cambridge: Cambridge University Press, 2013), 25.
13. Humboldt couldn't believe: *American Farmer*, 24 December 1824, 317.
14. Humphry Davy, *The Collected Works*, vol. 8, *Agricultural Lectures*, pt. 2, ed. John Davy (London: Smith, Elder, 1840), 26.
15. Composition of guano: *American Farmer*, 24 December 1824, 316.
16. Quoted in D. J. Browne, *The Field Book of Manures; or, the American Muck Book* (New York: A. O. Moore, 1858), 282.
17. Ben Perley Poore, "Biographical Notice of John S. Skinner," *The Plough, the Loom, and the Anvil* 7, no. 1 (1854): 11.
18. John James von Tschudi, *Travels in Peru, on the Coast, in the Sierra, Across the Cordilleras and the Andes, into the Primeval Forests*, trans. Thomasina Ross, new ed. (New York: A. S. Barnes, 1854), 169.
19. 44.7 meters: G. Evelyn Hutchinson, "The Biogeochemistry of Vertebrate Excretion (Survey of Contemporary Knowledge of Biogeochemistry)," *Bulletin of the American Museum of Natural History* 96: 1950. New York: By Order of the Trustees. (1950), 40.

20. Mann (2011), 4.
21. Middle of first millennium BCE: Hutchinson, "Biogeochemistry of Vertebrate Excretion," 70.
22. Justus Liebig, *Familiar Letters on Chemistry and Its Relation to Commerce, Physiology, and Agriculture*, ed. John Gardner (London: Taylor and Walton, 1844), letter 16.
23. Hutchinson, "Biogeochemistry of Vertebrate Excretion," 34, paraphrasing E. W. Middendorf, *Peru* (Berlin: R. Oppenheim, 1984), no page cited.
24. Thirteen million short tons, "a dazzling coat": Hutchinson, "Biogeochemistry of Vertebrate Excretion," 28.
25. George Washington Peck, quoted in ibid., 37
26. Late blight and guano: Nicholas Wade, "Testing Links Potato Famine to an Origin in the Andes," *New York Times*, 7 June 2001; J. B. Ristaino, "Tracking Historic Migrations of the Irish Potato Famine Pathogen, *Phytophthora infestans*," *Microbes and Infection* 4 (2002): 1369–76.
27. Quoted in Naomi Rogers, "Germs with Legs: Flies, Disease, and the New Public Health," *Bulletin of the History of Medicine* (Winter 1989): 602, n. 7.
28. Ibid., 601.
29. 95 percent of flies: Greene, *Horses at Work*, 249.
30. Frank J. Sprague, "The Solution of Municipal Rapid Transit," *American Institute of Electrical Engineers Transactions* 5 (1887): 177.
31. Brookline connected to central Boston: McShane and Tarr, *Horse in the City*, 171.
32. Number of city horses increased despite electric streetcars: Greene, *Horses at Work*, 175.
33. Ibid., 261–62.

FOURTEEN: PILLARS OF BLACK CLOUD

1. John W. Graham, *The Destruction of Daylight: A Study of the Smoke Problem* (London: George Allen, 1907), 1.
2. British firewood peak possibly c. 1750: Paul Warde, *Energy Consumption in England & Wales 1560–2000* (Naples, It.: Consiglio Nazionale delle Ricerche, Istituto di Studi sulle Società del Mediterraneo, 2007), 38.
3. Wood, coal peak uses: O'Connor, "Energy Transitions," 3.
4. Graham, *Destruction of Daylight*, 4.
5. Ian Douglas, Rob Hodgson, and Nigel Lawson, "Industry, Environment and Health Through 200 Years in Manchester," *Ecological Economics* 41 (2002): 235–55), 246.
6. Quoted in David Stradling, *Smokestacks and Progressives: Environmentalists,*

Engineers, and Air Quality in America, 1881–1951 (Baltimore: Johns Hopkins University Press, 1999), 6.

7. Quoted in A. E. Outerbridge Jr., "The Smoke Nuisance and Its Regulation, with Special Reference to the Condition Prevailing in Philadelphia," *Journal of the Franklin Institute* 143, no. 66 (1897): 396–97.
8. Cold-weather explanation of mortality increases during smoke fogs: H. R. Anderson, "Air Pollution and Mortality: A History," *Atmospheric Environment* 43 (2009): 143.
9. Stradling, *Smokestacks and Progressives*, 7.
10. Chicago 1893 fuel oil: Outerbridge Jr., "Smoke Nuisance and Its Regulation," 397–98.
11. Trollope (1862), vol. 2, 60.
12. Frederick Moore Binder, *Coal Age Empire: Pennsylvania Coal and Its Utilization to 1860* (Harrisburg: Pennsylvania Historical and Museum Commission, 1974), 22.
13. Trollope (1862), vol. 2, 60.
14. Quoted in Angela Gugliotta, "Class, Gender, and Coal Smoke: Gender Ideology and Environmental Justice in the City: A Theme for Urban Environmental History," *Environmental History* 5, no. 2 (2000): 165.
15. Robert Louis Stevenson, *Strange Case of Dr. Jekyll and Mr. Hyde* (London: Longmans, Green, 1886), 52.
16. 1900 US leading causes of death: National Office of Vital Statistics, cited in www.cdc.gov/nchs/data/dvs/lead1900_98.pdf.
17. 1900 U.S. homicide rate: *Crime and Justice Atlas 2000*, Justice Research and Statistics Association online, www.jrsa.org, 38.
18. David A. Waples, *The Natural Gas Industry in Appalachia: A History from the First Discovery to the Tapping of the Marcellus Shale*, 2nd ed. (Jefferson, NC: McFarland, 2012), 21.
19. Cast iron pipelines, wrought iron introduced c. 1890-1900: Tussing and Barlow (1984), 29.
20. Great Western Iron Company first to use gas: Waples, *Natural Gas Industry in Appalachia*, 45.
21. Five hundred miles of Pittsburgh NG pipeline: ibid., 48.
22. Ibid., 49.
23. Ibid., 52.
24. Quoted in Cliff I. Davidson, "Air Pollution in Pittsburgh: A Historical Perspective," *Journal of the Air Pollution Control Association* 29, no. 10 (1979): 1038.
25. Quoted in "The Costs and Benefits of Prevention," Editorial, *Journal of Public Health Policy* 1, no. 4 (1980): 286.

26. A United Nations study: *UN World Population Prospects: The 2015 Revision,* Table S.11: Ten Countries with the Highest and the Lowest Life Expectancy at Birth, p. 44 (online).

FIFTEEN: A GIFT OF GOD

1. Quadricycle details: Richard Snow, *I Invented the Modern Age: The Rise of Henry Ford* (New York: Scribner, 2013), 56–59. Gasoline fuel: some websites claim the quadricycle was alcohol fueled. The Henry Ford Research Center confirms that it was gasoline fueled. HFRC, Lauren S., personal communication, 25 January 2017.
2. Snow, *Rise of Henry Ford,* 39.
3. Hiram Percy Maxim, *Horseless Carriage Days* (New York: Harper & Brothers, 1937), 4.
4. 1896 automobiles: see images at Early American Automobiles, www.earlyamericanautomobiles.com.
5. Stanley Steamer 1898 best-seller: David A. Kirsch, "The Electric Car and the Burden of History: Studies in Automotive Systems Rivalry in America, 1890–1996" (PhD diss., Stanford University, 1996), 65; "of the 4,192": Rudy Volti, *Cars and Culture: The Life Story of a Technology* (Baltimore: Johns Hopkins University Press, 2004), 7.
6. Maxim, *Horseless Carriage Days,* 61.
7. Ibid., 113.
8. Stanley Steamer weights: Charles C. McLaughlin, "The Stanley Steamer: A Study in Unsuccessful Innovation," *Explorations in Environmental History* 7, no. 1 (1954): 40.
9. Volti, *Cars and Culture,* 667.
10. Steamer seventy-two moving parts: McLaughlin, "Stanley Steamer," 40.
11. Pedro G. Salom, "Automobile Vehicles," *Journal of the Franklin Institute* 141, no. 4 (1896): 290.
12. Ibid., 289.
13. 900 rpm, 3× for maximum efficiency: McLaughlin, "Stanley Steamer," 38.
14. Maxim, *Horseless Carriage Days,* 131.
15. 1914 New England hoof-and-mouth disease epidemic: Kirsch, "Electric Car and Burden of History," 50. Kirsch puts the outbreak "in the spring of 1914." Most sources date it from October. It lasted two years. See, for example, CQ Press CQ Researcher online, "The Foot and Mouth Disease."
16. 569,000 cars, about 1,000 Steamers: Kirsch, "Electric Car and Burden of History," 76.

17. 1912 cars outnumber horses: Eric Morris, "From Horse Power to Horsepower," *Access* 30 (Spring 2007): 8.
18. 1920 decline in horses: Paul G. Irwin, "Overview: The State of Animals in 2001," in *The State of the Animals 2001*, ed. D. J. Salem and A. N. Rowan (Washington, DC: Humane Society Press, 2001), 8.
19. H. E. Barnard, "Prospects for Industrial Uses of Farm Products," *Journal of Farm Economics* 20 (1938): 119.
20. Flex-fuel offered until 1931: David Blume, *Alcohol Can Be a Gas! Fueling an Ethanol Revolution for the 21st Century* (Santa Cruz, CA: International Institute for Ecological Agriculture, 2007), 11.
21. Model T flex-fuel features can be seen on a video hosted by David Blume at Fuel Freedom Foundation online, www.fuelfreedom.org, or on any museum Model T.
22. Standard Oil 85 percent of world market: Carolyn Dimitri and Anne Effland, "Fueling the Automobile: An Economic Exploration of Early Adoption of Gasoline over Ethanol," *Journal of Agricultural & Food Industrial Organization* 5, no. 2 (2007), 7.
23. 1913 knock problem: T. A. Boyd, *Professional Amateur: The Biography of Charles Franklin Kettering* (New York: E. P. Dutton, 1957), 98.
24. Charles F. Kettering, "More Efficient Utilization of Fuel" (SAE technical paper 190010, 1919), 204.
25. 13 percent gasoline: Alan W. Peters, William H. Flank, and Burtron H. Davis, "The History of Petroleum Cracking in the 20th Century," ch. 5 in *Innovations in Industrial and Engineering Chemistry,* ed. William H. Flank, Martin A. Abraham, and Michael A. Matthews. American Chemical Society Symposium Series (Washington, DC: American Chemical Society, 2009), 104.
26. Ibid., 105.
27. William Kovarik, "Charles F. Kettering and the Development of Tetraethyl Lead in the Context of Alternative Fuel Technologies" (SAE Technical Paper 941942, 1994), n.p.
28. High-percentage additive patent: US patent no. 1,296,832, filed 7 January 1918, awarded 11 March 1919 to Thomas Midgley Jr.
29. Corn prices: Ric Deverell and Martin Yu, *Reversing a 60-Year Trend, Exhibit 2: Real Corn and Wheat Prices, 1900–2010,* Credit Suisse Commodities Research, 2 February 2011, www.credit-suisse.com/researchandanalytics.
30. L. H. Bean and P. H. Bollinger, "The Base Period for Parity Prices," *Journal of Farm Economics* 21, no. 1 (1939): 253.
31. Quoted in Kovarik, "Charles F. Kettering," 6.
32. 1918 patent: see note 28 above. The patent actually discusses kerosene, but comprises any "hydrocarbon."

33. Thanks to Bill Kovarik for discovering this connection: Kovarik, "Charles F. Kettering," 6.
34. Hibbert inventor of antifreeze: Wolfrom (1958), 149.
35. Harold Hibbert, "The Role of the Chemist in Relation to the Future Supply of Liquid Fuel," *Journal of Industrial and Engineering Chemistry* 13 (1921): 841.
36. Ibid.
37. Ibid.
38. 28 February 1921 Associated Press story: "New Solvent Dissolves Rubber; Universal Solvent Sought by Scientists Through Ages May Be Found," Lincoln, NE, 28 February 1921.
39. Leo H. Baekeland, "The Synthesis, Constitution, and Uses of Bakelite," *Industrial and Engineering Chemistry* 1 (1909): 150. Bakelite patent: US 942,699, patented 7 December 1909.
40. Universal solvent joke: Boyd, *Professional Amateur*, 144.
41. Quoted in Alan P. Loeb, "Birth of the Kettering Doctrine: Fordism, Sloanism and the Discovery of Tetraethyl Lead," *Business and Economic History* 24, no. 1 (1995): 81; five times better, twenty times better: Boyd, *Professional Amateur*, 145.
42. Boyd, *Professional Amateur*, 145.
43. Kettering, "More Efficient Utilization of Fuel," 205.
44. Quoted in Stanton P. Nickerson, "Tetraethyl Lead: A Product of American Research," *Journal of Chemical Education* 31, no. 11 (1954): 562.
45. Quoted in Hélio Elael Bonini and Paulo Alves Porto, "Thomas Midgley, Jr., and the Development of New Substances: A Case Study for Chemical Educators," *Journal of Chemical Education* 90 (2013): 1634.
46. Boyd, *Professional Amateur*, 146, emphasis added.
47. 1000:1 dilution: Kovarik, "Charles F. Kettering," n.[8]; Kettering said day most dramatic: Boyd, *Professional Amateur*, 146.
48. Kettering chose "Ethyl": Boyd, *Professional Amateur*, 147.
49. Thomas Midgley Jr., "Tetraethyl Lead Poison Hazards," *Industrial and Engineering Chemistry* 17, no. 8 (1925): 828.
50. Quoted in Dietmar Seyferth, "The Rise and Fall of Tetraethyllead [*sic*] 2," *Organometallics* 22, no. 25 (2003): 5157.
51. Ethyl first public sale: Nickerson, "Tetraethyl Lead," 566.
52. Indianapolis 500: Terry Reed, *Indy: The Race and the Ritual of the Indianapolis 500* (Washington, DC: Potomac Books, 2005), 20.
53. Ethyl Corporation details: Loeb, "Birth of the Kettering Doctrine," 82.
54. Midgley flushing his eyes with mercury: Kovarik, "Charles F. Kettering," 10.

55. Midgley, "Tetraethyl Lead Poison Hazards," 827.
56. Alice Hamilton, Paul Reznikoff, and Grace M. Burnham, "Tetra-ethyl Lead," *Journal of the American Medical Association* 84, no. 20 (1925): 1482.
57. Alice Hamilton, *Exploring the Dangerous Trades: The Autobiography of Alice Hamilton, M.D.* (Boston: Little, Brown, 1943), 416.
58. Quoted in Kovarik, "Charles F. Kettering," 14.
59. Ibid., 11.
60. Midgley and Kettering approached Bureau of Mines, GM allowed to review results: Kovarik, "Charles F. Kettering," 11.
61. New York and New Jersey rescinded ban: Hamilton, Reznikoff, and Burnham, "Tetra-ethyl Lead," 1486.
62. Ibid., 1485.
63. Kettering and iron carbonyl patent: Kovarik, "Charles F. Kettering," 16.
64. Kettering in Europe and meeting surgeon general: ibid. Meeting in May 1925: Hamilton, *Exploring the Dangerous Trades,* 415–16.
65. Kettering among May 1925 conference speakers: Kovarik, "Charles F. Kettering," 17.
66. "Our continued development"; "it was no gift of God"; "to find something else": quoted in ibid., 18.
67. Conference's lead-only caveat: ibid., quoting a report on the conference in the *New York World.*
68. Quoted in ibid., 19.
69. Quoted in ibid., 11.
70. Why Ethyl wanted 1959 increase: US Department of Health, Education and Welfare Public Health Service Advisory Committee on Tetraethyl Lead to the Surgeon General (1959), 15.
71. Ibid., 10.
72. Hibbert, "Role of the Chemist," 841.

SIXTEEN: ONE-ARMED MEN DOING WELDING

1. Bahrain bitumen: see Jacques Connan et al., "The Archeological Bitumens of Bahrain from the Early Dilmun Period (c. 2200 BC) to the Sixteenth Century AD: A Problem of Sources and Trade," *Arabian Archeology and Epigraphy* 9, no. 2 (November 1998): 141–81.
2. Ralph Omer Rhoades memorial, *AAPG Bulletin* online, February 1962; "Dusty": Scott McMurray, *Energy to the World: The Story of Saudi Aramco* (Houston: Aramco Services, 2011), 26.
3. Fred Davies spotting *jabals*: McMurray, *Energy to the World,* 29.

4. Mecca pilgrims: Anthony Cave Brown, *Oil, God, and Gold: The Story of Aramco and the Saudi Kings* (Boston: Houghton Mifflin, 1999), 23.
5. Daniel Yergin, *The Prize: The Epic Quest for Oil, Money, and Power* (New York: Simon & Schuster, 1991), 286.
6. Quoted in Brown, *Oil, God, and Gold*, 23.
7. Global oil production: McMurray, *Energy to the World*, 23; automobile numbers: "State Motor Vehicle Registrations, by Years, 1900–1995," US Department of Transportation Federal Highway Administration online.
8. "Oklahoma Oil Prices Soar to Seventy Cents," *Chicago Tribune*, 23 August 1931, 1.
9. Wallace Stegner, *Discovery! The Search for Arabian Oil* (1971; repr., Vista, CA: Selwa Press, 2007), 10–11.
10. Socal benefited from lack of interest: K. S. Twitchell, *Saudi Arabia: With an Account of the Development of its Natural Resources* (Princeton, NJ: Princeton University Press, 1947), 149–50.
11. Negotiations concluded end of May 1933; on May 29: ibid., 151.
12. US Library of Congress Area Studies: Saudi Arabia Brief History, n.p. (online).
13. Dean Acheson: Stegner, *Discovery!*, 29.
14. Sovereigns with images of English kings: Yergin, *The Prize*, 292.
15. Socal-Saudi deal signed 14 July 1933: "New Oil-Field in Saudi Arabia," *Standard Oil Bulletin*, September 1–12, 1936, 3.
16. Stegner, *Discovery!*, 29.
17. Yergin, *The Prize*, 292.
18. Stegner, *Discovery!*, 30, says 320,00 square miles; Twitchell, *Saudi Arabia*, 142, says 140,000 square miles, presumably part of a larger area that Stegner cites.
19. Stegner, *Discovery!*, 31.
20. Dammam Dome, end September: ibid., 35; Twitchell, *Saudi Arabia*, 152; Robert Vitalis, *America's Kingdom: Mythmaking on the Saudi Oil Frontier* (Stanford, CA: Stanford University Press, 2007), 59.
21. For the heroic version, see Stegner, *Discovery!* For the realistic version, see Vitalis, *America's Kingdom.*
22. June 1934, November: Stegner, *Discovery!*, 74, 79–80.
23. Breaking rock, spudding in, 22.5-inch hole: ibid., 89. Stegner has "collar," presumably a typo; the cylindrical space around the base of an oil well, which allows access to the Christmas tree, is called a cellar.
24. 7 May 1935–4 January 1936 drilling notes: ibid., 93–95.
25. Ibid., 95–96.
26. Submarine spring water: ibid., 97.

27. Dammam Nos. 3 to 6: ibid., 102–3.
28. Ibid., 103.
29. Ibid., 115.
30. Dammam No. 7 flows: ibid., 118; Sultan Al-Sughair, "Well No. 7 That Established KSA on World Oil Map," *Arab News*, 1 August 2015, www.arabnews.com.
31. Dammam No. 7 lifetime production: ibid.
32. Wells Nos. 2 & 4, enlarged concession: McMurray, *Energy to the World*, 90.
33. Ibn Saud and retinue visiting port: ibid.
34. First US welding patent, no. 363,320, issued 17 May 1887.
35. Christopher James Tassava, "Launching a Thousand Ships: Entrepreneurs, War Workers, and the State of American Shipbuilding, 1940–1945" (PhD diss., Northwestern University, Evanston, IL, 2003), 32.
36. US ships built 1910–1914: see table, "Cramp Shipbuilding, Philadelphia, Pa.," at www.shipbuildinghistory.com/shipyards/large/cramp.htm.
37. *Dorothea M. Geary:* Shelly Terry, "New Arrival at Ashtabula Maritime and Surface Transportation Museum," *Ashtabula (OH) Star Beacon* online, 21 December 2014.
38. Twenty-seven German-owned vessels: William Lowell Putnam, *The Kaiser's Merchant Ships in World War I* (Jefferson, NC: McFarland), Kindle location (hereafter KL) 2277.
39. Report number four in James Caldwell, *Report to the United States Shipping Board Emergency Fleet Corporation on Electric Welding and its Application in the United States of America to Ship Construction* (Philadelphia: Emergency Fleet Corporation, 1918), 67.
40. Repairs took four months: Putnam, *Kaiser's Merchant Ships*, KL 2380.
41. Report number four in Caldwell, *Report to the United States Shipping Board*, 67.
42. More than a hundred ships repaired within eight months: André A. Odermatt, *Welding: A Journey to Explore Its Past* (Troy, OH: Hobart Institute of Welding Technology, 2010), 133.
43. "Secretary Daniels Urges the Need for More Ships," *Official Bulletin*, 19 February 1918, 2.
44. British experience with welding: Caldwell, *Report to the United States Shipping Board*, 66.
45. Captain James Caldwell, "welding investigations": ibid., 5.
46. Ibid., 26.
47. Ibid., 144.
48. "The Coming of the Rivetless Steel Ship," *American Marine Engineer* 13, no. 10 (October 1918): 12.

49. 1931: Robert W. Gilmer, "The History of Natural Gas Pipelines in the Southwest," *Texas Business Review* (May/June 1981): 131; improved ditching machines and compressors: Castaneda, *Invisible Fuel*, 85.
50. Natural gas disadvantages: Christopher J. Castaneda, *Regulated Enterprise: Natural Gas Pipelines and Northeastern Markets, 1938–1954* (Columbus: Ohio State University Press, 1993), discusses the disadvantages of natural gas in detail on 15–18.
51. Panhandle, Hugoton, 117 trillion cubic feet: Castaneda, *Invisible Fuel*, 84.
52. 1935 FTC report charted in ibid., 25, table 2.5.
53. Ibid., 104.
54. United States more than 60 percent world oil production: Arthur J. Brown, "World Sources of Petroleum," *Bulletin of International News* 17, no. 13 (1940): 769; > 1 million bbl./day surplus capacity: "Oil and World Power," Encyclopedia of the New American Nation online, www.americanforeignrelations.com/Oil-and-world-power.html.
55. Dönitz five U-boats, seventy-three hits and kills: *The Big Inch and Little Big Inch Pipelines*. Houston: Texas Eastern Transmission Corporation, 2000, 8.
56. Quoted in ibid., 5.
57. 1.4 million, 100,000: ibid., 9.
58. Petroleum Industry War Council Committee message: Homer H. Hickam Jr., *Torpedo Junction: U-Boat War Off America's East Coast, 1942* (Annapolis: Naval Institute Press, 1989), 122.
59. Quoted in ibid., 123.
60. Bucket Brigade: ibid., 124.
61. Petroleum pipeline history: *Big Inch and Little Big Inch Pipelines*, 12.
62. Ten thousand miles pipeline during 1930s: John F. Kiefner and Cheryl J. Trench, "Oil Pipeline Characteristics and Risk Factors: Illustrations from the Decade of Construction," *American Petroleum Pipeline Committee Publication* (Washington, DC: American Petroleum Institute, 2001), 24.
63. "War Emergency Pipelines," 25 June 1942: *Big Inch and Little Big Inch Pipelines*, 15.
64. "Men dug a ditch": ibid., 19; Mississippi River trench: ibid., 25.
65. Ibid., 35.
66. 26 January 1944, 2 March 1944; 185 million bbl: ibid., 28.
67. Sidney A. Swensrud, "Possibility of Converting the Large Diameter War Emergency Pipe Lines to Natural Gas Service After the War" (paper for presentation at the February 1944 meeting of the Petroleum Division of the American Institute of Mining and Metallurgical Engineers, New York City, 21 February 1944, *AIME*

Technical Publications & Contributions 1943–1944. http://library.aimehq.org/library, 2.

68. Ibid.
69. Natural gas price versus town gas: ibid., 12.
70. Ibid., 12.
71. Lewis Stark, "UMW Head Defiant," *New York Times,* 4 December 1946, 1ff.
72. Longhand note of President Harry S. Truman, 11 December 1946, Truman Papers—President's Secretary's File, Harry S. Truman Library & Museum online.
73. Cited in Castaneda, *Regulated Enterprise,* 83.

SEVENTEEN: FULL POWER IN FIFTY-SEVEN

1. 200 watts: Enrico Fermi, "Atomic Energy for Power," in *The Future of Atomic Energy: The George Westinghouse Centennial Forum, May 16, 17, and 18, 1946,* vol. 1 (Pittsburgh: Westinghouse Educational Foundation), 93.
2. Jeffries, Zay, Enrico Fermi et al., *Prospectus on Nucleonics* (Chicago: Metallurgical Laboratory MUC-RSM-234, 1944), 28 (online).
3. Ibid.
4. Sixty-four-thousand pounds of uranium oxide: Chenoweth (1988), 5: "Using a recovery factor of 70 percent at the mills, an estimated 64,000 pounds of U308 from Arizona went into the manufacture of the first atomic weapons. Although this represents only a small fraction of the total domestic production [during the war] of 2,698,000 pounds of U308 . . . Arizona is usually overlooked as contributing to the Manhattan project."
5. V. E. McKelvey, "Mineral Resource Estimates and Public Policy: Better Methods for Estimating the Magnitude of Potential Mineral Resources Are Needed to Provide the Knowledge That Should Guide the Design of Many Key Public Policies," *American Scientist* 60, no. 1 (1972): 32.
6. Zoellner (2000), 3.
7. McKelvey, "Mineral Resource Estimates and Public Policy," 13.
8. E. L. Berthoud, "On the Occurrence of Uranium, Silver, Iron, etc., in the Tertiary Formation of Colorado Territory," *Proceedings of the Academy of Natural Sciences of Philadelphia* 27(2), May–September 1875, p. 365.
9. Miller and Gill (1954), 36.
10. Chinese fly-ash uranium extraction, Central European and South African interest: "Sparton Produces First Yellowcake from Chinese Coal Ash," *World Nuclear News* online, 16 October 2007.
11. Rickover choosing commanders: Theodore Rockwell, personal communication.

12. Theodore Rockwell, *The Rickover Effect: How One Man Made a Difference* (Lincoln, NE: iUniverse, 2002), 27.
13. Kintner (1959), 2.
14. Rockwell, *The Rickover Effect*, 44–45.
15. Cited in Richard Rhodes, "A Demonstration at Shippingport: Coming On Line," *American Heritage* online 32, no. 4 (1981): NB: I no longer have a record of the references from this 30-plus-year-old source, an article I wrote for *American Heritage* magazine in 1981. It was based on interviews as well as document research, and was carefully checked by the magazine's editors at the time.
16. 170 bombs in 1949, 841 in 1952: Robert S. Norris and Hans M. Kristensen, "Global Nuclear Weapons Inventories, 1945–2010," *Bulletin of the Atomic Scientists* 66, no. 4 (2010), online.
17. John Foster Dulles, "The Evolution of Foreign Policy," before the Council of Foreign Relations, New York, NY, Department of State, press release no. 81 (January 12, 1954).
18. NSC 162/2, "A Report to the [US] National Security Council by the Executive Secretary on Basic National Security Policy, October 30, 1953," Washington, 39b. 1, 22 (online).
19. "normal uranium and fissionable": "Atoms for Peace Speech," December 8, 1953, online at www.iaea.org/about/history/atoms-for-peace-speech.
20. More than forty thousand kilograms of fuel uranium: Martin J. Medhurst, "Atoms for Peace and Nuclear Hegemony: The Rhetorical Structure of a Cold War Campaign," *Armed Forces & Society* 23, no. 4 (1997): 581.
21. Eisenhower message to Congress: 17 February 1954, quoted in ibid., 583–84.
22. Quoted in Richard Rhodes, *Nuclear Renewal: Common Sense About Energy* (New York: Whittle/Viking, 1993), 35.
23. L. A. Kotchetkov, "Obninsk: Number One," *Nuclear Engineering International:* July 13, 2004, 4, www.neimagazine.com/features.
24. 1950, 1953 large ship reactor: Rockwell, *The Rickover Effect*, 159–60.
25. Ibid., 196.
26. Fleger interview: Rhodes, "A Demonstration at Shippingport," q.v. fn. 15 above.
27. Medhurst, "Atoms for Peace," 580.
28. Thirty thousand megatons: ibid., 581, citing Ronald E. Powaski, *March to Armageddon: The United States and the Nuclear Arms Race, 1939 to the Present* (New York: Oxford University Press, 1987), 60.
29. Shippingport thorium core, decommissioning: J. C. Clayton, "The Shippingport Pressurized Water Reactor and Light Water Breeder Reactor," for presentation at

25th Central Regional Meeting, American Chemical Society, Pittsburgh, October 4–6, 1993, 2 (online).

30. May 1972 at Pierrelatte: George Cowan, "A Natural Fission Reactor," *Scientific American*, July 1976, passim.
31. G. W. Wetherill and M. G. Inghram. "Spontaneous Fission in Uranium and Thorium Ores," *Proceedings of the Conference on Nuclear Processes in Geologic Settings* (Williams Bay, WI.: National Research Council Committee on Nuclear Science, 1953), 31.
32. P. K. Kuroda, "On the Nuclear Physical Stability of the Uranium Minerals," *Journal of Chemical Physics* 25, no. 4 (1956): 782.
33. Gabon reactors' "nuclear waste": Rockwell (2010), 32. Cowan, "Natural Fission Reactor," 45; François Gauthier-Lefaye, "2 Billion Year Old Natural Analogs for Nuclear Waste Disposal: The Natural Nuclear Fission Reactors in Gabon (Africa)," *Comptes Rendus R. Physique* 3, nos. 7 and 8 (September/October 2002).

EIGHTEEN: AFFECTION FROM THE SMOG

1. Liege 1930 killer fog: Benoit Nemery, Peter H. M. Hoet, and Abderrahim Nemmar, "The Meuse Valley Fog of 1930: An Air Pollution Disaster," *Lancet* 357, no. 9257 (March 3, 2001): 704.
2. Twenty dead, six thousand sickened at Donora: H. H. Schrenk et al., *Air Pollution in Donora, Pa.: Epidemiology of the Unusual Smog Episode of October 1948, Preliminary Report,* Public Health Bulletin no. 306 (Washington, DC: Public Health Service, 1949), 12.
3. Berton Roueché, "Annals of Medicine: The Fog," *New Yorker*, 30 September 1950, 33.
4. Ibid.
5. Ibid.
6. Schrenk et al., *Air Pollution in Donora, Pa., iv.*
7. Roueché, "Annals of Medicine: The Fog," 38.
8. Ibid., 44.
9. USPHS staff numbers: ibid., 49.
10. Schrenk et al., *Air Pollution in Donora, Pa.,* 29.
11. Ibid., 161–62.
12. File, 13 February 1946, Lieutenant Colonel Cooper B. Rhodes, online at ibid.
13. "Industrial News: Fluorine Gases in Atmosphere as Industrial Waste Blamed for Death and Chronic Poisoning of Donora and Webster, PA, Inhabitants," *Chemical and Engineering News* 26, no. 50 (December 13, 1948): 3692.

14. Quoted in Chris Bryson, "The Donora Fluoride Fog: A Secret History of America's Worst Air Pollution Disaster," ActionPA.org, last modified 2 December 1998.
15. Class-Action Lawsuit, Public Health Service Report: "Donora: 'The Truth Was Concealed,'" *Pittsburgh Observer-Reporter,* 19 October 2008, reproduced at fluoridealert.org (online).
16. Three thousand excess deaths: Michelle L. Bell and Devra Lee Davis, "Reassessment of the Lethal London Fog of 1952: Novel Indicators of Acute and Chronic Consequences of Acute Exposure to Air Pollution," *Environmental Health Perspectives* 109, no. 3 (2001): 389.
17. Quoted in Christine L. Corton, *London Fog: The Biography* (Cambridge, MA: Harvard University Press, 2015), 280.
18. Ministry of Health attribution: Bell and Davis, "Reassessment of the Lethal London Fog of 1952," 389; unexpected versus flu deaths: 392, fig. 6.
19. "Great Killer Fog": Corton, *London Fog,* 284.
20. 1956 Clean Air Act: (4 and 5 Eliz. 2 ch. 52).
21. 80 percent reduction: Corton, *London Fog,* 305; 1962 killer fog, ibid.; 1956 Clean Air Act, ibid., 309.
22. *Smog* a 1905 coinage: "Smog," *Oxford English Dictionary.*
23. McGill et al. (1952), 286.
24. Quoted in Haagen-Smit (2000), 20.
25. Ibid., 23.
26. Ibid., 23–24.
27. Thirty thousand cubic feet: Chip Jacobs and William J. Kelly, *Smogtown: The Lung-Burning History of Pollution in Los Angeles* (New York: Overlook Press, 2013), 72.
28. Quoted in Haagen-Smit (2000), 24.
29. Haagen-Smit (1950), 10–11.
30. Ibid., 8.
31. Haagen-Smit (2000), 24–26.
32. Nixon creating EPA, Clean Air Act: Jeanne M. Logsdon, "Organizational Responses to Environmental Issues: Oil Refining Companies and Air Pollution," in *Research in Corporate Social Performance and Policy,* vol. 7, ed. L. E. Preston (Greenwich, CT: JAI Press, 1985), 52.
33. Unleaded gasoline: Richard G. Newell and Kristian Rogers, "The U.S. Experience with the Phasedown of Lead in Gasoline" (discussion paper, Washington, DC: Resources for the Future, 2003).
34. NYC soot pollution findings: Patrick Louchouarn et al., "Elemental and Molecular Evidence of Soot- and Char-Derived Black Carbon Inputs to New York City's

Atmosphere During the 20th Century," *Environmental Science & Technology* 41, no. 1 (2007): 82–87.

35. Princeton economists' findings: Gene M. Grossman and Alan B. Krueger, "Environmental Impacts of a North American Free Trade Agreement" (working paper 3914, Cambridge, MA: National Bureau of Economic Research, 1991), abstract.
36. EKC inconsistencies: Edward B. Barbier, "Introduction to the Environmental Kuznets Curve Special Issue," *Environment and Development Economics* 2, no. 4 (November 1997): 372.
37. "Environmental amenities," luxury model: J. Martínez-Alier, "The Environment as a Luxury Good or 'Too Poor to Be Green'?" *Ecological Economics* 13 (1995): 1–10.
38. The Great Leap Forward: Robert J. Gordon, *The Rise and Fall of American Growth: The U.S. Standard of Living Since the Civil War* (Princeton, NJ: Princeton University Press, 2016), 535.
39. Robert Higgs, "Wartime Prosperity? A Reassessment of the U.S. Economy in the 1940s," *Journal of Economic History* 52, no. 1 (1992): 57, quoted in Gordon, *Rise and Fall of American Growth*, 552.

NINETEEN: THE DARK AGE TO COME

1. Carson (1962), 1–2.
2. Quoted in William Souder, *On a Farther Shore: The Life and Legacy of Rachel Carson* (New York: Crown, 2012), 338.
3. Quoted in Jason Lemoine Churchill, "The Limits to Influence: The Club of Rome and Canada, 1968 to 1988" (PhD diss., University of Waterloo, Waterloo, Ont., 2006), 35.
4. Quoted in Souder, *On a Farther Shore*, 278.
5. Many veterans sought help only in retirement: clinical psychologist Ginger Rhodes, personal communication.
6. Quoted in Roger Revelle, "Harrison Brown 1917–1986," *Biographical Memoirs of the National Academy of Sciences* (Washington, DC: National Academy of Sciences, 1994), 55.
7. Harrison Brown, *The Challenge of Man's Future: An Inquiry Concerning the Condition of Man During the Years That Lie Ahead* (New York: Viking, 1954), 221.
8. Ibid., 104.
9. Ibid., 105.
10. "The postwar intellectual roots of the population bomb": Pierre Desrochers and Christine Hoffbauer, "The Postwar Intellectual Roots of the Population Bomb: Fairfield Osborn's 'Our Plundered Planet' and William Vogt's 'Road to Survival'

in Retrospect," *Electronic Journal of Sustainable Development* 1, no. 3 (2009): 37–61.

11. Thomas Malthus, *Essay on Population,* 6th ed., bk. 4, ch. 5 (London: John Murray, 1826), .300–1, quoted in Robert Zubrin, *Merchants of Despair: Radical Environmentalists, Criminal Pseudo-Scientists, and the Fatal Cult of Antihumanism* (New York: Encounter Books, 2013), 6.
12. Paul R. Ehrlich, *The Population Bomb* (San Francisco: Sierra Club, 1969), 12.
13. Ibid., 143.
14. Josué de Castro, *The Geography of Hunger* (Boston: Little, Brown, 1952), 312, quoted in Desrochers and Hoffbauer, "Postwar Intellectual Roots of the Population Bomb," 54.
15. David Brower, minutes, board meeting, Sierra Club, 17–18 September 1966, quoted in Thomas Raymond Wellock, *Critical Masses: Opposition to Nuclear Power in California, 1958–1978* (Madison: University of Wisconsin Press, 1998), 85.
16. Eleven types of reactors in 1958: Joseph G. Morone and Edward J. Woodhouse, *The Demise of Nuclear Energy? Lessons for Democratic Control of Technology* (New Haven, CT: Yale University Press, 1989), 53.
17. Quoted in Rhodes, *Nuclear Renewal,* 39. Quotations are not footnoted in this reference, but sources were carefully checked. Some language in this chapter is adapted from *Nuclear Renewal.*
18. Quoted in ibid., 40.
19. Alvin M. Weinberg, "Nuclear Energy and the Environment," *Bulletin of the Atomic Scientists* 26, no. 6 (1970): 73.
20. UN ten billion by 2100: Fran Willekens, "Demographic Transitions in Europe and the World" (working paper WP 2014-004, Max Planck Institute for Demographic Research, 2014), 2 (online).
21. Eighteen billion: Weinberg, "Nuclear Energy and Environment," 69.
22. UN growth rate to zero: Willekens, "Demographic Transitions," 2.
23. Weinberg, "Nuclear Energy and Environment," 79.
24. 30-megawatt reactor, surging to 100 megawatts: M. M. Werner, D. K. Meyers, and D. P. Morrison, "Follow-up of CRNL Employees Involved in the NRX Reactor Clean-up," paper presented at the Third Annual Meeting of the Canadian Radiation Protection Association, Vancouver, BC, 4 May 1982 (Chalk River, Ont.: Chalk River Nuclear Laboratories), 1; more than a thousand men, two women: ibid., 2, table 1.
25. Carter and damaged reactor: Arthur Milnes, "When Jimmy Carter Faced Radioactivity Head-On," *Ottawa Citizen,* 28 January 2009; "A team of three of us":

Jimmy Carter, *Why Not the Best? Jimmy Carter: The First Fifty Years* (Fayetteville: University of Arkansas Press, 1996; first published 1975), 54; Peter Jedicke, "The NRX Incident," Canadian Nuclear Society online, last modified 1989, www.cns-snc.ca/media/history/nrx.html; Werner, Meyers, and Morrison, "Follow-up of CRNL Employees."

26. Werner, Meyers, and Morrison, "Follow-up of CRNL Employees," title page abstract.
27. Quoted at blog.nader.org; "nearly a million": Helen Caldicott, Letter to *New York Times,* 30 October 2013.
28. Lewis L. Strauss, "The H-Bomb and World Opinion: Chairman Strauss's Statement on Pacific Tests," *Bulletin of the Atomic Scientists* 10, no. 5 (May 1954): 163–67.
29. Quoted in Richard G. Hewlett and Jack M. Holl, *Atoms for Peace and War, 1953–1961* (Berkeley: University of California Press, 1989), 177.
30. Sturtevant biography: Edward B. Lewis, "Alfred Henry Sturtevant," *Biographical Memoirs of the National Academy of Sciences,* vol. 73 (Washington, DC: National Academies Press, 1998).
31. A. H. Sturtevant, "Social Implications of the Genetics of Man," *Science* 120, no. 3115 (September 10, 1954): 407.
32. Muller in Spanish Civil War: Elof Axel Carlson, *Genes, Radiation, and Society: The Life and Work of H. J. Muller.* (Ithaca, NY: Cornell University Press, 1981), 237–40.
33. Quoted in Guido Pontecorvo, "Hermann Joseph Muller, 1890–1967," *Biographical Memoirs of Fellows of the Royal Society* 14 (November 1968): 356.
34. Quoted in Carlson, *Genes, Radiation, and Society,* 399.
35. Hermann J. Muller, "The Production of Mutations," Nobel Lecture (1946). *Nobelprize.org* online.
36. Ibid.
37. Ernest Caspari and Curt Stern, "The Influence of Chronic Irradiation with Gamma-Rays at Low Dosages on the Mutation Rate in *Drosophila Melanogaster,*" *Genetics* 33, no. 1 (1948): 81.
38. Muller's Nobel dilemma: this discussion follows Edward J. Calabrese, "On the Origin of the Linear No-Threshold (LNT) Dogma by Means of Untruths, Artful Dodges and Blind Faith," *Environmental Research* 142 (2015).
39. Quoted in ibid., 435, table 1.
40. Ibid., 435.
41. Ibid., 432 (title).
42. Jerry M. Cuttler, "Remedy for Radiation Fear: Discard the Politicized Science," *Dose Response* 12, no. 2 (2014): 171.

43. Jan Beyea, "Response to 'On the Origins of the Linear No-Threshold (LNT) Dogma by Means of Untruths, Artful Dodges and Blind Faith,'" *Environmental Research* 148 (2016): 531.
44. Muller (1950), 57.
45. Pushker A. Kharecha and James E. Hansen, "Prevented Mortality and Greenhouse Gas Emissions from Historical and Projected Nuclear Power," *Environmental Science and Technology* 47 (2013): 4889 (abstract).

TWENTY: ALL ABOARD

1. James Blythe: *Oxford DNB.*
2. Jacobs brothers: "History," Jacobs Wind Electric Company online, www.jacobswind.net; Paul Jacobs, personal communications. Paul Jacobs graciously corrected my text.
3. Selenium and silicon efficiencies: Morton B. Prince, "Early Work on Photovoltaic Devices at the Bell Telephone Laboratories," ch. 33 in Wolfgang Palz, *Power for the World: The Emergence of Electricity from the Sun* (Singapore: Pan Stanford, 2011), 497–98; D. M. Chapin, C. S. Fuller, and G. L. Pearson, "A New Silicon *P-N* Junction Photocell for Converting Solar Radiation into Electrical Power," *Journal of Applied Physics* 25 (1954): 676–77.
4. Georgia six-month trial: Prince, "Early Work on Photovoltaic Devices," 498.
5. Vanguard I information: "Vanguard I," NASA Science Data Coordinated Archive online; "Vanguard Project," US Naval Research Laboratory online; Constance McLaughlin Green and Milton Lomask, *Vanguard, A History,* ch. 7 (Washington, DC: National Aeronautics and Space Administration, 1970) (online).
6. Peter E. Varadi, "Terrestrial Photovoltaic Industry—The Beginning," in *Power for the World,* ed. Wolfgang Palz (Singapore: Pan Stanford Publishing, 2011), 558.
7. Adam Vaughan, "Solar Power Growth Leaps by 50% Worldwide Thanks to US and China," *Guardian,* 20 March 2017 (online); polymer thin films: Frederik C. Krebs, "Fabrication and Processing of Polymer Solar Cells: A Review of Printing and Coating Techniques," *Solar Energy Materials & Solar Cells* 93 (2009): 394–412.
8. Capacity factors: "US Capacity Factors by Fuel Type, 2016," Nuclear Energy Institute Knowledge Center, Nuclear Statistics (online).
9. NG eighty-four times as effective as CO_2: *Climate Change 2014 Synthesis Report,* Intergovernmental Panel on Climate Change (online), 2014, box 3.2, table 1, 87.
10. Vasilis M. Fthenakis and Hyung Chul Kim, "Greenhouse-Gas Emissions from Solar Electric- and Nuclear Power: A Life-Cycle Study," *Energy Policy* 35 (2007): 2549. "Lifetime GHG emissions from solar- and nuclear-fuel cycles in the United

States are comparable under actual production conditions and average solar irradiation."

11. CO_2 percentages compared with coal and natural gas: Eli Goldstein, "CO_2 Emissions from Nuclear Plants" (submitted as coursework for PH241, Introduction to Nuclear Energy, Stanford University, Winter 2012) (online).
12. 3.7 million pounds (1,700 tonnes) of graphite: Zhores Medvedev, *The Legacy of Chernobyl* (New York: W. W. Norton, 1990), 5.
13. Fukushima radiation exposure (updated February 2016), World Nuclear Association Information Library (online).
14. Three Mile Island radiation: Samuel J. Walker, *Three Mile Island: A Nuclear Crisis in Historical Perspective* (Berkeley: University of California Press, 2004), 204–8.
15. Fukushima accident: "Fukushima Accident," World Nuclear Association Information Library (online).
16. Anil Markandya and Paul Wilkinson, "Electricity Generation and Health," *Lancet* 370 (2007): 982.
17. Cesare Marchetti, "My CV as a Personal Story," Cesare Marchetti Web Archive online, 2003, 4–5.
18. Cesare Marchetti and N. Nakicenovic, "The Dynamics of Energy Systems and the Logistic Substitution Model," pt. 1, pt. 2, RR-79-13, IIASA, 1979, 1 (online). Italics in original.
19. Luis de Sousa, "Marchetti's Curves," The Oil Drum: *Europe,* 2007 (online).
20. W. Brian Arthur, *The Nature of Technology: What It Is and How It Evolves* (New York: Free Press, 2009), 131.
21. "In large measure": this and the following text from de Sousa, "Marchetti's Curves."
22. East and South Asia nuclear: "Asia's Nuclear Energy Growth," World Nuclear Association Information Library online.
23. Elaine Scarry, *The Body in Pain: The Making and Unmaking of the World* (New York: Oxford University Press, 1985), 288–89.
24. Improvements in mortality: Kevin M. White and Samuel H. Preston, "How Many Americans Are Alive Because of Twentieth-Century Improvements in Mortality?" *Population and Development Review* 22, no. 3 (1996): 415–29.

INDEX

Page numbers in *italics* refer to illustrations

IMAGE CREDITS

Uncredited images are public domain. Despite due diligence, the author was unable to locate all image sources and welcomes communications from copyright holders. Numbers refer to pages.

17: Spring pole: S. T. Pees and Associates.
22: Horse gin: source unknown.
25: Drebbel's submarine: source unknown.
43: Packhorse convoy: source unknown.
45: Mining cart: Zenit, Wikimedia Creative Commons.
56: Watt's separate condenser: UK Science & Society Picture Library.
65: Casting iron pigs: © 2010 Susan Campbell Kuo.
66: Mandrel: reprinted by permission of the estate of Aldren A. Watson from *The Village Blacksmith* (p. 36), written and illustrated by Aldren A. Watson, published by Thomas Y. Crowell, © 1968.
100: Stephenson's *Rocket*: John P. Glithero.
106: Common soft rush: iStock.
128: Cape Cod/Nantucket map: source unknown.
158: Freshet run accident: Reproduced courtesy of The Drake Well Museum, Pennsylvania Historical and Museum Commission.
164: Battling warships (Manet painting): Wikimedia.
222: Toronto harbor smoke: Courtesy City of Toronto Archives, City Engineer's Collection, Series 376, File 4, Item 40.
232: Stanley Steamer: Photo courtesy Early American Automobiles, www.earlyamericanautomobiles.com.
235: Tokheim fuel pump: US Patent Office.
250: Map of Saudi Arabia: Wikimedia.

251: Photograph of ibn Saud: Karl S. Twitchell Collection/Courtesy of the Fine Arts Library, Harvard College Library.

255: Oil well acidizer system: US Patent Office.

258: Pre-1925 US gas pipelines: Federal Trade Commission, US Temporary National Economic Committee: *Investigation of Concentration of Economic Power*, monograph no. 36, 76th Congress, 3rd session, 1940.

259: Arc-welding system: US Patent Office.

262: World ship launching record poster: Courtesy Smithsonian National Museum of American History.

264: 1940 US gas pipelines: Federal Trade Commission, US Temporary National Economic Committee: *Investigation of Concentration of Economic Power*, monograph no. 36, 76th Congress, 3rd session, 1940.

267: Big Inch pipe: U.S. government photo: John Vachon. NARA 208-LU-37C-50.

269: Big Inch and Little Big Inch pipelines: National Archives and Records Administration (NARA) 208-LU-37C-1.

273 (top): Assembling CP-1: US Department of Energy.

273 (bottom): CP-1: US Department of Energy.

280: "Super Sniffer" advertisement: Courtesy of Robert Goldstein, Nuclear Corp.

283: Pressurized water reactor: US Nuclear Regulatory Commission.

292: Gabon ore bed: David Mossman, Geological Society of America.

298: London Killer fog graph: Royal Meteorological Society.

303 (top): George Washington Bridge smog: US Environmental Protection Agency.

303 (bottom): SO-2 vs. GDP graph: Gene M. Grossman and Alan B. Krueger, "Environmental Impacts of a North American Free Trade Agreement" (Working Paper 3914, National Bureau of Economic Research, Cambridge, MA, 1991).

304: Environmental Kuznets curve: Wikipedia.

312: Population bomb: Reprinted by permission of the artist, John C. Holden.

318: Shrimp device: US Department of Energy.

319: Castle Bravo radioactivity: US Department of Energy.

328: Jacobs Wind Electric banner: Courtesy Paul Jacobs, Jacobs Wind Electric Company. Used with permission.

329: *Vanguard I*: US Navy.

331: Wind turbine: Source unknown.

334: Chernobyl accident: Source unknown.

339: Energy substitutions graph: *Visions of Technology* (New York: Touchstone, 1999), © Richard Rhodes, p. 273.

341: DeSousa graph revision: From TheOilDrum.com, reprinted with permission of Dr. Luis de Sousa.

ABOUT THE AUTHOR

RICHARD RHODES is the author or editor of twenty-six works of fiction, history, memoir, and theater, including *The Making of the Atomic Bomb,* which won a Pulitzer Prize in General Nonfiction, a National Book Award, and a National Book Critics Circle Award; *Dark Sun: The Making of the Hydrogen Bomb,* which was one of three finalists for a Pulitzer Prize in History; *Arsenals of Folly,* about the last years of the Cold War; and *The Twilight of the Bombs,* about the post–Cold War challenges of nuclear weapons and international policy. *Nuclear Renewal: Common Sense About Energy* assessed the development of and prospects for nuclear energy at the turn of the millennium. His play, *Reykjavik,* about the 1986 summit meeting between Ronald Reagan and Mikhail Gorbachev, has been read and performed nationwide.

Rhodes has received numerous fellowships for research and writing, including grants from the Ford Foundation, the Guggenheim Foundation, the MacArthur Foundation, and the Alfred P. Sloan Foundation. He has been a host and correspondent for documentaries on public television's *Frontline* and *American Experience* series. He has been a visiting scholar at the Massachusetts Institute of Technology, Harvard University, and Stanford University. He lives near San Francisco, on Half Moon Bay.